Jie Yang, Congfeng Liu

Random Signal Analysis

Information and Computer Engineering

Volume 2

Already published in the series:

Volume 1
Beijia Ning, Analog Electronic Circuit, 2018
ISBN 978-3-11-059540-6, e-ISBN 978-3-11-059386-0,
e-ISBN (EPUB) 978-3-11-059319-8

Jie Yang, Congfeng Liu

Random Signal Analysis

DE GRUYTER

Science Press
Beijing

Authors
Prof. Jie Yang
Associate professor, School of Communication and Information
Ai'an University of Post & Telecommunication
yangjie@xupt.edu.cn

Prof. Congfeng Liu
Associate professor, Research Institute of Electronic Countermeasure
Xidian University
cfliu@mail.xidian.edu.cn

ISBN 978-3-11-059536-9
e-ISBN (PDF) 978-3-11-059380-8
e-ISBN (EPUB) 978-3-11-059297-9
ISSN 2570-1614

Library of Congress Cataloging-in-Publication Data
Names: Liu, Congfeng, author.
Title: Random signal analysis / Congfeng Liu.
Description: Berlin ; Boston : De Gruyter, [2018] | Series: Information and computer engineering | Includes bibliographical references and index.
Identifiers: LCCN 2018026142 (print) | LCCN 2018029153 (ebook) | ISBN 9783110593808 (electronic Portable Document Format pdf) | ISBN 9783110595369 (alk. paper) | ISBN 9783110593808 (e-book pdf) | ISBN 9783110592979 (e-book epub)
Subjects: LCSH: Signal processing.
Classification: LCC TK5102.9 (ebook) | LCC TK5102.9 .L565 2018 (print) | DDC 621.382/2–dc23
LC record available at https://lccn.loc.gov/2018026142

Bibliographic information published by the Deutsche Nationalbibliothek
The Deutsche Nationalbibliothek lists this publication in the Deutsche Nationalbibliografie; detailed bibliographic data are available on the Internet at http://dnb.dnb.de.

Cover image: Prill/iStock/Getty Images Plus
Typesetting: le-tex publishing services GmbH, Leipzig
Printing and binding: CPI books GmbH, Leck

www.degruyter.com

Preface

Random signals (stochastic signals) are also known as random processes (stochastic processes). It is a quantitative description of the dynamic relationship of a series of random events. Random research and other branches of mathematics such as potential theory, differential equations, the mechanics and theory of complex functions, and so on, are closely linked in natural science, engineering science and social science research in every field of random phenomena is an important tool. Random signal research has been widely used in areas such as weather forecasting, statistical physics, astrophysics, management decision-making, economic mathematics, safety science, population theory, reliability and many fields such as computer science often use random process theory to establish mathematical models.

In the study of random processes, people accidently came to describe the inherent law of necessity and to describe these laws in probability form, realizing that the inevitable is the charm of this discipline.

The theoretical basis of the whole discipline of stochastic processes was laid by Kolmogorov and Dub. This discipline first originated from the study of physics, such as by Gibbs, Boltzmann, Poincare and others studying statistical mechanics, and later Einstein, Wiener, Levy and others with the pioneering work of the Brownian movement. Before and after 1907, Markov studied a series of random variables with specific dependencies, which were later called Markov chains. In 1923, Wiener gave the mathematical definition of Brown's movement, and this process is still an important research topic today.

The general theory of stochastic processes is generally considered to have begun in the 1930s. In 1931, Kolmogorov published the *Analytical Method of Probability Theory*. In 1934, Khintchine published "The theory of smooth process," which laid the theoretical basis of the Markov process and the stationary process. In 1953, Dub published the famous "random process theory," systematically and strictly describing the basic theory of random processes. At this point, the stochastic process developed into a systematic scientific theory.

In our daily lives, because of the presence of noise and interference, the signals we receive is no longer a clear signal, but a random process; usually we call this a random signal. A random signal is a kind of signal that is prevalent in the objective world. It is very important for college students in the information technology field to have a deep understanding of the statistical characteristics and to master the corresponding processing and analysis methods. Therefore, random signal analysis is an important basic course in the field of electronic information. Through the study of the course, students are taught to understand the basic concepts of random signals, to master the basic theory of random signals, statistical characteristics and analytical methods, to learn "statistical signal processing" or "signal detection and valuation," with other follow-up courses and future developments laying a solid foundation.

https://doi.org/10.1515/9783110593808-201

The book was written on the basis of the textbook *Random Signal Analysis* compiled by Professor Zhang Qianwu from Xidian University, which absorbed the experiences of similar teaching materials in brother colleges and universities, and which was finalized after a number of discussions in the project group. The textbook characteristics can be summarized as:

(1) Focus on the construction of the whole knowledge system in the field of signal processing.
From the point of view of the knowledge system, the mathematical basis of random signal analysis is "higher mathematics," "probability theory," and "linear algebra," and a professional background from "signals and systems," the following courses are "statistical signal processing," or "signal detection and estimation." Therefore, it continues to strengthen students' foundation of mathematical analysis and the known basic concept of signal analysis, basic principles and basic analysis and processing methods, and at the same time helps students to understand the application of random signal analysis methods to signal detection and parameter estimation with noise in the background. The textbook emphasizes the knowledge system and the structure of signal processing in its entirety, so as to avoid students learning and understanding random signal processing in isolation.

(2) Continuous random signal and discrete random sequence analysis.
Traditional random signal analysis materials mostly focus on the description, characterization and analysis of continuous stochastic process, often ignoring the introduction of discrete random sequences, so that the students doing this course can only carry out theoretical analysis and derivation and cannot use computers for simulation and emulation. However, making full use of computers to process and analyze random signals, on the one hand, is beneficial for students to obtain an intuitive understanding, and, on the other hand is helpful for students to apply their knowledge, truly combining theoretical research and practical applications. Therefore, in the course of compiling the textbook, the analysis of the discrete random sequence was also taken into consideration in detail while the continuous random process is analyzed.

(3) The combination of theoretical analysis and experimental practice.
Random signal analysis is a practical course, and most current textbooks only focus on theoretical teaching instead of experimental practice. This textbook will design the corresponding experimental content for each chapter, so that students can understand and grasp basic concepts, basic principles and basic methods through computer simulation experiments.

(4) Introduction of the latest research results.
Random signal analysis of existing teaching material is mainly limited to the characterization and analysis of stationary random processes lack a description of nonsta-

tionary random processes and related analysis of random processes after passing nonlinear systems. With the advancement of modern signal processing, nonstationary, aperiodic, non-Gaussian and nonlinear stochastic signal processing problems have led to a lot of research results; these results should be the basis of a preliminary understanding of today's undergraduates. Therefore, this textbook will devote a chapter to the introduction of time-frequency analysis and basic knowledge of wavelet analysis.

The book is divided into six chapters: Chapter 0 is an introduction, which reviews and summarizes the basic knowledge of probability theory and introduces random variables and their related digital features and characteristic functions, as well as other knowledge points. Chapter 1 introduces the basic concept of random signals. It discusses their basic characteristics and methods to describe them, complex stochastic process. The discrete-time stochastic process are also detailed, and the normal stochastic process and its spectral analysis and white noise process are introduced as well. Chapter 2 introduces the linear transformation of the stationary stochastic process, reviews the linear transformation and linear system. Moreover, the process of differential and integral of random process is also introduced therein. The stochastic process is analyzed by continuous and discrete-time systems. White noise is analyzed by a linear system and the method of solving the probability density after the linear transformation of the random process. In Chapter 3, we discuss the stationary narrow band stochastic process and first introduce the analytical process and Hilbert transform, narrow band stochastic representation, and the analytic complex stochastic process. We then discuss the probability density of the envelope and phase of the narrow band normal process and the probability density of the square of its envelope. Chapter 4 discusses the nonlinear transformation method of stationary random process, including the direct method, transformation and the analysis of the stochastic process through limiters and the method of calculating the signal-to-noise ratio at the output of the nonlinear system are also detailed. The characteristic description and analysis method of the nonstationary stochastic process is given in Chapter 5. First, the definition and description of the nonstationary stochastic process are introduced, and the correlation function and power spectral density are discussed. Finally, the analysis method of the nonstationary stochastic process in modern signal processing, such as Wigner–Ville distribution and wavelet analysis are introduced. The book incorporates a large number of examples and illustrations, and at the end of each chapter there are enough exercises for practice. It also provides some reference answers at the end of the Chinese book. The book also provides the derivation and proof of some important formulas, the English–Chinese terminology, and other appendices.

The book was completed by associate Professor Yang Jie and Liu Congfeng. Ongbwa Ollomo Armelonas, an international student of Xidian University, has made great efforts in the translation process of this book. The authors express their appreciation to Fu Panlong, Yin Chenyang, Yun Jinwei, Liu Chenchong, Sha Zhaoqun, Su Juan, Hou Junrong for translating and correcting Chapters 0, 1, 2, 3, 4, 5 and the

Appendix, respectively. The preparation process of this book was encouraged, helped and supported by the Xi'an University of Post & Telecommunication and Xidian University colleagues. The science publishers were strong support in the publication of this book; Pan Sisi and other editors dedicated a lot of energy to the book and the authors wish to express their heartfelt appreciation of this.

Limited to the knowledge of editors, the book fallacies and omissions are inevitable. Readers are encouraged to offer criticism and suggest corrections.

The Editors
2017. 08

Contents

0 Introduction

0.1 Probability space

In the probabilistic part of engineering mathematics, probabilities have been defined for both classical and geometric profiles. In the classical model, the possible results of the test are limited and have the same probability; for the geometric generalization, although the possible outcome of the experiment is infinite, it is still required to have some of the same probability. However, a large number of random test results in practical problems do not belong to these two types, so it is necessary to give a definite probability definition to the general stochastic phenomenon. In 1933, the Russian mathematician Kolmogorov combined his predecessors' research results, gave the axiomatic system of probability theory, and defined the basic concepts of events and probabilities, and probability theory became a rigorous branch of mathematics.

0.1.1 Randomized trials

In probability theory, a random test is a test with randomness under given conditions; E is generally used to represent randomized trails. Several random trials are presented below.

E_1: toss a coin, observe the positive H and the negative T appears.
E_2: throw a die, observe the points that appear.
E_3: a point is arbitrarily thrown on the $(0, 1)$ interval of the real axis.
E_4: pick one out of a batch of bulbs and test its lifespan.

The above examples of several experiments have common characteristics. For example, the test E_1 has two possible results, H or T, but we do not know H or T before throwing. This test can be repeated under the same conditions. There are six possible outcomes for the test E_2, but it is not possible to determine which outcome will occur before throwing the die, and this test can be repeated under the same conditions. Another example is the test E_4; we know the lamp life (in hours) $t \geq 0$ but cannot determine how long its life is before the test. This test can also be repeated under the same conditions. To sum up, these tests have the following characteristics:

(1) They can be repeated under the same conditions.
(2) There is more than one possible outcome of each trial, and all possible results of the test can be identified in advance.
(3) No outcome can be determined until each trial.

In probability theory, the experiment with these three characteristics is called a random experiment.

https://doi.org/10.1515/9783110593808-001

0.1.2 Sample space

For randomized trials, although the results of the tests cannot be predicted before each test, the set of all possible outcomes of the test is known. We refer to the set of all possible outcomes of stochastic test E as the sample space of random test E, and each possible test result is called a basic event, showing that the sample space consists of all the basic events of stochastic test E.

For example, in the random test E_1, "positive H" and "negative T" are the basic events. These two basic events constitute a sample space.

In the random experiment E_2, the respective points "1", "2", "3", "4", "5" and "6" are the basic events. These six basic events form a sample space.

In the random experiment E_3, each point in the (0, 1) interval is a basic event, and the set of all points (i.e., the (0, 1) interval) constitutes a sample space.

Abstractly, the sample space is a collection of points, each of which is called a sample point. The sample space is denoted by $\Omega = \{\omega\}$, where ω represents the sample point, as defined below.

Definition 0.1.1. Set the sample space $\Omega = \{\omega\}$, a set of some subsets specified $\mathscr{F}$, if $\mathscr{F}$ satisfy the following properties:
(1) $\Omega \in \mathscr{F}$.
(2) if $A \in \mathscr{F}$, then $\overline{A} = \Omega - A \in \mathscr{F}$.
(3) if $A_k \in \mathscr{F}$, $k = 1, 2, \ldots$, then $\bigcup_{k=1}^{\infty} A_k \in \mathscr{F}$.

That said, $\mathscr{F}$ is a Boral event domain or a σ event domain. A subset of each sample space Ω in the Boral event domain is called an event.

In particular, the sample Ω is called a certain event, and empty $\varnothing$ is called an impossible event.

In the example of the three sample spaces above, each sample point is a basic event. However, generally it is not required that sample points be basic events.

0.1.3 Probability space

The statistical definition of probability and the classical probability definition have been mentioned in probability theory. The following describes the axiomatic definition of probability. This definition is abstracted from the specific probability definition described above, while retaining some of the characteristics of the specific probability definition. The probability of an event is a number corresponding to a subset of Ω in Borel field F, which can be considered as the aggregation function.

Definition 0.1.2 (the axiomatic definition of probability). Set a set function in the Borel field $\mathscr{F}$ of the sample space Ω. If $P(A)$ satisfies the following conditions:
(1) non negativity: $\forall A \in \mathscr{F}$, we have $P(A) \geq 0$;
(2) polarity: $P(\Omega) = 1$;

(3) countable additivity: if the $\forall n = 1, 2, \ldots, A_n \in \mathscr{F}$ and on $\forall i \neq j$, $A_iA_j = \varnothing$, then $P(\bigcup_{n=1}^{\infty} A_n) = \sum_{n=1}^{\infty} P(A_n)$.

P is defined as the probability on the Borel event field, called $P(A)$ as the probability of events.

So far, we have introduced three basic concepts in probability theory: sample space Ω, event domain $\mathscr{F}$ and probability P. They are the three basic components of a random experiment. For the random trial E, the sample space Ω gives all of its test results; $\mathscr{F}$ gives a variety of events composed of these possible results, while P gives the probabilities of each event. Combining the three, we call the ordered three elements $(\Omega, \mathscr{F}, P)$ a probability space.

0.2 Conditional probability space

0.2.1 Conditional probability

Conditional probability is an important and practical concept in probability theory. Because in many practical problems, in addition to the probability that the event A occurs, it is sometimes necessary to find the probability that the event A will occur in the condition that the event B has occurred, which is the conditional probability, denoted as $P(A|B)$. Let us look at an example.

Example 0.2.1. Toss a coin twice and observe both the occurrences. Set B events as "at least once for the positive," event A is "two throws with one side." Now, find the conditional probability of event A under the condition that "event B has occurred."

Solution: It is easy to see that the sample space $\Omega = \{HH, HT, TH, TT\}$ contains four sample points, $B = \{HH, HT, TH\}$, $A = \{HH, TT\}$; the known event B has occurred, and all that is known to test that the possible results of collection should be B, and there are three elements in B, of which only $HH \in A$. Therefore, we can know from the same possibility that $P(A|B) = 1/3$.

In addition, it is easy to know that $P(A) = 2/4$; apparently it does not equal $P(A|B)$.

Besides, it is also found that $P(AB) = 1/4$, $P(B) = 3/4$, which is $P(A|B)$ exactly equal to the $P(AB)$ and $P(B)$ ratio.

The general definition of conditional probability is given below.

Definition 0.2.1. Set $A, B \in \mathscr{F}$ and $P(A) > 0$ and remember that

$$P(B|A) = \frac{P(AB)}{P(A)} \tag{0.1}$$

As under the condition of a known event A occurs, the conditional probability of event B occurs.

Remember that $P_A(B) \triangleq P(B|A)$, call $(\Omega, \mathscr{F}, P_A)$ the conditional probability space of the given event A, referred to as conditional probability space.

0.2.2 Multiplication formula

From the above eq. (0.1), we have

$$P(AB) = P(A)P(B|A)\,, \quad P(A) > 0 \tag{0.2}$$

Call this the multiplication formula of conditional probability.

If $P(B) > 0$, similar to the above, we can define the conditional probability of the occurrence of even A under known B occurrence conditions,

$$P(A|B) = \frac{P(AB)}{P(B)}\,, \quad P(B) > 0 \tag{0.3}$$

and the corresponding multiplication formula

$$P(AB) = P(B)P(A|B)\,, \quad P(B) > 0 \tag{0.4}$$

The multiplication formula can be extended to the case of n events.

The general multiplication formula is: set $A_i \in \mathscr{F}, i = 1, 2, \dots, n$, and $P(A_1A_2 \dots A_{n-1}) > 0$ and then

$$P(A_1A_2 \dots A_n) = P(A_1)P(A_2|A_1)P(A_3|A_1A_2) \dots P(A_n|A_1A_2 \dots A_{n-1}) \tag{0.5}$$

0.2.3 Total probability formula

The formula of total probability is an important formula for calculating probability. Before introducing the formula of total probability, the concept of the division of the sample space is introduced.

Definition 0.2.2. Set Ω as the sample space of the test E; $B_1, B_2, \dots, B_n$ are a set of events of E. If

(1) $B_iB_j = \varnothing, i \neq j, i, j = 1, 2, \dots, n$;

(2) $B_1 \cup B_2 \cup \dots \cup B_n = \Omega$.

It is said that $B_1, B_2, \dots, B_n$ is a division of the sample space Ω.

If $B_1, B_2, \dots, B_n$ is a division of the sample space, then, for each test, there must be one and only one occurrence in the events of $B_1, B_2, \dots, B_n$.

For example, set the test E to "throw a die and observe its points." Its sample space is $\Omega = \{1, 2, 3, 4, 5, 6\}$. A set of events of E is $B_1 = \{1, 2, 3\}$, $B_2 = \{4, 5\}$, $B_3 = \{6\}$, which is a division of Ω; an event group $C_1 = \{1, 2, 3\}$, $C_2 = \{3, 4\}$, $C_3 = \{5, 6\}$ is not a division of Ω.

Theorem 0.2.1. *Set the sample space of test E as Ω, B_i for Ω a partition, $B_i \in \mathscr{F}$ and $P(B_i) > 0$, $i = 1, 2, \ldots, n$. Set $A \in \mathscr{F}$, and then*

$$P(A) = \sum_{i=1}^{n} P(B_i)P(A|B_i) \tag{0.6}$$

Which is called the full probability formula.

In many practical problems, $P(A)$ is not easy to obtain straightforwardly, but it is easy to find a partition $B_1, B_2, \ldots, B_n$ of Ω, and $P(B_i)$ with $P(A|B_i)$ either known or easy to obtain, and then we can get $P(A)$ based on formula (0.6).

0.2.4 The Bayesian formula

The Bayesian formula can be obtained by the definition of conditional probability, the multiplication formula and the full probability formula.

Theorem 0.2.2. *Set the sample space of the test E as Ω, B_i for Ω a partition, $B_i \in \mathscr{F}$ and $P(B_i) > 0$, $i = 1, 2, \ldots, n$. Set up $A \in \mathscr{F}$ and $P(A) > 0$, and then*

$$P(B_i|A) = \frac{P(A|B_i)P(B_i)}{\sum_{j=1}^{n} P(A|B_j)P(B_j)}, \quad i = 1, 2, \ldots, n \tag{0.7}$$

which is called the Bayesian formula.

In eqs. (0.6) and (0.7) take $n = 2$ and set B_1 as B; then B_2 is $\overline{B}$, and the total probability formula and the Bayesian formula are, respectively,

$$P(A) = P(A|B)P(B) + P(A|\overline{B})P(\overline{B}) \tag{0.8}$$

$$P(B|A) = \frac{P(AB)}{P(A)} = \frac{P(A|B)P(B)}{P(A|B)P(B) + P(A|\overline{B})P(\overline{B})} \tag{0.9}$$

These two formulas are commonly used.

Example 0.2.2. Analysis of previous data shows that when the machine is well adjusted, the product pass rate is 98% and when the machine has some kind of fault, the passing rate is 55%. Every morning when the machine starts, the probability of the machine's adjustment is 95%. What is the probability of a good machine adjustment when the first product is known to be a qualified product on some morning?

Solution: Set the event A as "qualified products," B for the event "the machine adjustment is good." Through the question, we have $P(A|B) = 0.98$, $P(A|\overline{B}) = 0.55$, $P(B) = 0.95$, and $P(\overline{B}) = 0.05$. The required probability is $P(B|A)$. By the Bayesian formula we have

$$P(B|A) = \frac{P(A|B)P(B)}{P(A|B)P(B) + P(A|\overline{B})P(\overline{B})} = \frac{0.98 \times 0.95}{0.98 \times 0.95 + 0.55 \times 0.05} = 0.97$$

This means that the first production is a qualified product; the probability of good machine adjustment is 0.97. Here, the probability of 0.95 is obtained from the previous data analysis, called the prior probability. The probability of re-correction (i.e., 0.97) after the information (i.e., the first product produced is the qualifying product) is called the posterior probability. With the posterior probability, we will be able to have a better understanding of the situation of the machine.

0.3 Random variables

0.3.1 The concept of random variables

Another important concept in probability theory is the concept of random variables. A random variable is the size of a different value that can be obtained in the result of the experiment; the value depending on the test results, because as the result of the test is random, so its value is random. Random variables are defined as follows.

Theorem 0.3.1. *Given the probability space $(\Omega, \mathscr{F}, P)$. Set $X = X(\omega)$ as a function that defines the domain as Ω. If for any real number x, the subset of Ω is*

$$\{\omega : X(\omega) \leq x\} \in \mathscr{F}$$

then, call $X(\omega)$ a random variable, abbreviated as X.

When throw a coin as an example, $\Omega = \{\omega\} = \{\text{head, tail}\}$ is the sample space of this randomized trial, the regulation function of the $X(\omega)$ value: $X\{\text{head}\} = 1$, $X\{\text{tail}\} = 0$. This way, $X(\omega)$ is a random variable.

Taking the die as an example, $\Omega = \{1 \text{ point}, 2 \text{ points}, 3 \text{ points}, 4 \text{ points}, 5 \text{ points}, 6 \text{ points}\}$. The prescribed function $X(k \text{ points}) = k$, $k = 1, 2, 3, 4, 5, 6$, is $X(\omega) = \omega$ and then $X(\omega)$ is a random variable.

According to the possible values of a random variable, they can be divided into two basic types, namely, discrete random variables and continuous random variables.

Discrete random variables may only be obtained by finite or countable infinite numbers, i.e., sets of such numbers; all of these can be arranged in a certain order, which are thus expressed as sequences $x_1, x_2, \ldots, x_n, \ldots$. In engineering, it is often possible to obtain only random variables with non-negative integer values. For example, the number of defective products in a batch of products, the number of calls made by telephone users over a given period of time, etc.

Continuous random variables can take any value within a range. For example, the size of lathe machined parts and the specified size of the deviation, the shooting hit point and the target center deviation.

0.3.2 Discrete random variables

The probability distribution of discrete random variables is usually described by the probability distribution table (also known as the distribution law). All possible values obtained by setting a discrete random variable X are $x_1, x_2, \ldots, x_n, \ldots$ and the probabilities of obtaining these values are $p(x_1), p(x_2), \ldots, p(x_n), \ldots$, abbreviated as $p_1, p_2, \ldots, p_n, \ldots$; the probability distribution is shown in Tab. 0.1.

Tab. 0.1: Probability distribution table.

X	x_1	x_2	$\ldots$	x_n	$\ldots$
P	p_1	p_2	$\ldots$	p_n	$\ldots$

Usually the function

$$p(x_i) = P(X = x_i)\,, \quad i = 1, 2, \ldots, n, \ldots$$

Is called the probability function of discrete random variables X. The probability function has the following properties.

(1) The probability function is a non-negative function

$$p(x_i) \geq 0\,, \quad i = 1, 2, \ldots, n, \ldots$$

(2) The sum of the probability that all possible values obtained by the random variable X is equal to 1

$$\sum_{i-1}^{n} p(x_i) = 1 \tag{0.10}$$

If the random variable may yield countless infinitely many values, then eq. (0.10) becomes

$$\sum_{i=1}^{\infty} p(x_i) = 1 \tag{0.11}$$

Sometimes the distribution of probabilities can be expressed directly in terms of a series of equations,

$$p(x_i) = P(X = x_i)\,, \quad i = 1, 2, \ldots, n, \ldots \tag{0.12}$$

The above is called the probability distribution of X.

0.3.3 Continuous random variables

A continuous random variable is characterized by the fact that it may take all the values in a range, for example, the distance between the bombardment point and the target can be any one of the values $[0, +\infty)$. For a continuous random variable, it is impossible and meaningless to enumerate all its values and their corresponding probabilities. Normally, only the probability of the occurrence of event "$a < X \leq b$" is considered for continuous random variables. The concept of the distribution function of random variables is introduced for this purpose.

Theorem 0.3.2. *Set a random variable X, for any real number $x \in (-\infty, +\infty)$, calling $F(x) = P\{X \leq x\}$ the distribution function for random variables X.*

Considering the probability of the occurrence of the above events "$a < X \leq b$," there is

$$P\{a < X \leq b\} = P\{X \leq b\} - P\{X \leq a\} = F(b) - F(a)$$

therefore, if the distribution function $F(x)$ is known, we will know the probability that X falls into any interval $(a, b]$ and in this sense that the distribution function completely describes the statistical regularity of random variables.

It is through the distribution function that we can use the method of mathematical analysis to study random variables. If X is regarded as a random point on the coordinate axis, then distribution function $F(x)$ at x represents the probability that X falls on the interval $(-\infty, x]$.

Theorem 0.3.3. *For the random variable X, the distribution function F(x) has a non-negative function f(x), so that for any real number x there is*

$$F(x) = \int_{-\infty}^{x} f(t)\,\mathrm{d}t \tag{0.13}$$

X is called a continuous random variable, in which the function f(x) is called the probability density function of X, referred to as the probability density. By definition, the probability density f(x) has the following properties.

(1) *Non-negativity:* $f(x) \geq 0$.
(2) *Normative:* $\int_{-\infty}^{+\infty} f(x)\,\mathrm{d}x = 1$.
(3) *For any real number* $x_1, x_2 (x_1 \leq x_2)$,

$$P\{x_1 < X \leq x_2\} = F(x_2) - F(x_1) = \int_{x_1}^{x_2} f(x)\,\mathrm{d}x\,.$$

(4) *If $f(x)$ is continuous in point x, there are $F'(x) = f(x)$.*

Example 0.3.1. Set the probability density function of continuous random variable X as

$$f(x) = A\mathrm{e}^{-|x|}\,, \quad -\infty < x < +\infty$$

Find the following:

(1) the coefficient of A;

(2) $P\{x \in (0, 1)\}$;
(3) the distribution function $F(x)$ of X.

Solution: (1) By the basic properties of the probability density $\int_{-\infty}^{+\infty} f(x)\,dx = 1$, we obtain

$$A\left[\int_{-\infty}^{0} e^x\,dx + \int_{0}^{+\infty} e^{-x}\,dx\right] = 1$$

and then

$$A = \frac{1}{2}$$

(2) $P\{x \in (0, 1)\} = P\{0 < x < 1\} = \int_0^1 1/2e^{-x}\,dx = 1/2(1 - 1/e)$
(3) $F(x) = \int_{-\infty}^{x} f(t)\,dt$.
When $x < 0$,

$$F(x) = \frac{1}{2}\int_{-\infty}^{x} e^t\,dt = \frac{1}{2}e^x$$

When $x \geq 0$,

$$F(x) = \frac{1}{2}\left[\int_{-\infty}^{0} e^t\,dt + \int_{0}^{x} e^{-t}\,dt\right] = 1 - \frac{1}{2}e^{-x}$$

thus,

$$F(x) = \begin{cases} \frac{1}{2}e^x, & x < 0 \\ 1 - \frac{1}{2}e^{-x}, & x \geq 0. \end{cases}$$

0.3.4 Multidimensional random variables

What we discussed earlier is only the case of a single random variable, but in practice we often encounter several random variables at the same time in order to have a better description of an experiment or phenomenon. For example, the position of the explosion point of the shell on the ground is composed of a pair of random variables (X, Y). We call the overall $X = (X_1, X_2, \ldots, X_n)$ of n random variables $X_1, X_2, \ldots, X_n$ an n-dimensional random variable or an n-dimensional random vector. Because there is no principal difference between two-dimensional and n-dimensional, for simplicity and ease of understanding, we focus on the case of two-dimensional random variables.

0.3.4.1 The joint distribution function

Definition 0.3.1. Set X and Y as the two random variables defined in the same probability space $(\Omega, \mathscr{F}, P)$; then (X, Y) are called two-dimensional random variables, for any $x, y \in R$, of order

$$F(x, y) = P\{X \leq x, Y \leq y\}$$

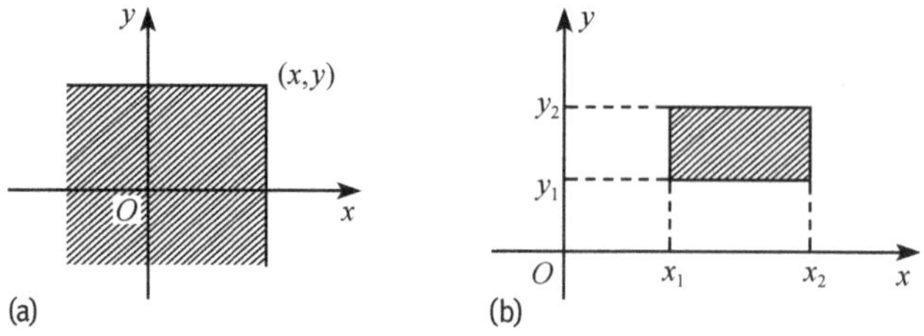

Fig. 0.1: The distribution function of two random variables.

$F(x, y)$ is called the joint distribution function or the two-dimensional distribution function of (X, Y).

If the two-dimensional random variable (X, Y) is regarded as the coordinate of the random point on the plane, then the function value of the distribution function $F(x, y)$ at (x, y) is the random point (X, Y) falling into the area shown in Fig. 0.1 (a), the probability of the infinite rectangular area with the point as the vertex (x, y).

Then, the probability that is the random point (X, Y) falling into the rectangular region shown in Fig. 0.1 (b) can be given by the distribution function, which probability is

$$P[x_1 < X \le x_2, y_1 < Y \le y_2] = F(x_2, y_2) - F(x_2, y_1) - F(x_1, y_2) + F(x_1, y_1)$$

The distribution function $F(x, y)$ has the following basic properties:

(1) $0 \le F(x, y) \le 1$;

(2) $F(x, y)$, both x and y are monotonous nondecreasing and right continuous;

(3)

$$F(-\infty, y) = \lim_{x \to -\infty} F(x, y) = 0$$

$$F(x, -\infty) = \lim_{y \to -\infty} F(x, y) = 0$$

$$F(-\infty, -\infty) = \lim_{(x,y) \to (-\infty,-\infty)} F(x, y) = 0$$

$$F(+\infty, +\infty) = \lim_{(x,y) \to (+\infty,+\infty)} F(x, y) = 1$$

In addition, if the distribution functions $F(x, y)$ and (X, Y) are known, then $F(x, y)$ can be used to derive the respective distribution functions $F_X(x)$ and $F_Y(y)$ of X and Y,

$$F_X(x) = P\{X \le x\} = P\{X \le x, Y \le +\infty\} = F(x, +\infty) \tag{0.14}$$

In the same way,

$$F_Y(y) = P\{Y \le y\} = P\{X \le +\infty, Y \le y\} = F(+\infty, y) \tag{0.15}$$

usually called $F_X(x)$ and $F_Y(y)$, are the marginal distribution functions of the joint distribution function $F(x, y)$.

0.3.4.2 The joint probability distribution of discrete random variables

Definition 0.3.2. When the random variables X and Y can only take finite or countable values, they are called (X, Y) two-dimensional discrete random variables (vectors). Setting (X, Y), all possible values are (x_i, y_j), $i, j = 1, 2, \ldots$, if

$$P\{X = x_i, Y = y_j\} = p_{ij}\,, \quad i, j = 1, 2, \ldots \tag{0.16}$$

are known.

Then (0.16) is called the probability distribution of the two-dimensional discrete random variables (X, Y), or the joint probability distribution of X and Y. It is easy to see that p_{ij} satisfies the following properties:
(1) $p_{ij} \geq 0$, $i, j = 1, 2, \ldots$.
(2) $\sum_i \sum_j p_{ij} = 1$.

For the sake of intuition, the joint probability distribution is sometimes expressed in table form (Tab. 0.2), which is called joint probability distribution table.

By the joint probability distribution of X and Y, we can calculate the probability distribution of X and Y, respectively, p_i^X, $i = 1, 2, \ldots$; p_j^Y, $j = 1, 2, \ldots$.

$$p_i^X = P\{X = x_i\} = \sum_j P\{X = x_i, Y = y_j\} = \sum_j p_{ij}\,, \quad i = 1, 2, \ldots \tag{0.17}$$

$$p_j^Y = P\{Y = y_j\} = \sum_i P\{X = x_i, Y = y_j\} = \sum_i p_{ij}\,, \quad j = 1, 2, \ldots \tag{0.18}$$

Usually, p_i^X and p_j^Y are called the marginal probability distributions of joint probability distributions $P\{X = x_i, Y = y_j\} = p_{ij}$, $i, j = 1, 2, \ldots$. In Tab. 0.2, the edge distributions are listed in the last line and the last columns, respectively. They are equal to the sum of the row and column of the joint probability distribution, respectively.

For discrete random variables, the joint probability distribution is not only more intuitive than the joint distribution function, but it can be more easily determined by

Tab. 0.2: Joint probability distribution.

Y / X	y_1	y_2	$\cdots$	y_j	$\cdots$	$P\{X = x_i\}$
x_1	p_{11}	p_{21}	$\cdots$	p_{i1}	$\cdots$	$\sum_j p_{1j}$
x_2	p_{12}	p_{22}	$\cdots$	p_{i2}	$\cdots$	$\sum_j p_{2j}$
$\vdots$	$\vdots$	$\vdots$		$\vdots$		$\vdots$
x_i	p_{1j}	p_{2j}	$\cdots$	p_{ij}	$\cdots$	$\sum_j p_{ij}$
$\vdots$	$\vdots$	$\vdots$		$\vdots$		$\vdots$
$P\{Y = y_j\}$	$\sum_i p_{i1}$	$\sum_i p_{i2}$	$\cdots$	$\sum_i p_{ij}$	$\cdots$	

the probability of (X, Y) on any region D, in fact,

$$P\{(X, Y) \in D\} = \sum_{(x_i, y_j) \in D} p_{ij} \tag{0.19}$$

In particular, the joint distribution function can be determined by the joint probability distribution:

$$F(x, y) = P\{X \le x, Y \le y\} = \sum_{x_i \le x, y_j \le y} p_{ij} \tag{0.20}$$

0.3.4.3 Joint probability density of continuous random variables

Definition 0.3.3. For the distribution function $F(x, y)$ of the two-dimensional random variable (X, Y), if there is a non-negative function $f(x, y)$, so for any real number x, y has

$$F(x, y) = \int_{-\infty}^{y} \int_{-\infty}^{x} f(u, v)\,du\,dv \tag{0.21}$$

(X, Y) is called a two-dimensional continuous random variables (vector), and the function $f(x, y)$ is called the probability density of the two-dimensional random variable (X, Y), or the joint probability density of random variables X and Y.

According to the definition, the probability density function $f(x, y)$ has the following properties:
(1) $f(x, y) \ge 0$;
(2) $\int_{-\infty}^{+\infty} \int_{-\infty}^{+\infty} f(x, y)\,dx\,dy = 1$;
(3) If D is an area in the plane, then

$$P\{(X, Y) \in D\} = \iint_D f(x, y)\,dx\,dy$$

(4) If $f(x, y)$ is continuous at point (x, y), we have

$$\frac{\partial^2 F(x, y)}{\partial x \partial y} = f(x, y)$$

In particular, the marginal distribution function $F_X(x)$ can be represented as

$$F_X(x) = P\{X \le x\} = P\{X \le x, Y \le +\infty\} = \int_{-\infty}^{x} \int_{-\infty}^{+\infty} f(u, v)\,du\,dv = \int_{-\infty}^{x} \left[\int_{-\infty}^{+\infty} f(u, v)\,dv \right] du$$

From the above formula, we can see that X is a continuous random variable, and the density function is as follows:

$$f_X(x) = \int_{-\infty}^{+\infty} f(x, y)\,dy \tag{0.22}$$

Similarly, if Y is a continuous random variable, the density function is

$$f_Y(y) = \int_{-\infty}^{+\infty} f(x, y)\,\mathrm{d}x \tag{0.23}$$

$f_X(x)$ and $f_Y(y)$ are usually called the marginal density functions of (X, Y) or joint density functions $f(x, y)$.

0.3.4.4 Independence of multidimensional random variables

Definition 0.3.4. Set the joint distribution function of the random variables X and Y as $F(x, y)$; the marginal distribution functions are $F_X(x)$ and $F_Y(y)$, respectively. If for any real number x and y, there is constantly

$$F(x, y) = F_X(x)F_Y(y) \tag{0.24}$$

then the random variables X and Y are independent of each other.

For discrete random variables, if the following relations is true for all i, j, namely:

$$P\{X = x_i, Y = y_j\} = P\{X = x_i\}P\{Y = y_j\} \tag{0.25}$$

then X and Y are independent random variables.

If (X, Y) is a continuous random variable, the density functions $f(x, y)$, $f_X(x)$ and $f_Y(y)$ are continuous and at the point (x, y), we have

$$f(x, y) = f_X(x)f_Y(y) \tag{0.26}$$

And then X and Y are independent random variables. The above definition can be further extended to n dimensional conditions.

There are n random variables $X_1, X_2, \ldots, X_n$, which joint distribution function is $F(x_1, x_2, \ldots, x_n)$ and the marginal distribution function is $F_i(x_i)$, $i = 1, 2, \ldots, n$, if for any real numbers $x_1, x_2, \ldots, x_n$, there are constantly

$$F(x_1, x_2, \ldots, x_n) = F_1(x_1)F_2(x_2)\ldots F_n(x_n) \tag{0.27}$$

So that $X_1, X_2, \ldots, X_n$ are independent of each other.

0.4 Distribution of random variable functions

Let $g(x)$ be a function defined on the set of all possible values x of the random variable X. The function of the so-called random variable X refers to a random variable Y that takes: $y = g(x)$, whenever the variable X is the value x. Remember that

$$Y = g(X)$$

In practical problems, we often need to discuss the distribution of random variable functions. This is because, in some experiments, the random variables we care about cannot be obtained by direct observation, but it is another function of a random variable that can be directly observed. Therefore, it is necessary to discuss how to find the distribution of the random variable function $Y = g(X)$ based on the distribution of known random variables X.

0.4.1 Distribution of discrete random variable functions

In $Y = g(X)$, if X is a discrete random variable, then Y is also a discrete random variable which distribution law can be obtained by the distribution law of X.

Example 0.4.1. Set the distribution law of X, as shown in Tab. 0.3.

Tab. 0.3: X distribution law.

X	−2	−1	0	1	2	3
P	0.1	0.15	0.2	0.25	0.21	0.09

Find the distribution law of $Y = X^2$.

Solution: When $X = -2$ and $X = 2$, we have $Y = X^2 = 4$. According to the additive property of probability:

$$P\{Y = 4\} = P\{X = -2\} + P\{X = 2\} = 0.1 + 0.21 = 0.31$$

we similarly have

$$P\{Y = 1\} = P\{X = -1\} + P\{X = 1\} = 0.15 + 0.25 = 0.4$$

The distribution law of $Y = X^2$ is shown in Tab. 0.4.

Tab. 0.4: Y distribution law.

Y	0	1	4	9
P	0.2	0.4	0.31	0.09

0.4.2 Distribution of continuous random variable functions

In $Y = g(X)$, if X is a continuous random variable, then Y is a continuous random variable, and the distribution of the function of continuous random variables has the following theorem.

Theorem 0.4.1. *The continuous random variable X is given by its probability density function, which is $f_X(x)$, the function $y = g(x)$ is strictly monotonic, and its inverse function $g^{-1}(y)$ is also a continuous random variable, and then $Y = g(X)$ is also a continuous random variable. Its probability density is*

$$f_Y(y) = \begin{cases} f_X[g^{-1}(y)]|[g^{-1}(y)]'|, & \alpha < y < \beta \\ 0, & \textit{otherwise} \end{cases} \tag{0.28}$$

and $\alpha = \min[g(-\infty), g(+\infty)]$, $\beta = \max[g(-\infty), g(+\infty)]$.

Example 0.4.2. Let a random variable be $X \sim N(\mu, \sigma^2)$, find the probability density function of $Y = aX + b$ (a, b are constants and $a \neq 0$).

Solution: The probability density of X is

$$f_X(x) = \frac{1}{\sqrt{2\pi}\sigma} e^{-\frac{(x-\mu)^2}{2\sigma^2}}, \quad -\infty < x < +\infty$$

From $y = g(x) = ax + b$, the inverse function is

$$x = g^{-1}(y) = \frac{y-b}{a}$$

Then

$$[g^{-1}(y)]' = \frac{1}{a}$$

The theorem shows that $Y = aX + b$ for probability density is

$$f_Y(y) = \frac{1}{|a|} f_X\left(\frac{y-b}{a}\right), \quad -\infty < y < +\infty$$

Namely,

$$f_Y(y) = \frac{1}{|a|} \frac{1}{\sqrt{2\pi}\sigma} e^{-\frac{\left(\frac{y-b}{a}-\mu\right)^2}{2\sigma^2}} = \frac{1}{\sqrt{2\pi}|a|\sigma} e^{-\frac{[y-(a\mu+b)]^2}{2(a\sigma)^2}}, \quad -\infty < y < +\infty$$

From the results of this example, we can see that the linear function $Y = aX+b$ of the normal random variable X is still a normal random variable, $Y \sim N(a\mu + b, (a\sigma)^2)$.

0.5 Numerical characteristics of random variables

Although the distribution function can describe the statistical properties of random variables completely, sometimes the distribution function of the random variable is not easy to obtain; in fact, in some practical problems, changes do not need to go to a comprehensive study of random variables, we only need to know some of the characteristics of random variables. For example, when assessing the level of grain yield in a particular area, on many occasions the average yield in the area is known. As a group, checking the quality of cotton, on the one hand we need to pay attention to the average fiber length, on the other hand, we need to pay attention to the degree of deviation from the fiber length and the average length; when the average length is larger and there is less deviation, the quality is better. As can be seen from the above examples, some values related to random variables cannot describe the random variables completely, but they can describe some important characteristics of random variables in some respects. These digital characteristics are of great significance, both in theory and in practice.

0.5.1 Mathematical expectations

Definition 0.5.1. The distribution law of discrete random variables is established as

$$P\{X = x_k\} = p_k\,, \quad k = 1, 2, \ldots$$

If the series

$$\sum_{k=1}^{\infty} x_k p_k$$

there is absolute convergence, and we then call the sum of the series the mathematical expectation of the random variable X, denoted as $E(X)$. That is,

$$E(X) = \sum_{k=1}^{\infty} x_k p_k \tag{0.29}$$

Let the probability density of continuous random variable X be $f(x)$. If integral

$$\int_{-\infty}^{\infty} x f(x)\,\mathrm{d}x$$

is the absolute convergence, then the value of the integral is called the mathematical expectation of the random variable X, denoted as $E(X)$. That is,

$$E(X) = \int_{-\infty}^{\infty} x f(x)\,\mathrm{d}x \tag{0.30}$$

The mathematical expectation is also called a mean value. Intuitively, the mathematical expectation of a random variable reflects the average value of random variables.

The mathematical expectation has the following properties:
(1) $E(c) = c$ (c is a constant).
(2) $E(kX) = kE(X)$ (k is a constant).
(3) $E(X \pm Y) = E(X) \pm E(Y)$.
(4) When X and Y are independent of each other, $E(XY) = E(X)E(Y)$.

For n-dimensional random variables $(X_1, X_2, \ldots, X_n)$, we call $(EX_1, EX_2, \ldots, EX_n)$ the mathematical expectation vector (or mean vector) of $(X_1, X_2, \ldots, X_n)$.

Sometimes, we also need the mathematical expectation of the function of the random variable. For example, the pressure W of the airplane wing is the function $W = kV^2$ ($k > 0$ constant) of the wind speed V, and the mathematical expectation of W can be obtained by the following theorem.

Theorem 0.5.1. *Let Y be a function of the random variable X*

$$Y = g(X)$$

Let X be a discrete random variable, which distribution law is $P\{X = x_k\} = p_k$, $k = 1, 2, \ldots$. If $\sum_{k=1}^{\infty} g(x_k)p_k$ is absolutely convergent, then we have

$$E(Y) = E[g(X)] = \sum_{k=1}^{\infty} g(x_k)p_k \tag{0.31}$$

Let X be a continuous random variable whose probability is $f(x)$. If $\int_{-\infty}^{\infty} g(x)f(x)\,dx$ converges absolutely, then we have

$$E(Y) = E[g(X)] = \int_{-\infty}^{\infty} g(x)f(x)\,dx \tag{0.32}$$

0.5.2 Variance and standard deviation

Intuitively speaking, the variance and the standard deviation of random variables reflect the average discrete degree of the random variable relative to its mathematical expectation.

Definition 0.5.2. The mathematical expectation of the function of the random variable X, $g(X) = (X - EX)^2$ is called the variance of the random variable X, which is denoted as $D(X)$, that is,

$$D(X) = E(X - EX)^2 \tag{0.33}$$

The arithmetic square root $\sqrt{D(X)}$ of the variance $D(X)$ of X is called the standard deviation (or root variance) of the random variable X.

The following formula is commonly used to calculate variance

$$D(X) = E(X^2) - (EX)^2 \tag{0.34}$$

The variance has the following properties:

(1) $D(X) \geq 0$.
(2) $D(c) = 0$ (c is a constant).
(3) $D(kX) = k^2 D(X)$ (k is a constant).
(4) When X and Y are independent of each other, $D(X \pm Y) = D(X) + D(Y)$.

0.5.3 Covariance and correlation coefficients

Intuitively, the covariance and correlation coefficients reflect the degree of tightness in the linear relation between the two random variables.

Definition 0.5.3. Given that a two-dimensional random variable (X, Y), called

$$\operatorname{cov}(X, Y) = E[(X - EX)(Y - EY)] \tag{0.35}$$

the covariance of random variables X and Y, and called

$$\rho(X, Y) = \frac{\operatorname{cov}(X, Y)}{\sqrt{D(X)D(Y)}} \tag{0.36}$$

the correlation coefficient between the random variable X and Y.

The following formula is commonly used to calculate

$$\operatorname{cov}(X, Y) = E(XY) - E(X)E(Y) \tag{0.37}$$

Covariance has the following properties:

(1) $\operatorname{cov}(X, X) = D(X)$;
(2) $\operatorname{cov}(X, Y) = \operatorname{cov}(Y, X)$;
(3) $\operatorname{cov}(X, c) = 0$ (c is constant);
(4) $\operatorname{cov}(kX, lY) = kl \operatorname{cov}(X, Y)$ (k, l are constants);
(5) $\operatorname{cov}(X, Y \pm Z) = \operatorname{cov}(X, Y) \pm \operatorname{cov}(X, Z)$.

The correlation coefficient has the following properties:

(1) $\rho(X, X) = 1$;
(2) $\rho(X, Y) = \rho(Y, X)$;
(3) $|\rho(X, Y)| \le 1$;
(4) when X and Y are independent of each other, $\rho(X, Y) = 0$; when $\rho(X, Y) = 0$, X and Y are uncorrelated;
(5) when $\rho(X, Y) = 1$, X and Y are in positive linear correlation; when $\rho(X, Y) = -1$, X and Y are in negative linear correlation.

0.5.4 The moment of random variables

Intuitively speaking, moments are the general form of numerical characteristics such as the mathematical expectation, variance and so on.

Let X be a random variable. The moment of $E(X^k)$ is the k-th-order origin moment of the random variable X, and $E(X - EX)^k$ is the k-th-order central moment of the random variable X. It is easy to see that the mathematical expectation is a first-order origin moment and the variance is the second-order central moment.

Let (X, Y) be a two-dimensional random variable; $E(X^kY^l)$ is a mixed origin moment of $k+l$ order of the random variable X and Y; $E[(X-EX)^k(Y-EY)^l]$ is the mixing central moment of the $k+l$-th order of the random variables X and Y. Obviously, the covariance $\mathrm{cov}(X, Y)$ is the second-order mixed central moment of X and Y.

From the above concept, we can see that the two-dimensional random variables (X_1, X_2) have four second-order central moments, respectively,

$$\begin{aligned} c_{11} &= E\left[(X_1 - EX_1)^2\right] \\ c_{12} &= E[(X_1 - EX_1)(X_2 - EX_2)] \\ c_{21} &= E[(X_2 - EX_2)(X_1 - EX_1)] \\ c_{22} &= E\left[(X_2 - EX_2)^2\right] \end{aligned}$$

Placing them in the matrix form:

$$\begin{bmatrix} c_{11} & c_{12} \\ c_{21} & c_{22} \end{bmatrix}$$

this matrix is called the covariance matrix of the random variables (X_1, X_2).

In fact, c_{11} is the second-order central moment of the random variable X_1, that is, the variance $D(X_1)$ of the random variable X_1; c_{22} is the variance $D(X_2)$ of the random variable X_2; c_{12} and c_{21} are the covariance of two random variables; $c_{12} = c_{21}$, so the covariance matrix is actually a symmetric matrix.

On the basis of the covariance matrices of the two-dimensional random variables, we can extend them to the case of n-dimensional random variables. The second-order mixed central moments of n-dimensional random variables $(X_1, X_2, \ldots, X_n)$ are

$$c_{ij} = \mathrm{cov}(X_i, X_j) = E[(X_i - EX_i)(X_j - EX_j)]\,, \quad i, j = 1, 2, \ldots, n$$

All of them exist and are the matrices

$$C = \begin{bmatrix} c_{11} & c_{12} & \cdots & c_{1n} \\ c_{21} & c_{22} & \cdots & c_{2n} \\ \vdots & \vdots & & \vdots \\ c_{n1} & c_{n2} & \cdots & c_{nn} \end{bmatrix} \tag{0.38}$$

being the covariance matrices of n-dimensional random variables $(X_1, X_2, \ldots, X_n)$. Since $c_{ij} = c_{ji}$, $(i \neq j, i, j = 1, 2, \ldots, n)$, the matrix is a symmetric matrix.

In general, the distribution of n-dimensional random variables is difficult to obtain, or too complex, which it is not easy to deal with in mathematics, so the covariance matrix is particularly important in practical applications.

Example 0.5.1. Let the mean vector of the two-dimensional random variable (X_1, X_2) be $(0, 1)$ and the covariance matrix

$$C = \begin{bmatrix} 1 & 0.5 \\ 0.5 & 1 \end{bmatrix}$$

Calculate:
(1) $D(2X_1 - X_2)$
(2) $E(X_1^2 - X_1X_2 + X_2^2)$

Solution: From the definition of the mean vector, $E(X_1) = 0$, $E(X_2) = 1$; by the definition of the covariance matrix, $DX_1 = DX_2 = 1$, $\text{cov}(X_1, X_2) = \text{cov}(X_2, X_1) = 0.5$. Then

(1)
$$\begin{aligned} D(2X_1 - X_2) &= D(2X_1) + DX_2 - 2\,\text{cov}(2X_1, X_2) \\ &= 4DX_1 + DX_2 - 4\,\text{cov}(X_1, X_2) = 3 \end{aligned}$$

(2) by

$$\begin{aligned} EX_1^2 &= DX_1 + (EX_1)^2 = 1 \\ EX_2^2 &= DX_2 + (EX_2)^2 = 2 \\ E(X_1X_2) &= \text{cov}(X_1, X_2) + EX_1EX_2 = 0.5 \end{aligned}$$

Therefore, $E(X_1^2 - X_1X_2 + X_2^2) = 2.5$.

Finally, we introduce the expression of the probability density of n-dimensional normal random variables that are often used in stochastic processes. Using the covariance matrix of n-dimensional random variables, this can be greatly simplified. The following starts with the probability density of the two-dimensional normal random variable and is then extended to the case where the n-dimensional normal random variables are applied.

The probability density of two-dimensional normal random variables (X_1, X_2) is

$$\begin{aligned} f(x_1, x_2) = &\frac{1}{2\pi\sigma_1\sigma_2\sqrt{1-\rho^2}} \\ &\exp\left\{\frac{-1}{2(1-\rho^2)}\left[\frac{(x_1-\mu_1)^2}{\sigma_1^2} - 2\rho\frac{(x_1-\mu_1)(x_2-\mu_2)}{\sigma_1\sigma_2} + \frac{(x_2-\mu_2)^2}{\sigma_2^2}\right]\right\} \end{aligned} \tag{0.39}$$

The part within the curly brackets is the type expressed with a matrix, so we introduce the following column matrices:

$$X = \begin{bmatrix} x_1 \\ x_2 \end{bmatrix}, \quad \mu = \begin{bmatrix} \mu_1 \\ \mu_2 \end{bmatrix}$$

The covariance matrix of (X_1, X_2):

$$C = \begin{bmatrix} c_{11} & c_{12} \\ c_{21} & c_{22} \end{bmatrix} = \begin{bmatrix} \sigma_1^2 & \rho\sigma_1\sigma_2 \\ \rho\sigma_1\sigma_2 & \sigma_2^2 \end{bmatrix}$$

The determinant is $\det C = \sigma_1^2\sigma_2^2(1-\rho^2)$, so the inverse matrix of C is

$$C^{-1} = \frac{1}{\det C}\begin{bmatrix} \sigma_2^2 & -\rho\sigma_1\sigma_2 \\ -\rho\sigma_1\sigma_2 & \sigma_1^2 \end{bmatrix}$$

Calculation shows that

$$(X-\mu)^{\mathrm{T}}C^{-1}(X-\mu)=\frac{1}{\det C}(x_1-\mu_1 x_2-\mu_2)\begin{bmatrix}\sigma_2^2 & -\rho\sigma_1\sigma_2\\ -\rho\sigma_1\sigma_2 & \sigma_1^2\end{bmatrix}\begin{bmatrix}x_1-\mu_1\\ x_2-\mu_2\end{bmatrix}$$
$$=\frac{1}{1-\rho^2}\left[\frac{(x_1-\mu_1)^2}{\sigma_1^2}-2\rho\frac{(x_1-\mu_1)(x_2-\mu_2)}{\sigma_1\sigma_2}+\frac{(x_2-\mu_2)^2}{\sigma_2^2}\right]$$

Thus, the probability of (X_1, X_2) can be written as

$$f(x_1,x_2)=\frac{1}{2\pi(\det C)^{1/2}}\exp\left\{-\frac{1}{2}(X-\mu)^{\mathrm{T}}C^{-1}(X-\mu)\right\} \tag{0.40}$$

Equation (0.40) is easily extended to the n-dimensional normal random variables $(X_1, X_2, \ldots, X_n)$.

Column matrices

$$X=\begin{bmatrix}x_1\\ x_2\\ \vdots\\ x_n\end{bmatrix},\quad \mu=\begin{bmatrix}\mu_1\\ \mu_2\\ \vdots\\ \mu_n\end{bmatrix}=\begin{bmatrix}EX_1\\ EX_2\\ \vdots\\ EX_n\end{bmatrix}$$

The probability density of the n-dimensional normal random variables $(X_1, X_2, \ldots, X_n)$ is

$$f(x_1,x_2,\ldots,x_n)=\frac{1}{(2\pi)^{n/2}(\det C)^{1/2}}\exp\left\{-\frac{1}{2}(X-\mu)^{\mathrm{T}}C^{-1}(X-\mu)\right\} \tag{0.41}$$

where C is the covariance matrix of $(X_1, X_2, \ldots, X_n)$.

Then n-dimensional normal random variables have the following four important properties.

(1) n-dimensional normal random variables $(X_1, X_2, \ldots, X_n)$ of each component of $X_i (i = 1, 2, \ldots, n)$ are normal random variables; on the contrary, if $X_1, X_2, \ldots, X_n$ are normal variables and independent, then $(X_1, X_2, \ldots, X_n)$ are n-dimensional normal random variables.

(2) The necessary and sufficient conditions for the n-dimensional random variables $(X_1, X_2, \ldots, X_n)$ obeying n-dimensional normal distribution are the arbitrary linear combination $l_1X_1 + l_2X_2 + \cdots + l_nX_n$ of $X_1, X_2, \ldots, X_n$ obeying a one-dimensional normal distribution (where $l_1, l_2, \ldots, l_n$ are not all zero).

(3) If $(X_1, X_2, \ldots, X_n)$ obeys the n-dimensional normal distribution, setting $Y_1, Y_2, \ldots, Y_k$ as the linear function of $X_i (i = 1, 2, \ldots, n)$, then $(Y_1, Y_2, \ldots, Y_k)$ also obeys the multidimensional normal distribution. This property is called the linear transformation invariance of normal variables.

(4) Setting $(X_1, X_2, \ldots, X_n)$ as obeying the n-dimensional normal distribution, the "$X_1, X_2, \ldots, X_n$ independent of each other" and "$X_1, X_2, \ldots, X_n$ not correlated to each other" are equivalent.

0.6 Characteristic functions of random variables

The characteristic functions are another important form of characterizing the distribution of random variables. It brings a lot of convenience to the study of random variables (especially multidimensional random variables). In order to define the feature function, we first introduce the concept of complex random variables.

0.6.1 Complex random variables

Definition 0.6.1. Set X and Y as the real value random variables in probability space $(\Omega, \mathscr{F}, P)$, then call

$$Z = X + \mathrm{j}Y$$

a complex random variable, where $\mathrm{j} = \sqrt{-1}$ is the imaginary unit. The distribution of Z is defined as the joint distribution function of (X, Y).

From the above definition, we can see that the study of the complex random variable $Z = X + \mathrm{j}Y$ is essentially the study of the real two-dimensional random vector (X, Y).

To extend the mathematical expectation, variance and covariance equal the moment of real random variables to complex random variables, we need the following. (1) When the variable $Y = 0$ (that is, Z is a real random variable), the moments of the complex variable Z should be equal to the moments of the real random variable X. (2) The characteristics of moments of random variables should be kept, for example, variance should be nonnegative.

(1) Definition of the mathematical expectation for the complex random variable Z:

$$m_Z = E[Z] = E[X] + \mathrm{j}E[Y] = m_X + \mathrm{j}m_Y \tag{0.42}$$

If $Y = 0$, then $m_Z = m_X$, which is in line with the aforementioned requirements.

(2) The variance of the complex random variable Z is defined as

$$\sigma_Z^2 = D[Z] = E[|\mathring{Z}|^2] \tag{0.43}$$

where $\mathring{Z} = Z - m_Z$, for the centralized complex random variable.

Because

$$\mathring{Z} = X + \mathrm{j}Y - (m_X + \mathrm{j}m_Y) = \mathring{X} + \mathrm{j}\mathring{Y}$$

then

$$D[Z] = E[\mathring{Z}^* \mathring{Z}] = E[\mathring{X}^2 + \mathring{Y}^2] = E[\mathring{X}^2] + E[\mathring{Y}^2] = D[X] + D[Y]$$

If $Y = 0$, then $D[Z] = D[X]$, which is in line with the aforementioned requirements. However, if we define $D[Z] = E[\mathring{Z}^2]$, the variance will be a complex quantity, which does not meet its characteristic requirement.

(3) If there are two complex random variables $Z_1 = X_1 + jY_1$, $Z_2 = X_2 + jY_2$, then the covariance of the complex random variables Z_1 and Z_2 is defined as

$$C_{Z_1Z_2} = E[\mathring{Z}_1^* \mathring{Z}_2] \tag{0.44}$$

where $\mathring{Z}_1^*$ is the complex conjugate of $\mathring{Z}_1$. Thus,

$$C_{Z_1Z_2} = E[(\mathring{X}_1 - j\mathring{Y}_1)(\mathring{X}_2 + j\mathring{Y}_2)] = C_{X_1X_2} + C_{Y_1Y_2} + j[C_{X_1Y_2} - C_{Y_1X_2}]$$

If $Y_1 = Y_2 = 0$, then $C_{Z_1Z_2} = C_{X_1X_2}$, which is in line with the aforementioned requirements. However, if we define $C_{Z_1Z_2} = E[\mathring{Z}_1\mathring{Z}_2]$, then when $Z_1 = Z_2 = Z$, the variance will be the complex quantity, which does not meet its characteristic requirement.

Below we introduce the noncorrelation, orthogonal and statistical independence of two complex random variables.

(1) If the covariance of complex random variables Z_1 and Z_2 is

$$C_{Z_1Z_2} = E[\mathring{Z}_1^* \mathring{Z}_2] = 0$$

the complex variables, Z_1 and Z_2 are uncorrelated.

(2) If the complex random variables Z_1 and Z_2 have

$$E[Z_1^* Z_2] = 0$$

then we say that the complex Z_1 and Z_2 are orthogonal.

(3) If the complex random variable $Z_1 = X_1 + jY_1$, $Z_2 = X_2 + jY_2$, we have

$$p(x_1, y_1; x_2, y_2) = p(x_1, y_1)p(x_2, y_2)$$

as the complex variables, and Z_1 and Z_2 are statistically independent.

0.6.2 Characteristic functions of random variables

With the aid of the concept of complex random variables, we introduce the characteristic functions of random variables.

Definition 0.6.2. Given probability spaces $(\Omega, \mathscr{F}, P)$ and random variable X, with

$$\varphi(\lambda) \triangleq E(e^{j\lambda X}), \quad -\infty < \lambda < +\infty \tag{0.45}$$

Called the characteristic function of the random variable X.

The characteristic function $\varphi(\lambda)$ is the complex function of the real variable λ, and since $|e^{j\lambda x}| = 1$, the characteristic function of the random variable must exist.

For discrete random variable X, when the probability function is $P\{X = x_k\} = p_k$ $(k = 1, 2, \ldots)$, the characteristic function of X is

$$\varphi(\lambda) = \sum_k p_k e^{j\lambda x_k} \tag{0.46}$$

For continuous random variable X, when the density function is $f(x)$, the characteristic function of X is

$$\varphi(\lambda) = \int_{-\infty}^{\infty} f(x) e^{j\lambda x}\, dx \tag{0.47}$$

For readers familiar with the Fourier transform it is not difficult to find that the introduction of the feature function is essentially the Fourier transform of the probability density function. Therefore, when the characteristic function is known, the probability density function can be obtained by inverse Fourier transform

$$f(x) = \frac{1}{2\pi} \int_{-\infty}^{\infty} \varphi(\lambda) e^{-j\lambda x}\, d\lambda \tag{0.48}$$

It can be seen that the characteristic function and the probability density function have a one-to-one correspondence, so the characteristic function is also a form of the description of the distribution of the random variables.

Example 0.6.1. Let the random variable obey the standard normal distribution and find its characteristic function.

Solution:

$$\begin{aligned}\varphi(\lambda) &= \int_{-\infty}^{\infty} f(x) e^{j\lambda x}\, dx = \int_{-\infty}^{\infty} \frac{1}{\sqrt{2\pi}} e^{-\frac{x^2}{2}} e^{j\lambda x}\, dx \\ &= \frac{1}{\sqrt{2\pi}} \int_{-\infty}^{\infty} e^{-\frac{x^2}{2}+j\lambda x}\, dx = \frac{1}{\sqrt{2\pi}} e^{-\frac{\lambda^2}{2}} \int_{-\infty}^{\infty} e^{-\frac{(x-j\lambda)^2}{2}}\, dx\end{aligned}$$

As

$$\int_{-\infty}^{\infty} e^{-\frac{(x-j\lambda)^2}{2}}\, dx = \sqrt{2\pi}$$

then

$$\varphi(\lambda) = e^{-\frac{\lambda^2}{2}}$$

0.6.3 Properties of characteristic functions

(1) The characteristic function $\varphi(\lambda)$ of the random variable X satisfies

$$\varphi(0) = 1,\ |\varphi(\lambda)| \le 1,\ \varphi(-\lambda) = \overline{\varphi(\lambda)}$$

Proof. By the characteristic function definition, we can obtain:

$$\varphi(0) = E[e^{j0X}] = E(1) = 1$$
$$|\varphi(\lambda)| = |E[e^{j\lambda X}]| \le E[|e^{j\lambda X}|] = 1$$
$$\varphi(-\lambda) = E[e^{j(-\lambda)X}] = E[e^{-j\lambda X}] = E[\overline{e^{j\lambda X}}] = \overline{E[e^{j\lambda X}]} = \overline{\varphi(\lambda)}$$ □

(2) The characteristic function of the random variable X is $\varphi_X(\lambda)$, and the characteristic function $Y = aX + b$ is

$$\varphi_Y(\lambda) = e^{jb\lambda}\varphi_X(a\lambda)$$

where a and b are constants.

Proof. $\varphi_Y(\lambda) = E[e^{j\lambda Y}] = E[e^{j\lambda(aX+b)}] = e^{jb\lambda}E[e^{ja\lambda X}] = e^{jb\lambda}\varphi_X(a\lambda)$ □

Example 0.6.2. Set X as obeying $N(\mu, \sigma^2)$ and find the characteristic function.

Solution: Setting $Y = X - \mu/\sigma$, then $Y \sim N(0, 1)$ and $X = \sigma Y + \mu$ and therefore,

$$\varphi_X(\lambda) = e^{j\mu\lambda}\varphi_Y(\sigma\lambda) = e^{j\mu\lambda}e^{-\frac{\sigma^2\lambda^2}{2}} = e^{j\mu\lambda-\frac{\sigma^2\lambda^2}{2}}$$

(3) If the random variables X and Y are independent of each other, then

$$\varphi_{X+Y}(\lambda) = \varphi_X(\lambda)\varphi_Y(\lambda)$$

That is, the characteristic function of the sum of the independent random variables is equal to the product of each characteristic function.

Proof. Since X and Y are independent of each other, the random variables $e^{j\lambda X}$ and $e^{j\lambda Y}$ are independent of each other, so,

$$\varphi_{X+Y}(\lambda) = E[e^{j\lambda(X+Y)}] = E[e^{j\lambda X}e^{j\lambda Y}] = E[e^{j\lambda X}]E[e^{j\lambda Y}] = \varphi_X(\lambda)\varphi_Y(\lambda)$$ □

0.6.4 Relationship between characteristic functions and moments

The probability density of the random variable X is $f(x)$ and its characteristic function is

$$\varphi(\lambda) = \int_{-\infty}^{\infty} f(x)e^{j\lambda x}\,dx$$

The above formula on both sides of λ can be derived to obtain

$$\frac{d\varphi(\lambda)}{d\lambda} = \int_{-\infty}^{\infty} jxf(x)e^{j\lambda x}\,dx$$

In the above formula, set $\lambda = 0$, we have

$$\left.\frac{\mathrm{d}\varphi(\lambda)}{\mathrm{d}\lambda}\right|_{\lambda=0} = \mathrm{j}\int_{-\infty}^{\infty} xf(x)\,\mathrm{d}x = \mathrm{j}E[X]$$

Further, the derivative times n, we have

$$\left.\frac{\mathrm{d}^n\varphi(\lambda)}{\mathrm{d}\lambda^n}\right|_{\lambda=0} = \mathrm{j}^n\int_{-\infty}^{\infty} x^n f(x)\,\mathrm{d}x = \mathrm{j}^n E[X^n]$$

So, we obtain

$$E[X^n] = \mathrm{j}^{-n}\left.\frac{\mathrm{d}^n\varphi(\lambda)}{\mathrm{d}\lambda^n}\right|_{\lambda=0} \tag{0.49}$$

It can be seen that the moments of the random variables can be obtained by deriving the characteristic functions without the need for complex integral operations.

0.6.5 Characteristic functions of multidimensional random variables

Definition 0.6.3. Letting $\boldsymbol{X} = (X_1, X_2, \ldots, X_n)$ be the n-dimensional random variables on the probability space $(\Omega, \mathscr{F}, P)$ and its distribution function be $F(x_1, x_2, \ldots, x_n)$, then

$$\begin{aligned}\varphi(\lambda_1, \lambda_2, \ldots, \lambda_n) &= E[\mathrm{e}^{\mathrm{j}(\lambda_1 X_1 + \lambda_2 X_2 + \cdots + \lambda_n X_n)}] \\ &= \underbrace{\int_{-\infty}^{+\infty} \cdots \int_{-\infty}^{+\infty}}_{n} \mathrm{e}^{\mathrm{j}(\lambda_1 x_1 + \lambda_2 x_2 + \cdots + \lambda_n x_n)}\,\mathrm{d}F(x_1, x_2, \ldots, x_n)\end{aligned} \tag{0.50}$$

is the characteristic function of $\boldsymbol{X}$.

Using a vector representation can obtain the compact form. Remembering that $\boldsymbol{x} = (x_1, \ldots, x_n)$ and $\boldsymbol{\lambda} = (\lambda_1, \ldots, \lambda_n)$, eq. (0.50) can be expressed as

$$\varphi(\boldsymbol{\lambda}) = E\left[\mathrm{e}^{\mathrm{j}\boldsymbol{\lambda}\boldsymbol{X}^{\mathrm{T}}}\right] = \underbrace{\int_{-\infty}^{+\infty} \cdots \int_{-\infty}^{+\infty}}_{n} \mathrm{e}^{\mathrm{j}\boldsymbol{\lambda}\boldsymbol{x}^{\mathrm{T}}}\,\mathrm{d}F(\boldsymbol{x}) \tag{0.51}$$

If $\boldsymbol{X}$ is a discrete random variable, its distribution law is

$$\begin{bmatrix} x_1 & x_2 & \ldots & x_n & \ldots \\ p_1 & p_2 & \ldots & p_n & \ldots \end{bmatrix}$$

Then the characteristic function of $\boldsymbol{X}$ is

$$\varphi(\boldsymbol{\lambda}) = \sum_k \mathrm{e}^{\mathrm{j}\boldsymbol{\lambda}\boldsymbol{x}_k^{\mathrm{T}}} p_k \tag{0.52}$$

If $\boldsymbol{X}$ is a continuous random variable, and the distribution density is $f(\boldsymbol{x})$, then the characteristic function of $\boldsymbol{X}$ is

$$\varphi(\boldsymbol{\lambda}) = \underbrace{\int_{-\infty}^{+\infty} \cdots \int_{-\infty}^{+\infty}}_{n} \mathrm{e}^{\mathrm{j}\boldsymbol{\lambda}\boldsymbol{x}^{\mathrm{T}}} f(\boldsymbol{x})\,\mathrm{d}x_1 \ldots \mathrm{d}x_n \tag{0.53}$$

The properties of the characteristic functions of n-dimensional random variables are similar to those of the characteristic functions of a one-dimensional random variable. A few more important properties are given without proof.

(1) If $E[X_1^{k_1} X_2^{k_2} \ldots X_n^{k_n}]$ exists, we have

$$E\left[X_1^{k_1} X_2^{k_2} \ldots X_n^{k_n}\right] = \mathrm{j}^{-\sum_{i=1}^{n} k_i} \left[\frac{\partial^{k_1+\cdots+k_n} \varphi(\lambda_1, \ldots, \lambda_n)}{\partial \lambda_1^{k_1} \ldots \partial \lambda_n^{k_n}}\right]_{\lambda_1=\cdots=\lambda_n=0}$$

(2) Letting a_i, b_i be constants ($i = 1, 2, \ldots, n$), then the characteristic function of the n-dimensional random variable $Y = (a_1X_1 + b_1, \ldots, a_nX_n + b_n)$ is

$$\varphi_Y(\boldsymbol{\lambda}) = \mathrm{e}^{\mathrm{j}\sum_{i=1}^{n} b_i\lambda_i} \varphi_X(a_1\lambda_1, \ldots, a_n\lambda_n)$$

(3) The characteristic functions of $(X_1, X_2, \ldots, X_n)$ is $\varphi(\lambda_1, \lambda_2, \ldots, \lambda_n)$, while the characteristic function of X_i is $\varphi_{X_i}(\lambda_i)$, and the necessary and sufficient condition that $X_1, X_2, \ldots, X_n$ are independent of each other is

$$\varphi(\lambda_1, \lambda_2, \ldots, \lambda_n) = \prod_{i=1}^{n} \varphi_{X_i}(\lambda_i)$$

Example 0.6.3. Let the random variables, X, Y be independent of each other and $X \sim N(\mu_1, \sigma_1^2)$, $Y \sim N(\mu_2, \sigma_2^2)$.

(1) Use the characteristic function of X to find $E[X^4]$.

(2) Find the characteristic function of $Z = a_1X + a_2Y$.

Solution: (1) As is known from Example 0.6.2, the characteristic function of X is

$$\varphi_X(\lambda) = \mathrm{e}^{\mathrm{j}\mu_1\lambda - \frac{\sigma_1^2\lambda^2}{2}}$$

Therefore, by property (1)

$$E[X^4] = \mathrm{j}^{-4}\varphi_X^{(4)}(0) = 3\sigma_1^4$$

(2) Since X and Y are independent of each other, the characteristic function of Z can be obtained from property (2) and (3)

$$\begin{aligned}\varphi_Z(\lambda) &= \varphi_{(X,Y)}(a_1\lambda, a_2\lambda) = \varphi_X(a_1\lambda)\varphi_Y(a_2\lambda)\\ &= \exp\left[\mathrm{j}(a_1\mu_1 + a_2\mu_2)\lambda - \frac{\left(a_1^2\sigma_1^2 + a_2^2\sigma_2^2\right)\lambda^2}{2}\right]\end{aligned}$$

0.7 Chebyshev inequality and the limit theorem

0.7.1 Chebyshev inequality

Theorem 0.7.1. *Letting random variable X have a finite second moment, that is,* $E[|X^2|] < +\infty$, *the Chebyshev inequality is as follows:*

$$P\{|X - EX| \geq \varepsilon\} \leq \frac{DX}{\varepsilon^2} \tag{0.54}$$

Where any $\varepsilon > 0$ *are true.*

Proof. If X is a discrete random variable, its distribution law is

$$P\{X = x_i\} = p_i\,, \quad i = 1, 2, \ldots$$

We have

$$P\{|X - EX| \geq \varepsilon\} = \sum_{|x_i - EX| \geq \varepsilon} P\{X = x_i\} \leq \sum_{|x_i - EX| \geq \varepsilon} \frac{(x_i - EX)^2}{\varepsilon^2} P\{X = x_i\}$$
$$\leq \frac{1}{\varepsilon^2} \sum_i (x_i - EX)^2 P\{X = x_i\} = \frac{DX}{\varepsilon^2}$$

If X is a continuous random variable, its distribution density is $f(x)$, then we have

$$P\{|X - EX| \geq \varepsilon\} = \int_{|x - EX| \geq \varepsilon} f(x)\,\mathrm{d}x \leq \int_{|x - EX| \geq \varepsilon} \frac{(x - EX)^2}{\varepsilon^2} f(x)\,\mathrm{d}x$$
$$\leq \frac{1}{\varepsilon^2} \int_{-\infty}^{+\infty} (x - EX)^2 f(x)\,\mathrm{d}x = \frac{DX}{\varepsilon^2}$$

When only the expectation and variance of random variables are known, and the probability distribution is unknown, the estimation of the probabilities of events is given by the Chebyshev inequality. □

0.7.2 Central limit theorem

In objective reality, many random variables are formed by the combined effects of a large number of independent random factors. Moreover, each of these factors plays a minor role in the overall impact, and this random variable tends to approximately obey the normal distribution. The central limit theorem provides a theoretical basis for this phenomenon.

The central limit theorem states that if there is a large number of statistically independent random variables,

$$Y = \sum_{i=1}^{n} X_i$$

where the effect of each random variable X_i on the total variable Y is small enough, then under certain conditions, $n \to \infty$ and the random variable Y obeys the normal distribution, which is unrelated to the distribution law of each random variable. In electronic technology we often meet this random phenomenon. For example, complex radar targets (aircraft, ships, etc.) can be seen as being composed of many independent scatterers, and each radar echo signal is the result of the superposition of the reflected signal. When the target moves, the distance and angle of the scatterers relative to the radar change, so the amplitude and phase of the reflected signals of every scatterer randomly change, and the echo signals superimposed by them are also randomly changed. Since the scatterers are independent of each other, each scatterer has little effect on the total echo signals. Therefore, according to the central limit theorem, it can be concluded that the instantaneous value of the echo signal of this type obeys the normal distribution. Below three commonly used central limit theorems are given.

Theorem 0.7.2 (Central theorem of the independent identical distribution). *Setting random variables $X_1, X_2, \ldots, X_n, \ldots$ as independent, subject to the same distribution, with the mathematical expectation and variance: $EX_k = \mu$, $DX_k = \sigma^2 > 0$ $(k = 1, 2, \ldots)$, the standardized variables of the sum of the random variables $\sum_{k=1}^n X_k$*

$$Y_n = \frac{\sum_{k=1}^n X_k - E\left(\sum_{k=1}^n X_k\right)}{\sqrt{D\left(\sum_{k=1}^n X_k\right)}} = \frac{\sum_{k=1}^n X_k - n\mu}{\sqrt{n}\sigma}$$

The distribution function $F_n(x)$ satisfies, for any x,

$$\lim_{n\to\infty} F_n(x) = \lim_{n\to\infty} P\left\{\frac{\sum_{k=1}^n X_k - n\mu}{\sqrt{n}\sigma} \le x\right\} = \int_{-\infty}^{x} \frac{1}{\sqrt{2\pi}} \mathrm{e}^{-\frac{t^2}{2}}\,\mathrm{d}t = \Phi(x)$$

which is a slight proof.

That is to say, the sum of the random variables $X_1, X_2, \ldots, X_n$, with the mean μ, the variance $\sigma^2 > 0$ and the same distribution is $\sum_{k=1}^n X_k$, and its standardized variable, $(\sum_{k=1}^n X_k - n\mu)/(\sqrt{n}\sigma)$, when n is sufficiently large, approximately obeys the standard normal distribution $N(0, 1)$.

The above standardized variables are transformed into $(1/n \sum_{k=1}^n X_k - \mu)/(\sigma/\sqrt{n})$, considering that $1/n \sum_{k=1}^n X_k$ is the arithmetic mean $\overline{X}$ of the random variables $X_1, X_2, \ldots, X_n$ and when n is sufficiently large, it is approximated,

$$\frac{\overline{X} - \mu}{\sigma/\sqrt{n}} \sim N(0, 1)$$

which is

$$\overline{X} \sim N(\mu, \sigma^2/n)$$

This is another form of the independent distribution central limit theorem.

Theorem 0.7.3 (Yaupon's theorem). *Setting random variables $X_1, X_2, \ldots, X_n, \ldots$ independent of each other, with mathematical expectation and variance,*

$$EX_k = \mu\,, \quad DX_k = \sigma^2 > 0\,, \quad k = 1, 2, \ldots$$

Remembering that $B_n^2 = \sum_{k=1}^{n} \sigma_k^2$, if there is a positive number δ, so when $n \to \infty$ time,

$$\frac{1}{B_n^{2+\delta}} \sum_{k=1}^{n} E\left\{|X_k - \mu_k|^{2+\delta}\right\} \to 0$$

With sum of the random variables $\sum_{k=1}^{n} X_k$, the standardized variables

$$Z_n = \frac{\sum_{k=1}^{n} X_k - E\left(\sum_{k=1}^{n} X_k\right)}{\sqrt{D\left(\sum_{k=1}^{n} X_k\right)}} = \frac{\sum_{k=1}^{n} X_k - \sum_{k=1}^{n} \mu_k}{B_n}$$

and the distribution function $F_n(x)$ for any x, satisfies

$$\lim_{n\to\infty} F_n(x) = \lim_{n\to\infty} P\left\{\frac{\sum_{k=1}^{n} X_k - \sum_{k=1}^{n} \mu_k}{B_n} \le x\right\} = \int_{-\infty}^{x} \frac{1}{\sqrt{2\pi}} \mathrm{e}^{-\frac{t^2}{2}}\, \mathrm{d}t = \Phi(x)$$

which is a slight proof.

Theorem 0.7.4 (Laplace's theorem). *Set random variables η_n $(n = 1, 2, \ldots)$ as obeying the binomial distribution with the parameters n, p $(0 < p < 1)$. For any x, we have*

$$\lim_{n\to\infty} P\left\{\frac{\eta_n - np}{\sqrt{np(1-p)}} \le x\right\} = \int_{-\infty}^{x} \frac{1}{\sqrt{2\pi}} \mathrm{e}^{-\frac{t^2}{2}}\, \mathrm{d}t = \Phi(x)$$

Exercises

0.1 A random experiment is to roll a dice twice, observe the points obtained twice, and try to write out the sample space Ω. If the event that A indicates that the number of occurrences occurs twice is the same, B indicates that the sum of occurrences is greater than 5 twice, and the event C indicates that at least one count is not larger than 3. Try to express events A, B and C with subsets of Ω.

0.2 Let random variable X have the probability density

$$f(x) = \begin{cases} kx\,, & 0 \le x < 3 \\ 2 - \frac{x}{2}\,, & 3 \le x \le 4 \\ 0\,, & \text{elsewhere} \end{cases}$$

(1) Determine the constant k.
(2) Find distribution function $F(x)$ of X.
(3) Find $P\{1 < X \le 7/2\}$.

0.3 Let random variable X obey the Rayleigh distribution, which probability density is

$$p(x) = \begin{cases} \frac{x}{\sigma^2} e^{-\frac{x^2}{2\sigma^2}}, & x \le 0 \\ 0, & x > 0 \end{cases}$$

where the constant $\sigma > 0$. Find $E[X]$ and $D[X]$.

0.4 Let (X, Y) the distribution law be

X \ *Y*	1	2	3
−1	0.2	0.1	0.0
0	0.1	0.0	0.3
1	0.1	0.1	0.1

(1) Find $E[Y]$, $E[X]$.
(2) Letting $Z = Y/X$, find $E[Z]$.
(3) Letting $Z = (X - Y)^2$, find $E[Z]$.

0.5 The known random variable X obeys the normal distribution $N(m, \sigma^2)$. Let the random variable $Y = e^X$.
Find the probability density $f(y)$ of Y.

0.6 Suppose that the probability density of random variable X is

$$f(x, y) = \begin{cases} x^2 + \frac{1}{3}xy, & 0 \le x \le 1, 0 \le y \le 2 \\ 0, & \text{elsewhere} \end{cases}$$

Try to find the distribution function of (X, Y) and two marginal probability densities of (x, y).

0.7 Let the probability density of random variables (X, Y) be

$$f(x, y) = \begin{cases} be^{-(x+y)}, & 0 \le x \le 1,\ 0 < y < \infty \\ 0, & \text{elsewhere} \end{cases}$$

(1) Try to determine constant b.
(2) Find the sum of the marginal density function.

0.8 Let the probability density of random variables (X, Y) be

$$f(x, y) = \begin{cases} \frac{1}{2}(x + y)e^{-(x+y)}, & x > 0,\ y > 0 \\ 0, & \text{elsewhere} \end{cases}$$

(1) Are X and Y statistically independent?
(2) Find the probability density of $Z = X + Y$.

0.9 Let the random variable X obey geometric distribution; its distribution is

$$P\{X = k\} = q^{k-1}p, \quad k = 1, 2, \ldots$$

where $0 < p < 1, q = 1-p$. Try to find the characteristic function of X and find the mathematical expectation and variance by using the characteristic functions.

0.10 Let $(X, Y)^{\mathrm{T}}$ be a two-dimensional normal vector. Its mathematical expectation vector and covariance matrix are $(0, 1)^{\mathrm{T}}$ and $\left[\begin{smallmatrix} 4 & 3 \\ 3 & 9 \end{smallmatrix}\right]$. Try to find the two-dimensional distribution density of $(X, Y)^{\mathrm{T}}$.

0.11 Let the random variables $X_1, X_2, \ldots, X_n$ be independent of each other and have the same normal distribution $N(m, \sigma^2)$. Try to find the n-dimensional distribution density of the vector $\boldsymbol{X} = (X_1, X_2, \ldots, X_n)^{\mathrm{T}}$ and its mathematical expectation vector and covariance matrix. Then find the probability distribution density of $\boldsymbol{X} = 1/n \sum_{i=1}^{n} X_i$.

0.12 The probability density of the known random variable X is

$$p(x) = \frac{1}{\sqrt{2\pi}\sigma} \mathrm{e}^{-\frac{x^2}{2\sigma^2}}, \quad -\infty < x < \infty$$

Try to find the mathematical expectation and variance of X using the relationship between the characteristic function and the moment.

0.13 The relation between random variables Y and X is $Y = 2X+1$, and X obeys normal distribution $N(m, \sigma^2)$. Find the characteristic function $\varphi_Y(\lambda)$ and the probability density $f(y)$ of the random variable Y.

1 Random processes

1.1 Basic concepts of random processes

1.1.1 Definition of random processes

The process of change in nature can be divided into two broad categories – the ascertained process and the random process. The former has a definite rule of change; while the latter has no definite rule of change. If the change process resulting from each experiment is the same, it is the same function of time t, and it is the ascertained process. If the change process resulting from each experiment is different, it is a different function of time t, and it is a random process. The electrical signal is the process of changing voltage or current over time, and by the rules above, it is divided into two categories – the ascertained signal and the random signal. Let us start with two examples.

Example 1.1.1. Sine (-type) ascertained signals

$$s(t) = A\cos(\omega_0 t + \varphi_0) \tag{1.1}$$

where the amplitude A, angular frequency ω_0 and phase φ_0 are known constants. Each time a high-frequency oscillator is directed, its steady-state part is the signal. Every incentive is equivalent to one test. Because each time the signal is identical with the time according to the ascertained function, the signal is an ascertained process.

Example 1.1.2. Sine (-type) random initial phase signals

$$X(t) = A\cos(\omega_0 t + \varphi) \tag{1.2}$$

where amplitude A and angular frequency ω_0 are both constants, and phase φ is a random variable that is evenly distributed in the range $(0, 2\pi)$. The phase φ is a continuous random variable, and it has numerous values in the interval $(0, 2\pi)$. In other words, it can be any possible value φ_i in the interval $(0, 2\pi) 0 < \varphi_i < 2\pi$. Accordingly, there are different functions

$$x_i(t) = A\cos(\omega_0 t + \varphi_i)\,, \quad \varphi_i \in (0, 2\pi) \tag{1.3}$$

Obviously, eq. (1.2) actually represents a group of different time functions, as is shown in Fig. 1.1 (only three function curves are drawn in the figure). So, this is a random process.

When doing power-on incentives to a general high-frequency oscillator that does not adopt phase measures, the steady-state part is such a signal. After power-on incentives, by the influences of some accidental factors, the oscillator's initial phase will

https://doi.org/10.1515/9783110593808-002

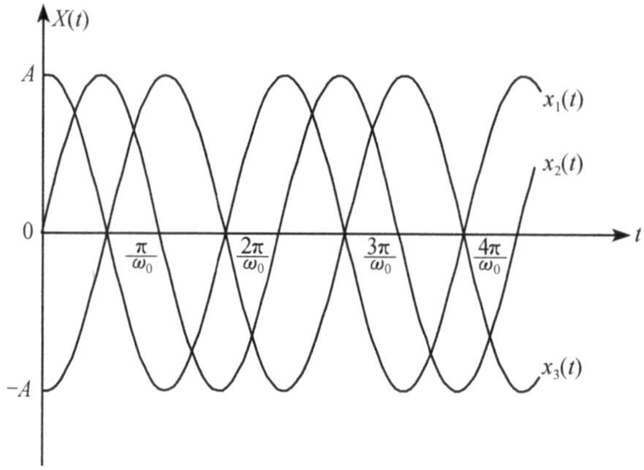

Fig. 1.1: Sine(-type) random initial confidence numbers.

be different each time. As a result, the high-frequency signal phase will have random changes. This is one of the most common random signals.

Similarly,

$$X(t) = A\cos(\omega t + \varphi) \tag{1.4}$$

In eq. (1.4), if the amplitude A is a random variable only, it will be a random amplitude signal. If the angular frequency ω is a random variable only, it will be a random frequency signal.

In the case of Example 1.1.2, the observation of each boot is equivalent to a random experiment. The observed and recorded result of each experiment is a determined time function, called the sample function, or sample or implementation for short. The total or set of all these sample functions constitutes a random process. Before each test, do not know the results of this experiment and should choose one sample of this set. Only after a large number of observations do we know their statistical regularity, namely the probability of exactly how to implement this one.

Definition 1.1.1. Set the random test E and sample space $S = \{e\}$, and for each of these elements e, determine a sample function $X(e, t)$ according to a certain rule. All elements e determine a group of sample functions $X(e, t)$ and this is called a random process. The usual practice of writing is not to write random factors e, and then the $X(e, t)$ is shortened to $X(t)$.

This definition regards the random process as taking a particular sample function in a random way (with a certain probability). Let us look at the meaning of random process $X(e, t)$ under certain circumstances.

In all kinds of situations, the following holds.

(1) If e fixed for e_i with only time t changes, we obtain a particular time function $X(e_i, t)$. This is a determined sample function, which is the record curve (implementation) of a certain observation [Fig. 1.1 $x_1(t)$, $x_2(t)$, $x_3(t)$]. In order to avoid

confusion, the random process is usually represented by capital letters, for example, $X(t)$ and $Y(t)$. The sample is in lowercase letters, for example, $x(t)$ and $y(t)$.
(2) If t is fixed for t_j and only random factors e change, the $X(e, t_j)$ turns into a random variable, abbreviated as $X(t_j)$. A random variable $X(t_j)$ is called the state of the random process $X(t)$ in the time $t = t_j$.
(3) If e is fixed for e_i with t fixed for t_j, the $X(e_i, t_j)$ turns into a certain value, abbreviated as $x_i(t_j)$.
(4) If e and t are both variables, the $\{X(e, t)\}$ stands for the collection of all samples or the population of all random variables. This is the random process $X(t)$.

Definition 1.1.2. If at each fixed time t_j ($j = 1, 2, \ldots$), $X(t_j)$ are all random variables define $X(t)$ as a random process.

This definition is to regard a random process as a group of random variables changing over time. The two kinds of different definitions above used different ways to describe the same thing, from different angles to help us understand the stochastic process. Thus, in essence the two definitions are consistent and complementary to each other. When making actual observations, we usually adopt Definition 1.1.1; while carrying out a theoretical analysis, we usually adopt Definition 1.1.2. According to Definition 1.1.2, a random process can be regarded as the generalization of random variables (n-dimensional random variables). If the time division is smaller, this means a larger dimension n, and the statistical regularity of the random process can be described in a more detailed way.

It should be pointed out that, in the process of general randomized trials, random variables will change with a parameter's change (such as time t or height h, etc.). In general, random variables that change with a parameter are called random functions, where the random function of time t is called a random process.

The random process can be divided into a continuous and a discrete one according to its state. It can also be divided into continuous a parametric random process (or a random process) and a discrete parametric random process (random sequence) by different time parameters t. Therefore, in total it can be divided into the following four categories, as shown in Fig. 1.2 (only one sample is shown in the figure, and samples are drawn for the usual interval sampling).
(1) A continuous random process: its state and time are both continuous [Fig. 1.2 (a)].
(2) A discrete random process: its state is discrete, but the time is continuous [Fig. 1.2 (b)]. Random sampling of continuous random processes is carried out, and the sampling values are maintained after quantization, then such random processes are obtained.
(3) A continuous random sequence: its state is continuous, but the time is discrete [Fig. 1.2 (c)]. Quantitative interval sampling of continuous random processes is carried out, and such random processes are obtained.

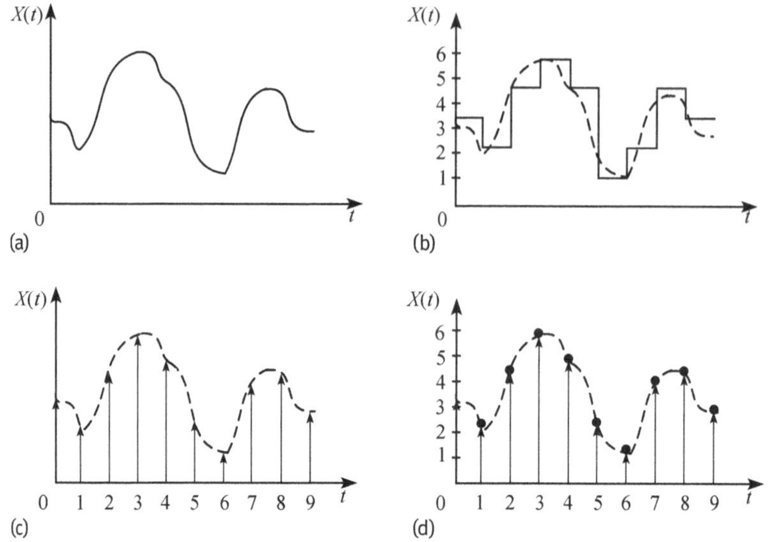

Fig. 1.2: Common classification of random processes.

(4) A discrete random sequence: its state and time are both discrete. This is shown in Fig. 1.2 (d). Carry out quantitative sampling on the continuous random process and change sampling values into some fixed values, such as the binary 0, 1, or $0 \sim 9$ in decimal, such random processes are obtained. In fact, this is a sequence of numbers or a digital signal.

In conclusion, the most basic process is the continuous random process, while the other three types just are applied discrete processing to it. Therefore, this book will mainly focus on continuous random processes.

According to its distribution function and probability density, the random process can be classified into independent random processes, Markov processes, independent increment processes, normal random processes, Rayleigh random processes, etc. In addition, according to the power spectrum of the random process, it can be divided into broadband or narrowband, white, or colored processes. In engineering technology, it can also be divided into stationary and nonstationary processes, based on its stability.

1.1.2 Probability distribution of random processes

Since the random process can be regarded as a group of random variables changing with time t, the concept of the probability distribution of random variables can be promoted for random processes, and the probability distribution of the random process can be obtained.

1.1.2.1 One-dimensional probability distribution

Random process $X(t)$ turns into a one-dimensional random variable $X(t_1)$ at any given moment t_1. The probability of $P\{X(t_1) \leq x_1\}$ is a function of value x_1 and the moment t_1. It can be written as:

$$F_1(x_1, t_1) = P\{X(t_1) \leq x_1\} \tag{1.5}$$

This is known as the one-dimensional distribution function of the process $X(t)$.

If it has a first-order derivative of x_1, define the one-dimensional probability density of the process $X(t)$ as

$$p_1(x_1, t_1) = \frac{\partial F_1(x_1, t_1)}{\partial x_1} \tag{1.6}$$

Usually we omit its footnote and write $p(x, t)$ for short.

The one-dimensional probability distribution can only describe the value statistics feature of the random process at any certain moment and it cannot reflect the correlation between the values of the random process at any certain moment.

1.1.2.2 Two-dimensional probability distribution

The values $X(t_1)$, $X(t_2)$ of the random process at two certain times become two-dimensional random variables $[X(t_1), X(t_2)]$. Mark

$$F_2(x_1, x_2; t_1, t_2) = P\{X(t_1) \leq x_1, X(t_2) \leq x_2\} \tag{1.7}$$

as the two-dimensional distribution function of the process $X(t)$.

For x_1 and x_2, if the second-order mixed partial derivative exists, define

$$p_2(x_1, x_1; t_1, t_2) = \frac{\partial^2 F_2(x_1, x_1; t_1, t_2)}{\partial x_1 \partial x_2} \tag{1.8}$$

as the two-dimensional probability density of the process $X(t)$.

The two-dimensional probability distribution can describe the correlation between the values of the random process at two times, and two one-dimensional probability densities can be obtained by integrating $p(x_1, t_1)$ and $p(x_2, t_2)$. Obviously, two-dimensional probability distribution contains more statistics information than the one-dimensional probability distribution and describes the random process with more details. However, it cannot reflect the correlation between the values of the random process at more than two moments.

1.1.2.3 Probability distribution of *n* dimensions

The values of the random process $X(t)$ at any n moments $t_1, t_2, \ldots, t_n$, $X(t_1), X(t_2), \ldots, X(t_n)$, constitute a n-dimensional random variable $[X(t_1), X(t_2), \ldots, X(t_n)]$. That is, the random vector $\boldsymbol{X}$ in n-dimensional space. Similarly, we can obtain the n-dimensional distribution function of the process $X(t)$:

$$F_n(x_1, x_2, \ldots, x_n; t_1, t_2, \ldots, t_n) = P(X(t_1) \le x_1, X(t_2) \le x_2, \ldots, X(t_n) \le x_n) \quad (1.9)$$

The n-dimensional probability density of the process $X(t)$ is:

$$p_n(x_1, x_2, \ldots, x_n; t_1, t_2, \ldots, t_n) = \frac{\partial^n F_n(x_1, x_2, \ldots, x_n; t_1, t_2, \ldots, t_n)}{\partial x_1 \partial x_2 \ldots \partial x_n} \quad (1.10)$$

The n-dimensional probability distribution can describe the correlation between the values of any n moments and has more statistical properties than its low-dimensional probability distribution, and the description of the random process is more subtle. Therefore, the statistical properties of random processes can be described more carefully if the number of observation points of the random processes is more (that is, a larger number of dimensions).

Theoretically, we must make n infinite if we want to fully describe the statistical characteristics of a random process, but for engineering practice, two dimensions are enough in many cases. According to Chapter 0 about the probability distribution of multidimensional random variables, the n-dimensional probability distribution of random processes has the following main properties:

(1) $$F_n(x_1, x_2, \ldots, -\infty, \ldots, x_n; t_1, t_2, \ldots, t_i, \ldots, t_n) = 0\,.$$

(2) $$F_n(\infty, \infty, \ldots, \infty; t_1, t_2, \ldots, t_n) = 1\,.$$

(3) $$p_n(x_1, x_2, \ldots, x_n; t_1, t_2, \ldots, t_n) \ge 0\,.$$

(4) $$\underbrace{\int_{-\infty}^{\infty} \cdots \int_{-\infty}^{\infty}}_{n \text{ times}} p_n(x_1, x_2, \ldots, x_n; t_1, t_2, \ldots, t_n)\, dx_1\, dx_2 \ldots dx_n = 1\,.$$

(5) $$\underbrace{\int_{-\infty}^{\infty} \cdots \int_{-\infty}^{\infty}}_{(n-m) \text{ times}} p_n(x_1, x_2, \ldots, x_n; t_1, t_2, \ldots, t_n)\, dx_{m+1}\, dx_{m+2} \ldots dx_n = p_m(x_1, x_2, \ldots, x_m; t_1, t_2, \ldots, t_m)\,.$$

(6) if $X(t_1), X(t_2), \ldots, X(t_n)$ statistical independence:

$$p_n(x_1, x_2, \ldots, x_n; t_1, t_2, \ldots, t_n) = p(x_1, t_1) p(x_2, t_2) \ldots p(x_n, t_n)$$

Example 1.1.3. Set a random amplitude signal $X(t) = X \cos \omega_0 t$, with ω_0 the constant and X the standard normal random variable.

Try to obtain the one-dimensional probability density of $X(t)$ at the moments when $t = 0$, $t = \pi/(3\omega_0)$, $t = \pi/(2\omega_0)$.

Solution: The one-dimensional probability density of the standard normal random variable is

$$p(x) = \frac{1}{\sqrt{2\pi}} \exp\left[-\frac{x^2}{2}\right], \qquad -\infty < x < \infty$$

The value of $X(t)$ at any moment is:

$$x_t = x \cos \omega_0 t$$

From probability density transformation of random variables, we can obtain the one-dimensional probability density of $X(t)$

$$\begin{aligned} p(x_t, t) &= p(x, t)\left|\frac{\mathrm{d}x}{\mathrm{d}x_t}\right| = p\left(\frac{x_t}{\cos \omega_0 t}, t\right)\left|\frac{1}{\cos \omega_0 t}\right| \\ &= \frac{1}{\sqrt{2\pi}\,|\cos \omega_0 t|} \exp\left[-\frac{x_t^2}{2\cos^2 \omega_0 t}\right], \qquad -\infty < x < \infty \end{aligned}$$

Therefore,

$$\text{when } t = t_1 = 0, \qquad p(x_1, t_1) = \frac{1}{\sqrt{2\pi}} \exp\left[-\frac{x_1^2}{2}\right]$$

$$\text{when } t = t_2 = \frac{\pi}{3\omega_0}, \qquad p(x_2, t_2) = \frac{1}{\sqrt{2\pi 0.5}} \exp\left[-2x_2^2\right]$$

$$\text{when } t = t_3 = \frac{\pi}{2\omega_0}, \qquad p(x_3, t_3) = \lim_{t \to t_3} \frac{1}{\sqrt{2\pi}\,|\cos \omega_0 t_3|} \exp\left[-\frac{x_3^2}{2\cos^2 \omega_0 t_3}\right] = \delta(x_3)$$

The above results are shown in Fig. 1.3, where we can see that the one-dimensional probability density of the random process $X(t)$ varies with time, and the values at any time are normal distributions, but the variance at each moment is different.

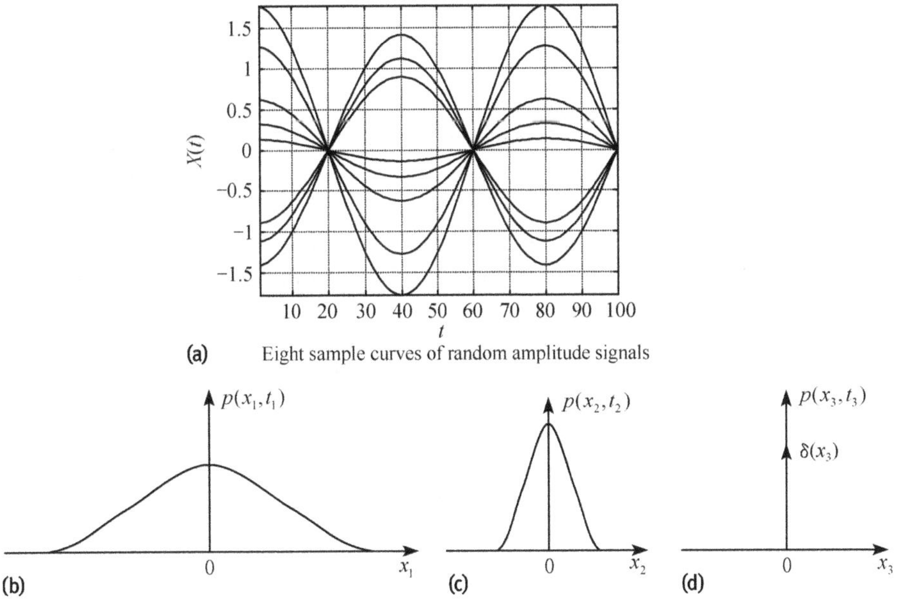

Fig. 1.3: Random amplitude signal and its probability density.

Example 1.1.4. Try to write the one-dimensional probability density of an ascertained signal $s(t) = A\cos(\omega_0 t + \varphi_0)$

Solution: In terms of classification, the ascertained signal $s(t)$ is not a random process, but it can be regarded as a special kind of random process $X(t)$. For any moment t, the process $X(t)$ can only evaluate $s(t)$. So, the one-dimensional probability density of the certain signal $s(t)$ can be written as

$$p(x, t) = \delta[x - s(t)]$$

1.1.3 The moment function of random processes

The average statistical parameters describing a random variable are the mathematical expectation, variance, covariance, correlation moment and so on. The most common numerical characteristics are the moment. Random processes can be regarded as a group of random variables changing over time, so the concept of statistical features of random variables is applied in the random process to obtain average statistical functions describing the random process (no longer a determined number, but a certain time function) and they are called the moment functions. The distribution functions and probability densities of random processes are their general statistical properties. They can describe the random processes completely, but they are not concise and are often difficult to obtain. In engineering techniques, a few moment functions that describe the main average statistical properties of the stochastic process are sufficient. Let us introduce the moment functions. Obviously, their definition and significance are just a generalization of random variable moments.

1.1.3.1 Mathematical expectations

At a particular time, the value of the random process $X(t)$ is a one-dimensional random variable, and its mathematical expectation is a definite value. At any time t, the value of the random process $X(t)$ is still a one-dimensional random variable $X(t)$ (note that t is fixed here, so $X(t)$ is no longer a random process). Shorten any of its values $x(t)$ to x. According to the definition of the mathematical expectation of a random variable, we obtain:

$$E[X(t)] = \int_{-\infty}^{\infty} xp(x, t)\,\mathrm{d}x = m_X(t) \tag{1.11}$$

This is a certain function of time and it is the mathematical expectation or statistical average of any moment t of the process $X(t)$, called the (instantaneous) mathematical expectation or statistical mean of the random process $X(t)$. We often use a special symbol $m_X(t)$ to mark it (the subscript X can be omitted if there is no confusion).

The statistical mean is the average value of all samples in the random process at any time t, so it is also called the mean of the set (it can be called mean value for short if there is no confusion).

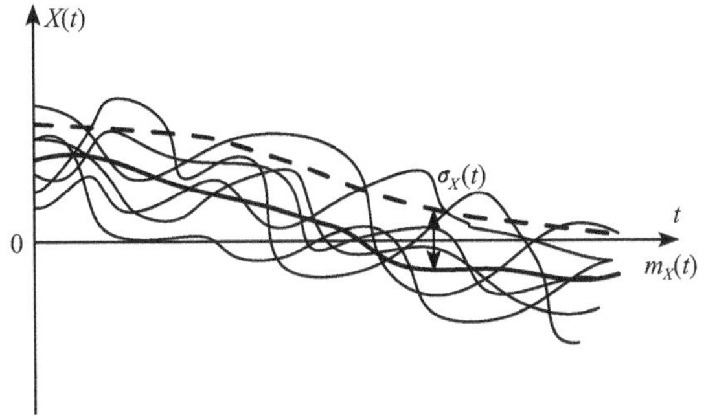

Fig. 1.4: Mathematical expectation and standard deviation of a stochastic process.

The mathematical expectation $m_X(t)$ of the random process $X(t)$ is shown in Fig. 1.4 with a thick line. It is the distribution center of the value (random variable) at any moment t of all the samples from the random process.

1.1.3.2 Variance

For the random process $X(t)$ and mathematical expectations $m_X(t)$, it is a definite time function. So, $\mathring{X}(t) = X(t) - m_X(t)$ is still a random process, called a centralizing random process, or the undulation of the process for short. The value of undulating $\mathring{X}(t)$ at any time is still a one-dimensional random variable, so similarly to the variance of the random variable, we obtain the following definition:

$$D[X(t)] = E\{[\mathring{X}(t)]^2\} = E\{[X(t) - m_X(t)]^2\} = \int_{-\infty}^{\infty} [x - m_X(t)]^2 p(x, t)\,dx = \sigma_X^2(t) \quad (1.12)$$

It is also a determined function of time, called the variance of the random process $X(t)$. Usually we denote it with the particular symbol $\sigma_X^2(t)$.

The variance $\sigma_X^2(t)$ must be a non-negative function; its square root $\sigma_X(t)$ is called the standard deviation or the variance root of the random processes $X(t)$. It is shown in Fig. 1.4 by the dotted line.

The variance $\sigma_X^2(t)$ shows the average discrete degree of the values of all samples at any given time t and the distribution center in the random process $X(t)$.

1.1.3.3 Autocorrelation function

The mathematical expectation and variance, respectively, are the first-order origin moment and the second-order central moment of one-dimensional random variables. They can only express the average statistical properties of the random process at each isolated time but cannot reflect the correlation between values of a random process at two moments. To show the correlation degree of two values at two moments of the random process, we need to use the second-order moments or the central moment of the two-dimensional variables. This is the autocorrelation function and the centralized autocorrelation function of the random process.

The values of random process $X(t)$ at two moments t_1, t_2 form a two-dimensional random variable. Denote two values of the variables $X(t_1)$ and $X(t_2)$ at two moments, $x(t_1)$ and $x(t_2)$ by x_1 and x_2 for short. Define the second-order mixed origin moment as

$$R_X(t_1, t_2) = E[X(t_1)X(t_2)] = \int_{-\infty}^{\infty}\int_{-\infty}^{\infty} x_1 x_2 p_2(x_1, x_1; t_1, t_2)\,\mathrm{d}x_1\,\mathrm{d}x_2 \tag{1.13}$$

This is called the autocorrelation function of the random process $X(t)$ or the correlation function if there is no confusion. It represents the average degree of correlation between the values at any two moments of the process $X(t)$.

Similarly, denote the second-order mixing central moment of variables $X(t_1)$ and $X(t_2)$ as:

$$\begin{aligned} C_X(t_1, t_2) &= E[\mathring{X}(t_1)\mathring{X}(t_2)] = E\{[X(t_1) - m_X(t_1)][X(t_2) - m_X(t_2)]\} \\ &= \int_{-\infty}^{\infty}\int_{-\infty}^{\infty} [x_1 - m_X(t_1)][x_2 - m_X(t_2)]p_2(x_1, x_1; t_1, t_2)\,\mathrm{d}x_1\,\mathrm{d}x_2 \end{aligned} \tag{1.14}$$

This is called the centralized autocorrelation function or autocovariance function, or covariance function for short, of the random process $X(t)$. It represents the average correlation degree between the ups and downs of the two periods of the process $X(t)$. These two kinds of autocorrelation function have the following relationship:

$$\begin{aligned} C_X(t_1, t_2) &= E\{[X(t_1) - m_X(t_1)]\,[X(t_2) - m_X(t_2)]\} \\ &= E\{X(t_1)X(t_2) - m_X(t_1)X(t_2) - m_X(t_2)X(t_1) + m_X(t_1)m_X(t_2)\} \\ &= R_X(t_1, t_2) - m_X(t_1)m_X(t_2) \end{aligned} \tag{1.15}$$

Because when $t_1 = t_2 = t$, we have

$$R_X(t, t) = E[X(t)X(t)] = E[X^2(t)] \tag{1.16}$$

and

$$C_X(t, t) = E\{[X(t) - m_X(t)]^2\} = \sigma_X^2(t) \tag{1.17}$$

Therefore, we obtain

$$\sigma_X^2(t) = E[X^2(t)] - m_X^2(t) \tag{1.18}$$

Obviously, when $t_1 = t_2 = t$, the covariance function turns into the variance, and the correlation function is the second-order moment of the one-dimensional random variable $E[X^2(t)]$, called the mean square value of the random process $X(t)$.

Example 1.1.5. Set random amplitude signal $X(t) = X\cos\omega_0 t$, with ω_0 the constant and X the standard normal random variable. Determine the mathematical expectation, variance, correlation function and covariance function of the random process $X(t)$.

Solution: X is the standard normal random variable, so its mathematical expectation $E[X] = 0$, the variance $D[X] = 1$ and the mean square value

$$E[X^2] = D[X] + E^2[X] = 1$$

According to the definition of the moment function of the random process, and using the property of digital features, we have

$$\begin{aligned}
m_X(t) &= E[X(t)] = E[X\cos\omega_0 t] = \cos\omega_0 t \cdot E[X] = 0 \\
\sigma_X^2(t) &= D[X(t)] = D[X\cos\omega_0 t] = \cos^2\omega_0 t \cdot D[X] = \cos^2\omega_0 t \\
R_X(t_1, t_2) &= E[X(t_1)X(t_2)] = E[X^2\cos\omega_0 t_1 \cos\omega_0 t_2] \\
&= \cos\omega_0 t_1 \cos\omega_0 t_2 \cdot E[X^2] = \cos\omega_0 t_1 \cos\omega_0 t_2 \\
C_X(t_1, t_2) &= E[\mathring{X}(t_1)\mathring{X}(t_2)] = E\{[X(t_1) - m_X(t_1)][X(t_2) - m_X(t_2)]\} \\
&= E\{X(t_1)X(t_2)\} = R_X(t_1, t_2) = \cos\omega_0 t_1 \cos\omega_0 t_2
\end{aligned}$$

1.1.4 Characteristic functions of random processes

We know that the probability density and characteristic function of a random variable is a pair of Fourier transforms, and the moment of the random variable is uniquely determined by the characteristic function. Therefore, when solving the probability density and the digital characteristics of random variables, such as normal distribution, the operation can be significantly simplified by using the characteristic functions. Similarly, when solving the probability density and the moment function of the random process, using the characteristic functions can also achieve that goal.

The one-dimensional characteristic function of the random process $X(t)$ is defined as:

$$\Phi_X(\lambda, t) = E[\mathrm{e}^{\mathrm{j}\lambda X(t)}] = \int_{-\infty}^{\infty} \mathrm{e}^{\mathrm{j}\lambda x} p(x, t)\,\mathrm{d}x \tag{1.19}$$

where x is the value of the random variable $X(t)$, $p(x, t)$ is the one-dimensional probability density of the process $X(t)$, which forms a pair of Fourier transforms with the one-dimensional characteristic function $\Phi_X(\lambda, t)$, namely:

$$p(x, t) = \frac{1}{2\pi}\int_{-\infty}^{\infty} \Phi_X(\lambda, t)\mathrm{e}^{-\mathrm{j}\lambda x}\,\mathrm{d}\lambda \tag{1.20}$$

With n times partial derivatives of the variable λ on both ends of the eq. (1.19), we have:

$$\frac{\partial^n}{\partial\lambda^n}\Phi_X(\lambda, t) = \mathrm{j}^n \int_{-\infty}^{\infty} x^n \mathrm{e}^{\mathrm{j}\lambda x} p(x, t)\,\mathrm{d}x \tag{1.21}$$

Thus, the n-order origin moment function of the process $X(t)$ is:

$$E[X^n(t)] = \int_{-\infty}^{\infty} x^n p(x,t)\,\mathrm{d}x = \mathrm{j}^{-n}\left[\frac{\partial^n}{\partial\lambda^n}\Phi_X(\lambda,t)\right]_{\lambda=0} \tag{1.22}$$

The mathematical expectation and mean square value of the random process can be obtained by using this formula.

The two-dimensional feature function of the random process $X(t)$ is defined as

$$\Phi_X(\lambda_1,\lambda_2;t_1,t_2) = E[\mathrm{e}^{\mathrm{j}\lambda_1 X(t_1)+\mathrm{j}\lambda_2 X(t_2)}] = \int_{-\infty}^{\infty}\int_{-\infty}^{\infty} \mathrm{e}^{\mathrm{j}\lambda_1 x_1+\mathrm{j}\lambda_2 x_2} p_2(x_1,x_2;t_1,t_2)\,\mathrm{d}x_1\,\mathrm{d}x_2 \tag{1.23}$$

where $x_1 = X(t_1)$, $x_2 = X(t_2)$, $p_2(x_1,x_2;t_1,t_2)$ is the two-dimensional probability density of the process $X(t)$. It forms a double Fourier transform with $\Phi_X(\lambda_1,\lambda_2;t_1,t_2)$, namely,

$$p_2(x_1,x_2;t_1,t_2) = \frac{1}{(2\pi)^2}\int_{-\infty}^{\infty}\int_{-\infty}^{\infty} \Phi_X(\lambda_1,\lambda_2;t_1,t_2)\mathrm{e}^{-\mathrm{j}\lambda_1 x_1-\mathrm{j}\lambda_2 x_2}\,\mathrm{d}\lambda_1\,\mathrm{d}\lambda_2 \tag{1.24}$$

With a partial derivative of both variable λ_1 and λ_2 on both ends of eq. (1.23), we have:

$$\frac{\partial^2}{\partial\lambda_1\partial\lambda_2}\Phi_X(\lambda_1,\lambda_2;t_1,t_2) = -\int_{-\infty}^{\infty}\int_{-\infty}^{\infty} x_1 x_2 \mathrm{e}^{\mathrm{j}\lambda_1 x_1+\mathrm{j}\lambda_2 x_2} p_2(x_1,x_2;t_1,t_2)\,\mathrm{d}x_1\,\mathrm{d}x_2 \tag{1.25}$$

Thus, the correlation function of the process $X(t)$ is:

$$\begin{aligned} R_X(t_1,t_2) &= \int_{-\infty}^{\infty}\int_{-\infty}^{\infty} x_1 x_2 p_2(x_1,x_2;t_1,t_2)\,\mathrm{d}x_1\,\mathrm{d}x_2 \\ &= -\left[\frac{\partial^2}{\partial\lambda_1\partial\lambda_2}\Phi_X(\lambda_1,\lambda_2;t_1,t_2)\right]_{\lambda_1=\lambda_2=0} \end{aligned} \tag{1.26}$$

By using the properties of characteristic functions, we can obtain the variance and covariance functions of the process $X(t)$:

$$D[X(t)] = -\left\{\frac{\partial^2}{\partial\lambda^2}e^{-\mathrm{j}\lambda E[X(t)]}\Phi_X(\lambda,t)\right\}_{\lambda=0} \tag{1.27}$$

$$C_X(t_1,t_2) = -\left\{\frac{\partial^2}{\partial\lambda_1\partial\lambda_2}\mathrm{e}^{-\mathrm{j}\lambda_1 E[X(t_1)]-\mathrm{j}\lambda_2 E[X(t_2)]}\Phi_X(\lambda_1,\lambda_2;t_1,t_2)\right\}_{\lambda_1=\lambda_2=0} \tag{1.28}$$

Example 1.1.6. The value of the random process $X(t)$ at the moment t obeys normal distribution, and its one-dimensional probability density is

$$p(x,t) = \frac{1}{\sqrt{2\pi}\sigma}\exp\left[-\frac{(x-m)^2}{2\sigma^2}\right]$$

So, let us try to solve the characteristic function of this random process $X(t)$, the mathematical expectation, the variance and the mean square value.

Solution: From the definition of one-dimensional characteristic functions we can obtain:

$$\Phi_X(\lambda, t) = E[e^{j\lambda X(t)}] = \int_{-\infty}^{\infty} e^{j\lambda x} p(x, t)\,dx$$

$$= \frac{1}{\sqrt{2\pi}\sigma} \int_{-\infty}^{\infty} \exp\left[j\lambda x - \frac{(x-m)^2}{2\sigma^2}\right] dx$$

Making the variable substitution $y = x - m/\sigma$, and then:

$$\Phi_X(\lambda, t) = \frac{1}{\sqrt{2\pi}} e^{j\lambda x} \int_{-\infty}^{\infty} \exp\left[-\frac{y^2}{2} + j\lambda\sigma y\right] dy = \exp\left[j\lambda m - \frac{1}{2}\lambda^2\sigma^2\right]$$

The mathematical expectation and the mean square value can be obtained from eq. (1.22).

$$E[X(t)] = j^{-1}\left[\frac{\partial}{\partial\lambda}\Phi_X(\lambda, t)\right]_{\lambda=0} = m$$

$$E[X^2(t)] = j^{-2}\left[\frac{\partial^2}{\partial\lambda^2}\Phi_X(\lambda, t)\right]_{\lambda=0} = \sigma^2 + m^2$$

So, the variance is

$$D[X(t)] = E[X^2(t)] - E^2[X(t)] = \sigma^2$$

1.2 Stationary random processes

Random processes can be mainly divided into stationary and nonstationary random processes. Strictly speaking, all random processes are nonstationary, but the analysis of the stationary random process is much easier, and usually most random processes met in radio technology are close to stationary, so this book will mainly focus on stationary random processes.

1.2.1 Characteristics and classification

The main characteristic of a stationary random process is that its statistical properties do not change with time. That is, its probability distribution or moment function has no relation with the time starting point in the observation. So, we can choose the time starting point in the observation. If a random process does not have the stability above, then it is a nonstationary random process.

According to the different levels of requirement for the condition of stability, the stationary random process is usually divided into two categories: strict-sense stationary processes (also called special stationary processes) and generalized stationary processes (also known as wide-sense stationary processes).

1.2.1.1 Strict-sense stationary processes

If any n dimension the probability distribution of the random process $X(t)$ does not vary with the selection of the time starting point, equally when the time is shifted by any constant ε, the n-dimensional probability density (or the distribution function) does not change, then we can call $X(t)$ a strict-sense stationary process. The following relations should be satisfied for the strict-sense stationary process:

$$p_n(x_1, x_2, \ldots, x_n; t_1, t_2, \ldots, t_n) = p_n(x_1, x_2, \ldots, x_n; t_1+\varepsilon, t_2+\varepsilon, \ldots, t_n+\varepsilon) \quad (1.29)$$

Now let us look at the characteristics of two-dimensional probability density and the moment function of strictly stationary processes.

Use eq. (1.29) in the case of one dimension. Letting $\varepsilon = -t_1$, then:

$$p_1(x_1, t_1) = p_1(x_1, 0) \quad (1.30)$$

This shows that one-dimensional probability density has no relation with time t. So it can be denoted by $p(x)$ for short. Therefore, the mathematical expectation and variance are also invariable constants with no relation to time t. Namely:

$$E[X(t)] = \int_{-\infty}^{\infty} xp(x,t)\,\mathrm{d}x = \int_{-\infty}^{\infty} xp(x)\,\mathrm{d}x = m_X \quad (1.31)$$

$$D[X(t)] = \int_{-\infty}^{\infty} [x - m_X(t)]^2\, p(x,t)\,\mathrm{d}x = \int_{-\infty}^{\infty} [x - m_X]^2\, p(x)\,\mathrm{d}x = \sigma_X^2 \quad (1.32)$$

Using eq. (1.29) in the case of two dimensions, make $\varepsilon = -t_1$ and $\tau = t_2 - t_1$, and then

$$p_2(x_1, x_2; t_1, t_2) = p_2(x_1, x_2; 0, \tau) \quad (1.33)$$

This indicates that the two-dimensional probability density only has a relation to the interval of time $\tau = t_2 - t_1$ but has no relation with the moments t_1 or t_2. So, it can be denoted by $p_2(x_1, x_2; \tau)$ for short. Therefore, the correlated function is a function of one single variable, namely:

$$R_X(t_1, t_2) = \int_{-\infty}^{\infty}\int_{-\infty}^{\infty} x_1 x_2 p_2(x_1, x_1; \tau)\,\mathrm{d}x_1\,\mathrm{d}x_2 = R_X(\tau) \quad (1.34)$$

Similarly, we can obtain the covariance function:

$$C_X(t_1, t_2) = C_X(\tau) = R_X(\tau) - m_X^2 \quad (1.35)$$

Therefore, when $t_1 = t_2 = t$ namely $\tau = 0$, we have

$$\sigma_X^2 = R_X(0) - m_X^2 \quad (1.36)$$

1.2.1.2 Wide-sense stationary processes

If the mathematical expectation of the random process $X(t)$ is a constant uncorrelated with time t, and correlation functions only have a relation to time intervals $\tau = t_2 - t_1$, namely:

$$\begin{cases} E[X(t)] = m_X \\ R_X(t_1, t_2) = R_X(\tau) \end{cases} \tag{1.37}$$

then $X(t)$ is a wide-sense stationary process.

From the moment functions of the strict-sense stationary process in the discussion above, we know that the wide-sense stationary process is just a special case when the strict-sense stationary process has relaxed its requirements of stationary conditions. Therefore, the wide-sense stationary process may not be strict-sense stationary.

After comparing the two requirements for stationarity, if one is strict-sense stationary, its second moment must exist. So, it must be wide-sense stationary. On the other hand, if it is wide-sense stationary, it does not have to be strict-sense stationary. In electronic information technology, generally we only study the wide-sense stationary process, which is suitable for engineering applications. So, unless specifically stated, in the following, stationary processes are referred to as wide-sense stationary processes.

Example 1.2.1. Setting random initial phase signal $X(t) = A\cos(\omega_0 t + \varphi)$ and denoting A and ω_0 as constants, φ is a random variable that is evenly distributed in $(0, 2\pi)$. Try to prove that $X(t)$ is a stationary process.

Solution: The probability density of random variables φ is

$$p(\varphi) = \begin{cases} \frac{1}{2\pi}, & 0 < \varphi < 2\pi \\ 0, & \text{otherwise} \end{cases}$$

At any time t, $X(t)$ is a function of the same random variable φ. So, we can calculate the mean statistic value of this function. The mathematical expectation is

$$\begin{aligned} m_X(t) = E[X(t)] &= \int_{-\infty}^{\infty} x(t)p(\varphi)\,\mathrm{d}\varphi \\ &= \int_0^{2\pi} A\cos(\omega_0 t + \varphi)\frac{1}{2\pi}\,\mathrm{d}\varphi = 0 \end{aligned}$$

Similarly, variables $X(t_1)X(t_2)$ are also functions of the same random variable φ, so the correlation function is

$$
\begin{aligned}
R_X(t_1, t_2) = E[X(t_1)X(t_2)] &= \int_{-\infty}^{\infty} x(t_1)x(t_2)p(\varphi)\,\mathrm{d}\varphi \\
&= \int_0^{2\pi} A^2 \cos(\omega_0 t_1 + \varphi)\cos(\omega_0 t_2 + \varphi)\frac{1}{2\pi}\,\mathrm{d}\varphi \\
&= \frac{A^2}{4\pi}\int_0^{2\pi} \{\cos \omega_0(t_2 - t_1) + \cos[\omega_0(t_1 + t_2) + 2\varphi]\}\,\mathrm{d}\varphi \\
&= \frac{A^2}{2}\cos \omega_0 \tau
\end{aligned}
$$

where $\tau = t_2 - t_1$.

The mathematical expectation $m_X(t)$ obtained is uncorrelated with time t, and the correlation functions $R_X(t_1, t_2)$ only relation with time interval τ, $X(t)$ is at least a wide-sense stationary process.

Similarly, we know that the random amplitude signal $X(t) = X \cos \omega_0 t$ is not a stationary process, while the random amplitude and phase signal $X(t) = X \cos(\omega_0 t + \varphi)$, which amplitude and phase are independent, is a stationary process.

Example 1.2.2. A typical sample of the random telegraph signals $X(t)$ is shown in Fig. 1.5. At any time t, the value of $X(t)$ can just be 0 or 1, while their probability is equal to 0.5. The transformation moment between the two values is random, and in the unit time the average number of transformations is λ. At any moment in the time interval $|\tau| = |t_2 - t_1|$, the probability of times of transformation obeys Poisson's distribution:

$$P_k(\lambda|\tau|) = \frac{(\lambda|\tau|)^k}{k!}\mathrm{e}^{-\lambda|\tau|}$$

Here, λ is the average number of transformations in a unit time. Try to prove that $X(t)$ is a stationary process.

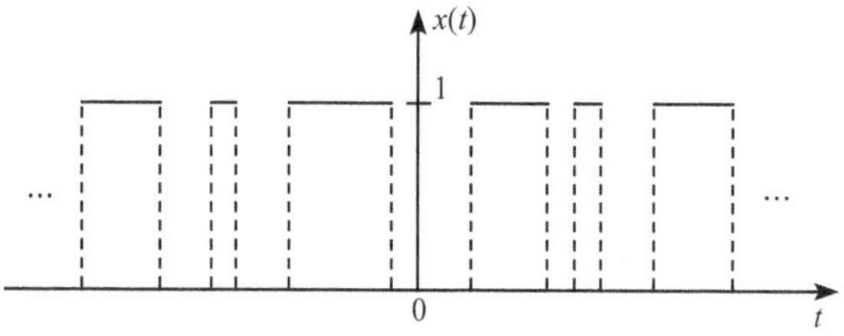

Fig. 1.5: A typical sample of random telegraph signals.

Solution:

$$m_X(t) = 1 \cdot P[X(t) = 1] + 0 \cdot P[X(t) = 0] = \frac{1}{2}$$

When trying to obtain the correlation function, if in a time interval $|\tau|$ the inner transformation changes an even number of times, $X(t_1)$ with $X(t_2)$ must be the same value, and the product is 1 or 0; if it changes an odd number of times, the two values are different and the product is 0. The probability of both $X(t_1)$ and $X(t_2)$ being equal to 1 (or equal to 0) is:

$$P_1 = \frac{1}{2} \sum_{k=0}^{\infty} P_k(\lambda|\tau|)\,, \quad k = 0, 2, 4, \dots$$

$$R_X(t_1, t_2) = E[X(t_1)X(t_2)] = 1 \cdot P_1$$

$$= \frac{1}{2} e^{-\lambda|\tau|} \left[1 + \frac{(\lambda|\tau|)^2}{2!} + \frac{(\lambda|\tau|)^4}{4!} + \cdots \right]$$

Then,

$$= \frac{1}{4} e^{-\lambda|\tau|} [e^{\lambda|\tau|} + e^{-\lambda|\tau|}]$$

$$= \frac{1}{4} + \frac{1}{4} e^{-2\lambda|\tau|}$$

Because it meets the conditions of the wide-sense stationary process, $X(t)$ is a stationary process.

1.2.2 Ergodic processes

There are two variables in the random process $X(e, t)$, so there are two methods to obtain the mean value: statistical average and time average. Let us start with the two approaches.

1.2.2.1 Statistical average

The average value in the discussion of the moment functions above is the statistical average. These moment functions are statistical average parameters. This method that obtains the value of all the samples at the same (or some) time in a statistical way is called the statistical average or the set average. We denote it by $E[\bullet]$ or $\overline{\bullet}$. For example, for stationary processes $X(t)$, the set average is

$$E(X(t)) = \overline{X(t)} = \int_{-\infty}^{\infty} xp(x)\,\mathrm{d}x = m_X \tag{1.38}$$

The set correlation function is

$$R_X(t_1, t_2) = \overline{X(t_1)X(t_2)} = \int_{-\infty}^{\infty} \int_{-\infty}^{\infty} x_1 x_2 p_2(x_1, x_1; \tau)\,\mathrm{d}x_1\,\mathrm{d}x_2 = R_X(\tau) \tag{1.39}$$

where, $\tau = t_2 - t_1$.

When using the two equations above to obtain the statistical average, the first step is to obtain a group of n samples of the process $X(t)$ (in theory we need to make n infinite), then use the statistical method on the values of one same moment or two same moments to get the one- or two-dimensional probability density. Until then, we can use the two equations above. So, this is a very complicated solution. The usual practice is to go through the experiment first and get a fairly large number of finite samples, and then work out the arithmetic average and approximate it as the statistical average. For example, when the accuracy requirement is not high, the following approximation formula can be adopted:

$$m_X(t) \approx \frac{1}{n}\sum_{i=1}^{n} x_i(t) \tag{1.40}$$

$$R_X(t, t+\tau) \approx \frac{1}{n}\sum_{i=1}^{n} x_i(t)x_i(t+\tau) \tag{1.41}$$

In the formula, n is the total number of samples. However, even so, we still need to make a large number of repeat observations on random processes to obtain enough samples $x_i(t)$. Obviously, the statistical average method experiments require a lot of work and complicated data processing. So, naturally, people will wonder: could we use time average instead of statistical average? If it works, what conditions do we need?

1.2.2.2 Time average

It has been pointed out that the random process $X(t)$ is a set of time functions, and every sample in this set is a function of time. For the values of a particular sample in the set at each moment, the general mathematical method is used to get the average.

So, we can define the time average of one particular sample $x_i(t)$ as:

$$\langle x_i(t)\rangle = \lim_{T\to\infty} \frac{1}{2T}\int_{-T}^{T} x_i(t)\,\mathrm{d}t \tag{1.42}$$

The time autocorrelation function of the sample is

$$\langle x_i(t)x_i(t+\tau)\rangle = \lim_{T\to\infty} \frac{1}{2T}\int_{-T}^{T} x_i(t)x_i(t+\tau)\,\mathrm{d}t \tag{1.43}$$

where $\tau = t_2 - t_1$. Obviously, we only need one sample to obtain the time average, and general mathematics methods for calculation are available for the definite time function. Even if the time function can be complex and difficult to calculate, we can still use integral average apparatus for measuring (the test condition should be kept unchanged). In theory, we need to make the average time infinite, that is, the sample should last forever. However, in fact, it only needs to be long enough to meet certain precision requirements of the project (estimating that the variance is small enough). The average time is much more practical than the statistical average, so actually it is the main method of measuring the random process.

Example 1.2.3. Set $x_i(t) = A\cos(\omega_0 t+\varphi_i)$ as one of the examples of the random initial phase signal $X(t) = A\cos(\omega_0 t+\varphi)$ shown in Example 1.2.1. Try to solve the time average and time autocorrelation function of $x_i(t)$ and compare these with the results obtained from the example.

Solution:

$$\begin{aligned}
\langle x_i(t)\rangle &= \lim_{T\to\infty}\frac{1}{2T}\int_{-T}^{T} A\cos(\omega_0 t+\varphi_i)\,\mathrm{d}t \\
&= \lim_{T\to\infty}\frac{A\cos\varphi_i}{2T}\int_{-T}^{T}\cos\omega_0 t\,\mathrm{d}t \\
&= \lim_{T\to\infty}\frac{A\cos\varphi_i\sin\omega_0 T}{\omega_0 T} = 0 \\
\langle x_i(t)x_i(t+\tau)\rangle &= \lim_{T\to\infty}\frac{1}{2T}\int_{-T}^{T} A^2\cos(\omega_0 t+\varphi_i)\cos[\omega_0(t+\tau)+\varphi_i]\,\mathrm{d}t \\
&\approx \lim_{T\to\infty}\frac{A^2}{4T}\int_{-T}^{T}\cos\omega_0\tau\,\mathrm{d}t \\
&= \frac{A^2}{2}\cos\omega_0\tau
\end{aligned}$$

Compared with the results of Example 1.2.1, we can obtain the equations of relation:

$$E[x_i(t)] = \langle x_i(t)\rangle$$
$$E[x_i(t)x_i(t+\tau)] = \langle x_i(t)x_i(t+\tau)\rangle$$

Obviously for this random initial phase signal, the time average of any sample is equal to the statistical average of the set, so we can use the simple time average to replace the complex statistical average.

1.2.2.3 Characteristic of ergodicity

If the time average of each sample of the random process $X(t)$ is equal to the statistical average of the set in probability, or more exactly, its time average has convergence in the statistical average of the set in probability 1, we say the process $X(t)$ has ergodicity. Namely, if:

$$\langle X(t)\rangle \overset{P}{=} E[X(t)] \tag{1.44}$$

we can say that the mean of the process $X(t)$ has ergodicity. The letter P on the middle of the medium means that they are equal in probability.

Similarly, if we have

$$\langle X(t)X(t+\tau)\rangle \overset{P}{=} E[X(t)X(t+\tau)] \tag{1.45}$$

we can say that the correlation function of the process $X(t)$ has ergodicity. Moreover, similarly, we can use the definition of ergodicity for the moment functions such as the variance and the mean square value. If both the mean value and the correlation function of the random process $X(t)$ have ergodicity, we can call $X(t)$ a wide-sense ergodic random process.

The physical meaning of ergodicity is that any sample of the random process has experienced all possible states of this random process over a long enough period of time. So, for the ergodic process, any one of its samples can be used as a fully representative typical sample. The time average obtained from this sample is equal to the statistical average of the set in probability, which can simplify the calculation and measurement.

1.2.2.4 Ergodicity conditions

By eqs. (1.42) and (1.43), the time average results are as follows: the time average value must be a constant that has no relation with the time t, and the time autocorrelation function must be a function of only the time interval τ. Obviously, the ergodic process must be stationary, but on the other hand, not all stationary processes have ergodicity. Thus, a stationary process is a necessary condition for the ergodic process, but not a sufficient condition. Namely, they have the following relationship:

$$\{\text{Ergodic Process}\} \subset \{\text{Stationary Process}\} \subset \{\text{Random Process}\}$$

Let us discuss the necessary and sufficient conditions for ergodicity.

(1) The necessary and sufficient condition of the ergodicity of the mean value m_X of the stationary process $X(t)$ is:

$$\lim_{T\to\infty}\frac{1}{2T}\int_{-2T}^{2T}\left(1-\frac{|\tau|}{2T}\right)C_X(\tau)\,\mathrm{d}\tau = 0 \tag{1.46}$$

where $C_X(\tau) = R_X(\tau) - m_X^2$ is the covariance function.

Proof. Because each sample of $X(t)$ is a different time function, the time average value of the process $\langle X(t)\rangle$ is a random variable that varies with the sample. Its mathematical expectation is a certain value, namely:

$$E[\langle X(t)\rangle] = E\left[\lim_{T\to\infty}\frac{1}{2T}\int_{-T}^{T}X(t)\,\mathrm{d}t\right] = \lim_{T\to\infty}\frac{1}{2T}\int_{-T}^{T}E[X(t)]\,\mathrm{d}t = m_X \tag{1.47}$$

□

One property is used in the equation: we can change the order of obtaining the statistic average value and doing the integration. What is more, the integral here is not a general integral (see Section 2.2 for details). According to the Chebyshev inequality:

$$P\{|X - m_X| < \varepsilon\} \geq 1 - \frac{\sigma_X^2}{\varepsilon^2} \tag{1.48}$$

We can obtain:

$$P\{|\langle X(t)\rangle - m_X| < \varepsilon\} \geq \lim_{T\to\infty}\left[1 - \frac{\sigma_{XT}^2}{\varepsilon^2}\right] \tag{1.49}$$

where ε stands for any positive number;

Variance:

$$\begin{aligned}
\sigma_{XT}^2 &= E\{[\langle \mathring{X}(t)\rangle]^2\} = E\left\{\left[\lim_{T\to\infty}\frac{1}{2T}\int_{-T}^{T}\mathring{X}(t)\,\mathrm{d}t\right]^2\right\} \\
&= E\left\{\lim_{T\to\infty}\frac{1}{4T^2}\int_{-T}^{T}\mathring{X}(t_1)\,\mathrm{d}t_1\int_{-T}^{T}\mathring{X}(t_2)\,\mathrm{d}t_2\right\} \\
&= \lim_{T\to\infty}\frac{1}{4T^2}\int_{-T}^{T}\int_{-T}^{T}E[\mathring{X}(t_1)\mathring{X}(t_2)]\,\mathrm{d}t_1\,\mathrm{d}t_2
\end{aligned}$$

That is,

$$\sigma_{XT}^2 = \lim_{T\to\infty}\frac{1}{4T^2}\int_{-T}^{T}\int_{-T}^{T}C_X(t_1 - t_2)\,\mathrm{d}t_1\,\mathrm{d}t_2 \tag{1.50}$$

Now, we substitute t, τ for the variables in the above equation t_1, t_2. With $t = -t_1$, $\tau - t_2 - t_1$, the Jacobian determinant of the transformation is

$$J = \frac{\partial(t_1, t_2)}{\partial(t, \tau)} = \begin{vmatrix} \frac{\partial t_1}{\partial t} & \frac{\partial t_1}{\partial \tau} \\ \frac{\partial t_2}{\partial t} & \frac{\partial t_2}{\partial \tau} \end{vmatrix} = \begin{vmatrix} \frac{\partial(-t)}{\partial t} & \frac{\partial(-t)}{\partial \tau} \\ \frac{\partial(\tau-t)}{\partial t} & \frac{\partial(\tau-t)}{\partial \tau} \end{vmatrix} = \begin{vmatrix} -1 & 0 \\ -1 & 1 \end{vmatrix} = -1$$

In this case, as is shown in Fig. 1.6, the integral range transforms from the square in solid lines to the parallelogram in dotted lines, so the equation above can be rewritten as

$$\begin{aligned}
\sigma_{XT}^2 &= \lim_{T\to\infty}\frac{1}{4T^2}\iint_{\diamond} C_X(t_1 - t_2)|J|\,\mathrm{d}\tau\,\mathrm{d}t \\
&= \lim_{T\to\infty}\frac{1}{4T^2}\left[\int_{0}^{2T}C_X(\tau)\,\mathrm{d}\tau\int_{\tau-T}^{T}\mathrm{d}t + \int_{-2T}^{0}C_X(\tau)\,\mathrm{d}\tau\int_{-T}^{\tau+T}\mathrm{d}t\right]
\end{aligned}$$

or

$$\sigma_{XT}^2 = \lim_{T\to\infty}\frac{1}{2T}\int_{-2T}^{2T}\left(1 - \frac{|\tau|}{2T}\right)C_X(\tau)\,\mathrm{d}\tau \tag{1.51}$$

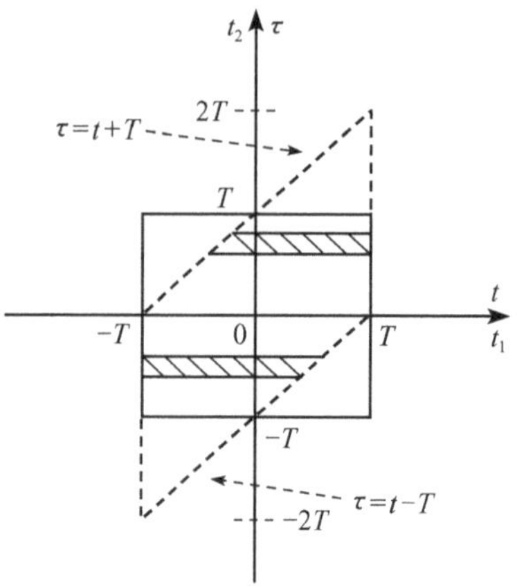

Fig. 1.6: Transformation of the integral range.

If $\lim_{T\to\infty} \sigma_{XT}^2 \to 0$, we learn from eq. (1.49) that:

$$P\{|\langle X(t)\rangle - m_X| < \varepsilon\} = 1 \tag{1.52}$$

That is,

$$\langle X(t)\rangle \overset{P}{=} E[X(t)]$$

Therefore, when eq. (1.51) is equal to zero, it is possible to obtain the conclusion that is to be proved.

(2) The necessary and sufficient condition of the ergodicity of the correlation function $R_X(\tau)$ of the stationary process $X(t)$ is

$$\lim_{T\to\infty} \frac{1}{2T} \int_{-2T}^{2T} \left(1 - \frac{|\alpha|}{2T}\right) C_\varphi(\alpha)\, d\alpha = 0 \tag{1.53}$$

where

$$C_\varphi(\alpha) = R_\varphi(\alpha) - E^2[\varphi(t)] = R_\varphi(\alpha) - R_X^2(\tau)$$
$$R_\varphi(\alpha) = E\left[\varphi(t)\varphi(t+\alpha)\right] = E\left[X(t)X(t+\tau)X(t+\alpha)X(t+\tau+\alpha)\right]$$

In the certification of (1) above, we used $\varphi(t) = X(t)X(t+\tau)$ instead of $X(t)$ to obtain this equation.

(3) The zero mean stationary normal random process $X(t)$ is the most common. The necessary and sufficient conditions of the ergodicity of the process can be proved as follows:

$$\lim_{T\to\infty} \frac{1}{T} \int_0^T |C_X(\alpha)|^2\, d\alpha = 0 \tag{1.54}$$

or

$$\lim_{|\tau|\to\infty} C_X(\tau) = 0 \tag{1.55}$$

Obviously, when $C_X(\tau)$ contains periodic components, the upper two equations will not be true.

It can be seen from above that it is very difficult to strictly test whether the mean value and the correlation functions of a random process have ergodicity because its second- and fourth-order moment functions must be calculated separately first. Therefore, it is usually based on experience that the stationary process is assumed as an the ergodic process first, and then this assumption is tested according to the experimental results. It should be pointed out that the conditions of ergodicity are relatively loose, and most stationary processes encountered in actual engineering are ergodic processes.

Example 1.2.4. Known is a stationary normal noise $X(t)$, its mean value is zero and the correlation function is $R_X(\tau) = \sigma^2 e^{-\tau^2} \cos \omega_0 \tau$, where σ and ω_0 are constants. Try to determine whether the process is ergodic.

Solution: $X(t)$ is a stationary normal noise with zero mean,

$$C_X(\tau) = R_X(\tau) = \sigma^2 e^{-\tau^2} \cos \omega_0 \tau$$

Due to $\lim_{|\tau| \to \infty} C_X(\tau) = 0$.

This meets the conditions of judgment in eq. (1.55). Now, we know that $X(t)$ is an ergodic process.

1.2.2.5 Engineering significance of moment functions of the ergodic process

If a stationary process $X(t)$ is ergodic, and its statistic average can be replaced by its time average, all its moment functions can be explained by the respective time average parameters. For example, when $X(t)$ means the voltage (or the current), every moment function has its engineering significance as stated below.

(1) Because the mean value of the set can be replaced by the time average, its mathematical expectation m_X expresses the DC voltage (or current) component of the process $X(t)$, while m_X^2 expresses the $X(t)$ process' DC component power consumed on a 1-ohm resistor.
(2) Because the set variance can be replaced by the time variance, the set variance σ_X^2 expresses the average undulating power of the process $X(t)$ on a 1-ohm resistor (i.e., alternating power), while the standard deviation σ_X expresses the effective value of the fluctuation voltage (or current).
(3) When $\tau = 0$, the correlation function of the set $R_X(0)$ is equal to the mean square value of the set $E\{[X(t)]^2\}$. From eq. (1.18) we know that:

$$R_X(0) = \sigma_X^2 + m_X^2 \tag{1.56}$$

Obviously, the mean square value of the set expresses the total average power on a 1-ohm resistor of the process $X(t)$, and it is the sum of the DC component power and the average fluctuation power.

1.2.3 Properties of correlation functions

To make the discussion more general, we make the random process have a nonzero mean. The correlation function $R_X(\tau)$ of the stationary random processes $X(t)$ has the following properties:

(1)
$$R_X(\tau) = R_X(-\tau) \tag{1.57}$$

Namely, the correlation function is an even function of the variable τ, as is shown in Fig. 1.7 (a).

Proof. $R_X(\tau) = E[X(t)X(t+\tau)] = E[X(t+\tau)X(t)] = R_X(-\tau)$. □

Similarly, we have
$$C_X(\tau) = C_X(-\tau) \tag{1.58}$$

as shown in Fig. 1.7 (b).

(2)
$$R_X(0) \geq 0 \tag{1.59}$$

This means that the correlation function is non-negative when $\tau = 0$, as is shown in Fig. 1.7 (a).

Proof. When $\tau = 0$. From eq. (1.56) we know: $R_X(0) = \sigma_X^2 + m_X^2 \geq 0$. □

Similarly, we have

$$C_X(0) = \sigma_X^2 \geq 0 \tag{1.60}$$

(3)
$$R_X(0) \geq |R_X(\tau)| \tag{1.61}$$

This means the correlation function reaches its maximum value when $\tau = 0$, as is shown in Fig. 1.7 (a). This indicates that two variables whose interval is zero in the process $X(t)$ have the biggest statistical correlation degree.

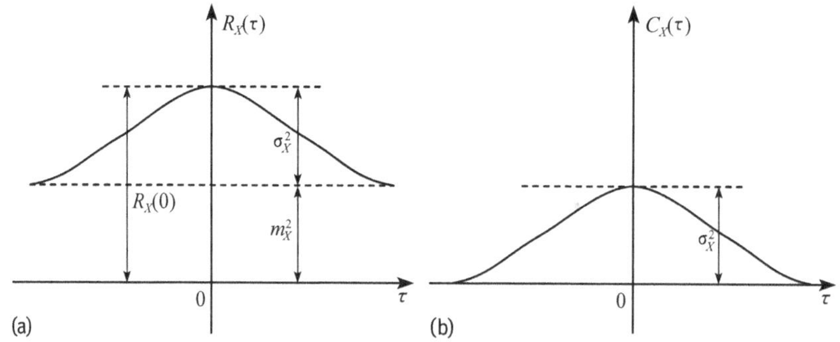

Fig. 1.7: Correlation function and covariance function of a stationary random process.

The mathematical expectation of any positive function is non-negative, so,

$$E\{[X(t) \pm X(t+\tau)]^2\} \geq 0 \quad \text{or} \quad E\{X^2(t) \pm 2X(t)X(t+\tau) + X^2(t+\tau)\} \geq 0$$

If process is smooth, then

$$E[X^2(t)] = E[X^2(t+\tau)] = R_X(0)$$

Thus,

$$2R_X(0) \pm 2R_X(\tau) \geq 0 \quad \text{or} \quad R_X(0) \geq |R_X(\tau)|$$

Similarly, there are:

$$C_X(0) \geq |C_X(\tau)|$$

(4) If the periodic component is not included in the stationary random process, then we have

$$R_X(\infty) = \lim_{|\tau|\to\infty} R_X(\tau) = m_X^2 \tag{1.62}$$

Proof. For this kind of random process, when $|\tau|$ increases, the correlation degree of the variables $X(t)$ and $X(t+\tau)$ declines. In the limit case of $|\tau| \to \infty$, the two random variables will show independence, so we have

$$\lim_{|\tau|\to\infty} R_X(\tau) = \lim_{|\tau|\to\infty} E\left[X(t)X(t+\tau)\right] = \lim_{|\tau|\to\infty} E[X(t)] \cdot E\left[X(t+\tau)\right] = m_X^2 \qquad \square$$

Similarly, it is not hard to find:

$$C_X(\infty) = \lim_{|\tau|\to\infty} C_X(\tau) = 0 \tag{1.63}$$

For this kind of random process, the mean value and variance can be expressed as follows:

$$\begin{cases} m_X = \pm\sqrt{R_X(\infty)} \\ \sigma_X^2 = C_X(0) = R_X(0) - R_X(\infty) \end{cases} \tag{1.64}$$

(5) For periodic random processes, we have

$$R_X(\tau) = R_X(\tau + T) \tag{1.65}$$

The stationary functions are also periodic, and the period remains the same.

Proof. Periodic processes have properties $X(t) = X(t+T)$, therefore, are

$$R_X(t) = E[X(t)X(t+\tau)] = E[X(t)X(t+\tau+T)] = R_X(\tau+T) \qquad \square$$

(6) $R_X(\tau)$ is a non-negative function, so to any arbitrary array $t_1, t_2, \ldots, t_n$ or any function $f(t)$, we have

$$\sum_{i,j=1}^{n} R_X(t_i - t_j)f(t_i)f(t_j) \geq 0 \tag{1.66}$$

Proof.

$$\sum_{i,j=1}^{n} R_X(t_i - t_j)f(t_i)f(t_j) = \sum_{i,j=1}^{n} E[X(t_i)X(t_j)]f(t_i)f(t_j)$$

$$= E\left\{\sum_{i,j=1}^{n} X(t_i)X(t_j)f(t_i)f(t_j)\right\} = E\left\{\left[\sum_{i=1}^{n} X(t_i)f(t_i)\right]^2\right\} \geq 0 \quad \square$$

Note: This is an important property of correlation functions, which guarantees the power spectral density $G(\omega)$ is a non-negative function of ω, and it shows that not any function can be a correlation function.

Example 1.2.5. Suppose the correlation functions of each stationary random process shown in Fig. 1.8. Try to determine if they can be formed and describe the characteristics of these processes (whether there is a DC component, if it is periodic, how fast or slow the wave fluctuation is.)

Solution: Fig. 1.8 (a) violates property (3), therefore it cannot be formed.

Figure 1.8 (b) can be formed, there is a DC component, which is not periodic and the fluctuation in process is slow.

Figure 1.8 (c) can be formed, there is no DC component, it is periodic and the fluctuation in the process is very fast (it is a δ-function whose width is very narrow).

Figure 1.8 (d) can be formed, there is no DC component and it is not periodic, the fluctuation in the process is fast (this refers to its carrier wave, while the envelope wave has a slower fluctuation), and it changes from positive to negative alternately.

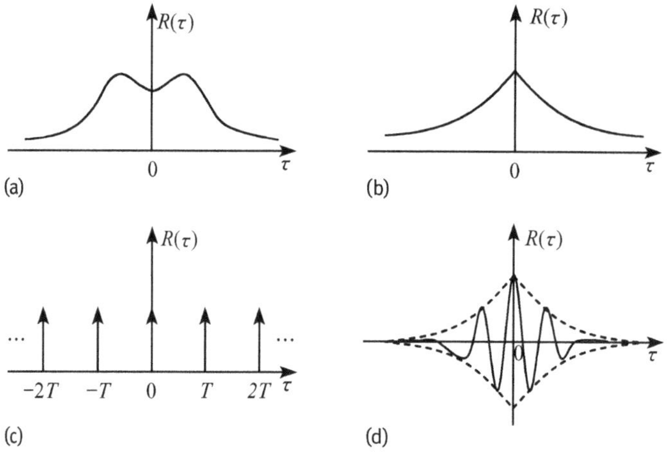

Fig. 1.8: Relative functions of stationary processes.

1.2.4 Correlation coefficient and correlation time

1.2.4.1 Correlation coefficient

For the stationary process $X(t)$, the correlation degree of two fluctuation variables with interval τ [variable $\mathring{X}(t)$ and $\mathring{X}(t+\tau)$] can be expressed by the covariance function $C_X(\tau) = E[\mathring{X}(t)\mathring{X}(t+\tau)]$. However, $C_X(\tau)$ also have a relation to two intensities of fluctuation. If $\mathring{X}(t)$ and $\mathring{X}(t+\tau)$ are very small, even if the degree of correlation is strong (namely a small τ), $C_X(\tau)$ will not be big. So, the covariance function cannot exactly express the degree of the association. In order to show the degree of correlation with certainty, the effect of the fluctuation intensity should be removed. The covariance function should be normalized, and we introduce a dimensionless ratio:

$$r_X(\tau) = \frac{C_X(\tau)}{C_X(0)} = \frac{C_X(\tau)}{\sigma_X^2} \tag{1.67}$$

This is called the autocorrelation coefficient of a random process $X(t)$ (also known as the normalized correlation function or the standard covariance function), short for correlation coefficient. It can accurately represent the degree of (linear) correlation between two fluctuations in a random process.

Obviously, the correlation coefficient also has similar properties to the correlation function. Figure 1.9 shows two typical curves of $r_X(\tau)$. The figure shows that $r_X(\tau)$ could be positive, zero, or negative. The plus sign means positive correlation, indicating that it is more likely that variable $\mathring{X}(t)$ has the same sign as $\mathring{X}(t+\tau)$; the minus sign means negative correlation, indicating that it is more likely that they have the opposite sign. The waveform of Fig. 1.9 (b) shows an alternating change between positive and negative, indicating that the waveform of this process changes rapidly. The absolute value $|r_X(\tau)|$ represents the degree of correlation; $r_X(\tau) = 1$, it is completely correlated; $r_X(\tau) = 0$, it is uncorrelated.

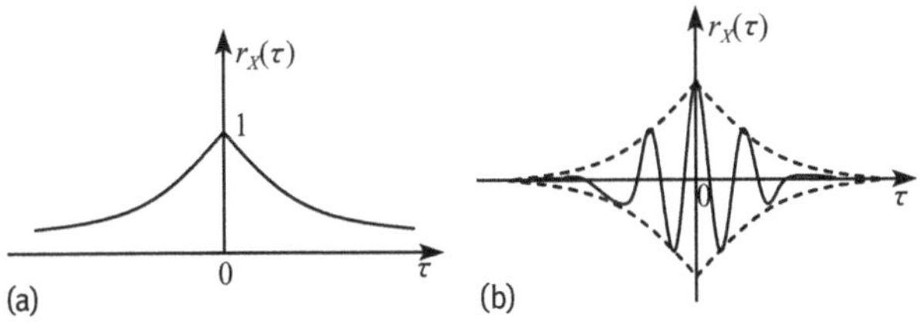

Fig. 1.9: Typical correlation coefficient curves.

1.2.4.2 Correlation time

For general random processes, the degree of correlation between $\mathring{X}(t)$ and $\mathring{X}(t+\tau)$ will decline with the increase of interval τ. When $\tau \to \infty$, $r_X(\tau) \to 0$, at this moment $\mathring{X}(t)$ and $\mathring{X}(t+\tau)$ are no longer correlated. In fact, as long as $r_X(\tau)$ is less than an approximate zero value, they can be considered uncorrelated in engineering. Therefore, we

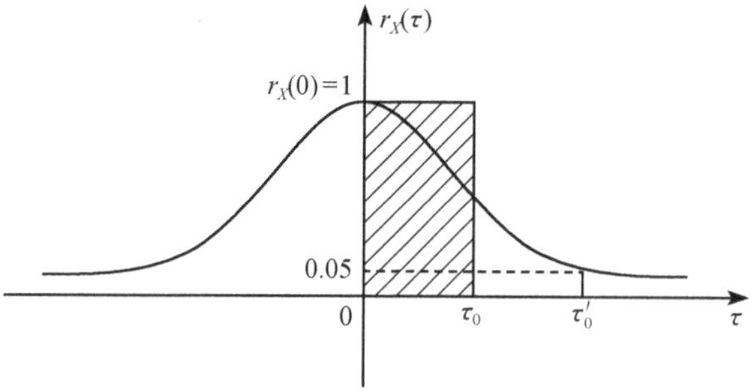

Fig. 1.10: Two definitions of correlation coefficients.

often call the time interval it experienced as $r_X(\tau)$ decreases from its maximum value $r_X(\tau) = 1$ to a value near zero (usually we make it 0.05) the correlation time and denote it by τ_0'. When $\tau \geq \tau_0'$, we can regard $\mathring{X}(t)$ and $\mathring{X}(t+\tau)$ as uncorrelated.

In general, the area enclosed by the curve $r_X(\tau)$ and the horizontal axis τ is equivalent to the rectangular area (Fig. 1.10). Its high is $r_X(0)$ and the bottom width τ_0 is called the correlation time, which is

$$\tau_0 = \int_0^\infty r_X(\tau)\,\mathrm{d}\tau \tag{1.68}$$

The correlation time τ_0 (or τ_0') only has a relation to the speed of the decrease of the curve $r_X(\tau)$ and it is a parameter, which is uncorrelated with the time interval τ It can simply reflect the average fluctuation speed of the random process. The smaller correlation time τ_0 indicates a faster speed of the average fluctuation of the process. On the other hand, the bigger τ_0 indicates a slower speed of the average fluctuation of the process.

Example 1.2.6. Given two random processes $X(t)$ and $Y(t)$, the covariance functions are

$$C_X(\tau) = \frac{1}{4}\mathrm{e}^{-2\lambda|\tau|}, \quad C_Y(\tau) = \frac{\sin\lambda\tau}{\lambda\tau}$$

(1) Compare the speed of fluctuation of two processes.
(2) Compare the degree of correlation between the two processes when $\tau = \pi/\lambda$.
(3) Compare the degree of correlation of the process $Y(t)$ when $\tau = 0$ and $\tau = \pi/\lambda$.

Solution: (1)

$$\sigma_X^2 = C_X(0) = \frac{1}{4}$$

$$r_X(\tau) = \frac{C_X(\tau)}{\sigma_X^2} = \mathrm{e}^{-2\lambda|\tau|}$$

$$\tau_{0X} = \int_0^\infty r_X(\tau)\,\mathrm{d}\tau = \int_0^\infty \mathrm{e}^{-2\lambda\tau}\,\mathrm{d}\tau = \frac{1}{2\lambda}$$

$$\sigma_Y^2 = C_Y(0) = 1$$

$$r_Y(\tau) = \frac{C_Y(\tau)}{\sigma_Y^2} = \frac{\sin \lambda\tau}{\lambda\tau}$$

$$\tau_{0Y} = \int_0^\infty \frac{\sin \lambda\tau}{\lambda\tau}\, d\tau = \frac{\pi}{2\lambda}$$

Because $\tau_{0X} < \tau_{0Y}$, the fluctuation speed of $X(t)$ is faster than $Y(t)$.
(2) When $\tau = \pi/\lambda$, we have

$$r_X\left(\frac{\pi}{\lambda}\right) = e^{-2\lambda|\frac{\pi}{\lambda}|} = e^{-2\pi}$$

$$r_Y\left(\frac{\pi}{\lambda}\right) = \frac{\sin \pi}{\pi} = 0$$

So, in this case, the process $X(t)$ is correlated, while the process $Y(t)$ is uncorrelated.
(3) When $\tau = 0$, we have $r_Y(0) = 1$. In this case, the process $Y(t)$ is completely correlated, while when $\tau = \pi/\lambda$ it becomes uncorrelated.

In this example, the random process $Y(t)$ is sometimes correlated and sometimes uncorrelated. This is because $\sin \lambda\tau/\lambda\tau$ is an orthogonal function.

This example illustrates that the correlation coefficient $r(\tau)$ can be used both to express the correlation degree of two fluctuation values in one process and the correlation degree of two fluctuation values of two processes in the interval τ, while the correlation time τ_0 can only be used to compare the two fluctuation speeds of two processes.

1.3 Joint stationary random processes

So far, we have only discussed the statistical properties of a single random process, while we actually often need to consider the statistical features of two or more random processes. For example, to detect signal from noise the background, we must study the two signal and noise random processes at the same time. In the following, the joint probability distribution and correlation function of two random processes are introduced.

1.3.1 Joint probability distribution and moment functions of two random processes

There are two random processes $X(t)$ and $Y(t)$. Their probability densities are, respectively,

$$p_n(x_1, x_2, \ldots, x_n; t_1, t_2, \ldots, t_n)$$
$$p_m(y_1, y_2, \ldots, y_m; t_1', t_2', \ldots, t_m')$$

Define the $n+m$ dimension joint distribution function of these two processes as:

$$\begin{aligned} &F_{n+m}(x_1, x_2, \dots, x_n; y_1, y_2, \dots, y_m; t_1, t_2, \dots, t_n; t_1', t_2', \dots, t_m') \\ &= P\{X(t_1) \le x_1, X(t_2) \le x_2, \dots, X(t_n) \le x_n; \\ &\quad Y(t_1') \le y_1, Y(t_2') \le y_2, \dots, Y(t_m') \le y_m\} \end{aligned} \tag{1.69}$$

Define the $n+m$ dimension joint distribution density of these two processes as:

$$\begin{aligned} &p_{n+m}(x_1, x_2, \dots, x_n; y_1, y_2, \dots, y_m; t_1, t_2, \dots, t_n; t_1', t_2', \dots, t_m') \\ &= \frac{\partial^{n+m} F_{n+m}(x_1, x_2, \dots, x_n; y_1, y_2, \dots, y_m; t_1, t_2, \dots, t_n; t_1', t_2', \dots, t_m')}{\partial x_1 \partial x_2 \dots \partial x_n \partial y_1 \partial y_2 \dots \partial y_m} \end{aligned} \tag{1.70}$$

If

$$\begin{aligned} &p_{n+m}(x_1, x_2, \dots, x_n; y_1, y_2, \dots, y_m; t_1, t_2, \dots, t_n; t_1', t_2', \dots, t_m') \\ &\quad = p_n(x_1, x_2, \dots, x_n; t_1, t_2, \dots, t_n) p_m(y_1, y_2, \dots, y_m; t_1', t_2', \dots, t_m') \end{aligned} \tag{1.71}$$

We can say that the processes $X(t)$ and $Y(t)$ are statistically independent.

If for any $n+m$ dimension, p_{n+m} or F_{n+m} does not vary with different selection of the starting point of the clock, that is, when the time is shifted by any constant ε, the probability density (or distribution function) will not change, and then we say that the processes $X(t)$ and $Y(t)$ have stationary dependency.

When analyzing two random processes at the same time, the most important moment function is the cross-correlation function. There are two random processes $X(t)$ and $Y(t)$; at any two moments t_1, t_2 their values are the random variables $X(t_1)$, $Y(t_2)$. We define their cross-correlation function as follows:

$$R_{XY}(t_1, t_2) = E[X(t_1)Y(t_2)] = \int_{-\infty}^{\infty}\int_{-\infty}^{\infty} xy p_2(x, y; t_1, t_2)\,\mathrm{d}x\,\mathrm{d}y \tag{1.72}$$

where $p_2(x, y; t_1, t_2)$ is the joint two-dimensional probability density of $X(t)$ and $Y(t)$.

Similarly, we define the centralized cross-correlation function (commonly known as the cross-covariance function) as

$$C_{XY}(t_1, t_2) = E\{[X(t_1) - m_X(t_1)][Y(t_2) - m_Y(t_2)]\} \tag{1.73}$$

where variables $m_X(t_1)$ and $m_Y(t_2)$ are the mean values of the variables $X(t_1)$ and $Y(t_2)$, respectively.

If both $X(t)$ and $Y(t)$ are wide-sense stationary processes, and their cross-correlation function is a single variable function about time intervals $\tau = t_2 - t_1$,

$$R_{XY}(t_1, t_2) = R_{XY}(\tau) \tag{1.74}$$

we say that the processes $X(t)$ and $Y(t)$ are wide-sense stationary correlated (or joint stationary for short). When the processes $X(t)$ and $Y(t)$ are joint stationary, the time cross-correlation is defined as

$$\langle X(t)Y(t+\tau)\rangle = \lim_{T\to\infty}\frac{1}{2T}\int_{-T}^{T} x(t)y(t+\tau)\,\mathrm{d}t \tag{1.75}$$

If they converge to the cross-correlation function of the set $R_{XY}(\tau)$ in the probability 1, that is, they are equal in the probability, we have:

$$\langle X(t)Y(t+\tau)\rangle \overset{P}{=} \overline{X(t)Y(t+\tau)} \tag{1.76}$$

We say that the processes $X(t)$ and $Y(t)$ have the joint ergodicity characteristic.

Example 1.3.1. Random process $X(t)$ is statistically independent of $Y(t)$. Try to solve their cross-correlation and cross-covariance functions.

Solution: $X(t)$ is statistically independent of $Y(t)$, so we can learn from eq. (1.71) that:

$$p_2(x_1, y_2; t_1, t_2) = p(x_1, t_1)p(y_2, t_2)$$

So, the cross-correlation function is:

$$\begin{aligned}
R_{XY}(t_1, t_2) &= E[X(t_1)Y(t_2)] \\
&= \int_{-\infty}^{\infty}\int_{-\infty}^{\infty} x_1 y_2 p_2(x_1, y_2; t_1, t_2)\,\mathrm{d}x_1\,\mathrm{d}y_2 \\
&= \int_{-\infty}^{\infty} x_1 p(x_1, t_1)\,\mathrm{d}x_1 \int_{-\infty}^{\infty} y_2 p(y_2, t_2)\,\mathrm{d}y_2 \\
&= m_X(t_1)m_Y(t_2)
\end{aligned}$$

The cross-covariance function is:

$$\begin{aligned}
C_{XY}(t_1, t_2) &= E\{[X(t_1) - m_X(t_1)]\,[Y(t_2) - m_Y(t_2)]\} \\
&= \int_{-\infty}^{\infty}\int_{-\infty}^{\infty} [x_1 - m_X(t_1)]\,[y_2 - m_Y(t_2)]\,p_2(x_1, y_2; t_1, t_2)\,\mathrm{d}x_1\,\mathrm{d}y_2 \\
&= \int_{-\infty}^{\infty} [x_1 - m_X(t_1)]\,p(x_1, t_1)\,\mathrm{d}x_1 \int_{-\infty}^{\infty} [y_2 - m_Y(t_2)]\,p(y_2, t_2)\,\mathrm{d}y_2 \\
&= [m_X(t_1) - m_X(t_1)]\,[m_Y(t_2) - m_Y(t_2)] \\
&= 0
\end{aligned}$$

1.3.2 Moment function of joint stationary random processes

When the random process $X(t)$ are joint stationary with $Y(t)$, their cross-correlation function is as shown eq. (1.74). Similarly, their covariance function is

$$C_{XY}(t_1, t_2) = C_{XY}(\tau) \tag{1.77}$$

where $\tau = t_2 - t_1$.

If the random process $X(t)$ is joint stationary with $Y(t)$, we can define their cross-correlation coefficient as follows:

$$r_{XY}(\tau) = \frac{C_{XY}(\tau)}{\sigma_X \sigma_Y} \tag{1.78}$$

where σ_X, σ_Y are the standard deviations of the two random processes, respectively.

For any value of τ, we have

$$r_{XY}(\tau) = 0 \quad \text{or} \quad C_{XY}(\tau) = 0 \tag{1.79}$$

We say that the process $X(t)$ is uncorrelated with $Y(t)$. The upper right eq. (1.79) can be equivalent to the following:

$$R_{XY}(\tau) = m_X(t) m_Y(t+\tau) \quad \text{or} \quad E[X(t)Y(t+\tau)] = E[X(t)]E[Y(t+\tau)]$$

For any value of τ, we have

$$E[X(t)Y(t+\tau)] = 0 \tag{1.80}$$

We say that the process $X(t)$ is orthogonal to $Y(t)$.

Similarly to the relation of two random variables, if two random processes are statically independent, they must be uncorrelated. On the other hand, if two random processes are uncorrelated, they may not be statistically independent.

Example 1.3.2. There are two stationary processes

$$X(t) = \cos(\omega_0 t + \varphi)\,, \qquad Y(t) = \sin(\omega_0 t + \varphi)$$

where ω_0 is a constant and φ is a random variable that is evenly distributed in $(0, 2\pi)$. Try to solve whether the two processes are jointly stationary, whether they are correlated, orthogonal and statistically independent.

Solution: Their cross-correlation function is

$$\begin{aligned} R_{XY}(t, t+\tau) &= E[X(t)Y(t+\tau)] \\ &= E\{\cos(\omega_0 t + \varphi)\sin[\omega_0(t+\tau) + \varphi]\} \\ &= \frac{1}{2} E\{\sin[\omega_0(2t+\tau) + 2\varphi] + \sin \omega_0 \tau\} \\ &= \frac{1}{2} \sin \omega_0 \tau \end{aligned}$$

because both $X(t)$ and $Y(t)$ are stationary, and the cross-correlation function is a single-valued function about the interval τ, so the processes $X(t)$ and $Y(t)$ are joint stationary.

Because:

$$\begin{aligned} m_X(t) &= E[X(t)] = E[\cos(\omega_0 t + \varphi)] = 0 \\ m_Y(t+\tau) &= E[Y(t+\tau)] = E\{\sin[\omega_0(t+\tau) + \varphi]\} = 0 \end{aligned}$$

the cross-covariance function is:

$$\begin{aligned} C_{XY}(t, t+\tau) &= E\{[X(t) - m_X(t)][Y(t+\tau) - m_Y(t+\tau)]\} \\ &= E\{X(t)Y(t+\tau)\} \\ &= R_{XY}(t, t+\tau) \\ &= R_{XY}(\tau) \end{aligned}$$

That is, $C_{XY}(\tau) = 1/2 \sin \omega_0 \tau$.

Because only a part of τ will make $C_{XY}(\tau) = 0$, and it cannot satisfy this condition for any τ. The process $X(t)$ is correlated with $Y(t)$ [only in the case of some values of τ are their values (random variables) uncorrelated]. So, the process $X(t)$ is not statistically independent of $Y(t)$. In fact, we know from the given conditions that they have nonlinear relations: $X^2(t) + Y^2(t) = 1$.

Because $E[X(t)Y(t+\tau)] = (\sin \omega_0 \tau)/2$, only with a part of the value τ can be equal to zero. Not all values of τ meet the condition (1.80), and thus the processes $X(t)$ and $Y(t)$ are not orthogonal [only in the case of a part of values of τ are their values (random variables) orthogonal].

It is important to note that the properties of the correlation functions are different from those of the autocorrelation function. When $X(t)$ and $Y(t)$ are joint stationary real random processes, we have the following properties:

(1) $$R_{XY}(\tau) = R_{YX}(-\tau)$$

Obviously, the cross-correlation functions generally are not the odd functions $R_{XY}(\tau) \neq -R_{YX}(-\tau)$ and are not the even functions $R_{XY}(\tau) \neq R_{YX}(-\tau)$ either. They have has the image relationship shown in Fig. 1.11. This is because

$$R_{XY}(\tau) = E[X(t)Y(t+\tau)] = E[X(t'-\tau)Y(t')] = E[Y(t')X(t'-\tau)] = R_{YX}(\tau)$$

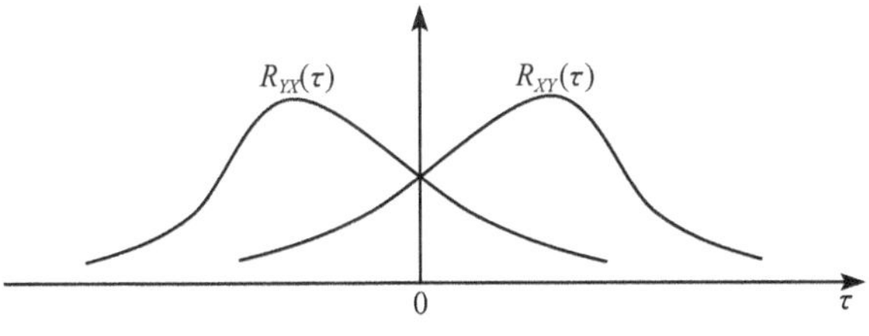

Fig. 1.11: Image relation of the cross-correlation functions.

Similarly, we have

$$C_{XY}(\tau) = C_{YX}(-\tau) \tag{1.81}$$

(2) $$|R_{XY}(\tau)|^2 \leq R_X(0)R_Y(0) \tag{1.82}$$

Similarly, we have

$$|C_{XY}(\tau)|^2 \le C_X(0)C_Y(0) \tag{1.83}$$

Note: The cross-correlation functions do not have the same properties as eqs. (1.59) and (1.61). That is, the cross-correlation functions may have a negative value when $\tau = 0$, so in this case, they may not get their maximum value.

1.4 Discrete-time random process

In the previous sections, we focused on continuous-time random processes. This section will focus on discrete-time random processes. The so-called discrete-time random process is the general term for continuous random sequences and discrete random sequences referred to in the classification of random processes. Next, we introduce the general concept of the discrete-time random process, and then we introduce the probability distribution and numerical characteristics of discrete-time random processes.

1.4.1 Definition of discrete-time random processes

As mentioned earlier, the continuous-time random process is the function of both time t and the random factor e. So, the random process has the characteristics of both random variables and time functions. It can be marked as $X(e, t)$ or, generally, $X(t)$ for short.

If the parameter t takes discrete values $t_1, t_2, \ldots, t_n$, this random process is called a discrete-time random process. In this case, $X(t)$ is a random sequence consisting of a group of variables $X(t)_1, X(t_2), \ldots, X(t_n)$, namely, a random sequence. Random sequences can also be denoted as $X_1, X_2, \ldots, X_n$, or we can use $X(n)$ or $\{X_n, n = 1, 2, \ldots, N\}$ to represent them. In addition, because the integer variable n in the random sequence $\{X_n\}$ represents the increase in time intervals, people often call random sequences time sequences. It can be said that time series is a sequence of random changes over time. Here is a practical example of a time sequence.

Example 1.4.1. The monthly precipitation data of a region from January 1950 to now is $19, 23, 0.47, 0, 0, 123, \ldots$ (mm). Monthly precipitations $X_1, X_2, \ldots$ form a time series. The data above is a sample function of the time series. The sample function can usually be represented by the following point graph (Fig. 1.12), where t is from January 1950.

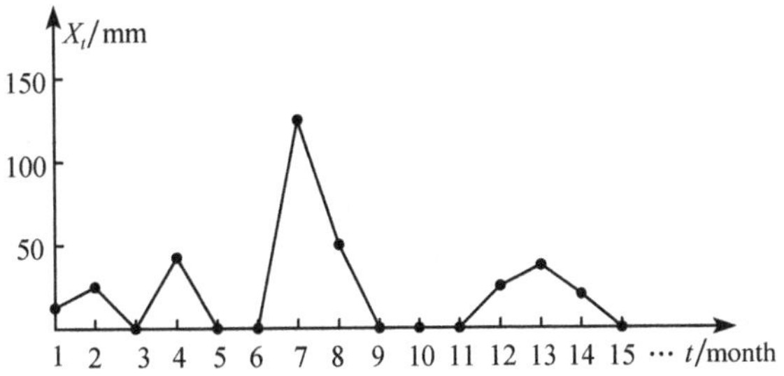

Fig. 1.12: January precipitation.

1.4.2 Probability distribution of discrete-time random processes

A discrete-time random process is a random sequence $X(n)$ varying with n. Since the random variable sequence can be described by the probability distribution function, the probability distribution of the random variable sequence $X(n)$ can be used to describe the discrete-time random process. For a single random variable X_n, we can use the following probability distribution function for the description:

$$F_X(x_n; n) = P\{X_n \le x_n\} \tag{1.84}$$

It is called the one-dimensional probability distribution function of the random variable sequence $X(n)$; x_n is one possible value of the random variable X_n. If X_n evaluate a value on a continuous range, and the partial derivative for x_n of $F_X(x_n; n)$ exists, then we have

$$f_X(x_n; n) = \frac{\partial F_X(x_n; n)}{\partial x_n} \tag{1.85}$$

or

$$F_X(x_n; n) = \int_{-\infty}^{x_n} f_X(x_n; n)\,\mathrm{d}x_n \tag{1.86}$$

We call $f_X(x_n; n)$ the one-dimensional probability distribution function of the random variable sequence $X(n)$.

For two different parameters n and m, the random variables sequence $X(n)$ corresponds to two different random variables X_n and X_m. Then, we use the following joint probability distribution functions to describe it:

$$F_X(x_n, x_m; n, m) = P\{X_n \le x_n, X_m \le x_m\} \tag{1.87}$$

It is called the two-dimensional probability density function of the random variable sequence $X(n)$, if X_n and X_m are continuous random variables, and the second-order mixed partial derivative for x_n, x_m of $F_X(x_n, x_m; n, m)$ exists; we have

$$f_X(x_n, x_m; n, m) = \frac{\partial^2 F_X(x_n, x_m; n, m)}{\partial x_n \partial x_m} \tag{1.88}$$

It is called the two-dimensional probability density function of the random variable sequence $X(n)$.

Similarly, as for the N random variables $X_1, X_2, \dots, X_N$ corresponding to N parameters, we have

$$F_X(x_1, x_2, \dots, x_N; 1, 2, \dots, N) = P\{X_1 \le x_1, X_2 \le x_2, \dots, X_N \le x_N\} \tag{1.89}$$

$$f_X(x_1, x_2, \dots, x_N; 1, 2, \dots, N) = \frac{\partial^N F_X(x_1, x_2, \dots, x_N; 1, 2, \dots, N)}{\partial x_1 \partial x_2 \dots \partial x_N} \tag{1.90}$$

We call $F_X(x_1, x_2, \dots, x_N; 1, 2, \dots, N)$ and $f_X(x_1, x_2, \dots, x_N; 1, 2, \dots, N)$ the N-dimensional probability distribution function and the N-dimensional probability density function of the random variable sequence $X(n)$, respectively.

If a discrete-time random process is shifted by time K (K as an integer), its probability distribution function stays same, namely:

$$F_X(x_{1+K}, x_{2+K}, \dots, x_{N+K}; 1 + K, 2 + K, \dots, N + K) = F_X(x_1, x_2, \dots, x_N; 1, 2, \dots, N)$$

The discrete-time process is called strictly stationary. It can be seen that the one-dimensional probability distribution function of a stationary discrete-time random process is uncorrelated with time

$$F_X(x_n; n) = F_X(x_n)$$

The two-dimensional probability distribution function only relates to the time difference

$$F_X(x_n, x_m; n, m) = F_X(x_n, x_m; n - m)$$

For two random sequences $X(n)$ and $Y(m)$, if they have the probability density functions

$$f_X(x_1, x_2, \dots, x_N; 1, 2, \dots, N)$$

$$f_Y(y_1, y_2, \dots, y_M; 1, 2, \dots, M)$$

we can define the $N + M$-dimensional joint distribution function of these two sequences as:

$$\begin{aligned} &F_{XY}(x_1, x_2, \dots, x_N; y_1, y_2, \dots, y_M; 1, 2, \dots, N; 1, 2, \dots, M) \\ &\quad = P\{X_1 \le x_1, X_2 \le x_2, \dots, X_N \le x_N; Y_1 \le y_1, Y_2 \le y_2, \dots, Y_M \le y_M\} \end{aligned} \tag{1.91}$$

If it has an $N + M$-order mixed derivative of $x_1, \dots, x_N, y_1, \dots, y_M$, the $N + M$ dimension joint probability density of the two sequences is

$$\begin{aligned} &f_{XY}(x_1, x_2, \dots, x_N; y_1, y_2, \dots, y_M; 1, 2, \dots, N; 1, 2, \dots, M) \\ &\quad = \frac{\partial^{N+M} F_{XY}(x_1, x_2, \dots, x_N; y_1, y_2, \dots, y_M; 1, 2, \dots, N; 1, 2, \dots, M)}{\partial x_1 \partial x_2 \dots \partial x_N \partial y_1 \partial y_2 \dots \partial y_M} \end{aligned} \tag{1.92}$$

If

$$\begin{aligned} &F_{XY}(x_1, x_2, \dots, x_N; y_1, y_2, \dots, y_M; 1, 2, \dots, N; 1, 2, \dots, M) \\ &\quad = F_X(x_1, x_2, \dots, x_N; 1, 2, \dots, N) F_Y(y_1, y_2, \dots, y_M; 1, 2, \dots, M) \end{aligned} \tag{1.93}$$

or

$$f_{XY}(x_1, x_2, \ldots, x_N; y_1, y_2, \ldots, y_M; 1, 2, \ldots, N; 1, 2, \ldots, M)$$
$$= f_X(x_1, x_2, \ldots, x_N; 1, 2, \ldots, N) f_Y(y_1, y_2, \ldots, y_M; 1, 2, \ldots, M) \quad (1.94)$$

we call the random sequence $\{X_n\}$ and $\{Y_m\}$ statistically independent.

If the joint probability distribution of two random sequences does not change with a shift of time, and it is uncorrelated to the time starting point, the two sequences are called the joint strict-sense stationary or strict-sense stationary dependent.

1.4.3 Digital characteristics of discrete-time random processes

1.4.3.1 Mathematical expectations

The mean value or the mathematical expectation of the discrete-time random process $\{X_n\}$ is defined as

$$m_{X_n} = E[X_n] = \int_{-\infty}^{+\infty} x f_X(x; n)\, \mathrm{d}x \quad (1.95)$$

If $g(\bullet)$ is a single-valued function, then $g(X_n)$ will form a new discrete-time random process. Its mathematical expectation is defined as

$$E[g(X_n)] = \int_{-\infty}^{+\infty} g(x) f_X(x; n)\, \mathrm{d}x \quad (1.96)$$

The mathematical expectation has the following properties:

(1) $E[X_n + Y_m] = E[X_n] + E[Y_m]$, namely the mean value of the sum is the sum of the mean values.

(2) $E[aX_n] = aE[X_n]$, namely the mean value of the discrete-time random process $\{X_n\}$ multiplied by a constant is equal to the mean value of $\{X_n\}$ multiplied by this constant.

(3) If $E[X_n Y_m] = E[X_n]E[Y_m]$, we call $\{X_n\}$ linear independent with $\{Y_m\}$; its sufficient condition is:

$$f_{XY}(x_n, y_m; n, m) = f_X(x_n; n) f_Y(y_m; m)$$

namely $\{X_n\}$ is linear independent of $\{Y_m\}$.

Statistically independent random processes must be linearly independent, but a linear independence process may not be statistically independent.

1.4.3.2 Mean square value and variance

The mean square value of discrete-time random processes is defined as

$$\psi_{X_n}^2 = E[X_n^2] = \int_{-\infty}^{+\infty} x^2 f(x; n)\, \mathrm{d}x \quad (1.97)$$

The variance of the discrete-time random processes is defined as

$$\sigma_{X_n}^2 = D[X_n] = E[(X_n - m_X)^2] \tag{1.98}$$

From the property of mathematical expectation, we can prove that the above formula can be expressed as

$$\sigma_{X_n}^2 = E[X_n^2] - E^2[X_n] = \psi_{X_n}^2 - m_X^2 \tag{1.99}$$

The square root of the variance is called the standard deviation or the root variance of the discrete-time random process

$$\sigma_{X_n} = \sqrt{\sigma_{X_n}^2} = \sqrt{D[X_n]} \tag{1.100}$$

In general, the mean value, mean square value and variance are all functions of parameter n. However, for the strict-sense stationary discrete-time random process, they are not dependent on n, but are all constants and we have

$$m_X = E[X_n] \tag{1.101}$$

$$\psi_X^2 = E[X_n^2] \tag{1.102}$$

$$\sigma_X^2 = E[(X_n - m_X)^2] \tag{1.103}$$

1.4.3.3 Correlation functions and covariance functions

1. Autocorrelation functions and autocovariance functions

The autocorrelation function is a metric that describes the degree of dependency between the values at different times in a random process, and it is defined as:

$$R_X(n_1, n_2) = E[X_{n_1} X_{n_2}] = \int_{-\infty}^{\infty}\int_{-\infty}^{\infty} x_1 x_2 f_X(x_1, x_2; n_1, n_2)\,dx_1\,dx_2 \tag{1.104}$$

The autocovariance function of the random process is defined as

$$K_X(n_1, n_2) = E[(X_{n_1} - m_{X_{n1}})(X_{n_2} - m_{X_{n2}})] \tag{1.105}$$

After some simple derivations, the above formula can be written as

$$K_X(n_1, n_2) = R_X(n_1, n_2) - m_{X_{n1}} m_{X_{n2}} \tag{1.106}$$

For the random process with zero mean, $m_{X_{n1}} = m_{X_{n2}} = 0$ and then,

$$K_X(n_1, n_2) = R_X(n_1, n_2) \tag{1.107}$$

That is, the autocovariance function is same as the autocorrelation function.

For stationary discrete-time random processes, similarly to stationary continuous random processes, the autocorrelation function is only correlated with the time interval, namely

$$R_X(m) = R_X(n_1, n_2) = R_X(n_1, n_1 + m) \tag{1.108}$$

where, m denotes the time interval, $m = n_2 - n_1$.

Similarly to the stationary continuous random process, if the mean value of the discrete-time random process is a constant, the autocorrelation function is only correlated with the time interval $m = n_2 - n_1$; the mean square value is limited. Namely:

$$\begin{cases} m_{X_n} = E[X_n] = m_X \\ R_X(n_1, n_2) = E[X_{n_1} X_{n_2}] = R_X(m) \\ \psi_{X_n}^2 = E[X_n^2] = \psi_X^2 < \infty \end{cases} \tag{1.109}$$

We call such discrete-time random processes wide-sense stationary, or stationary for short. For the strict stationary random process with mean value and variance, it must be a wide-sense stationary random process. However, conversely, it may not be true.

2. Cross-correlation functions and cross-covariance functions

The cross-correlation function is a metric that describes the degree of dependence between two different random processes. It is defined as

$$R_{XY}(n_1, n_2) = E[X_{n_1} Y_{n_2}] = \int_{-\infty}^{\infty}\int_{-\infty}^{\infty} xy f_{XY}(x, y; n_1, n_2)\, \mathrm{d}x\, \mathrm{d}y \tag{1.110}$$

The cross-covariance function is defined as

$$K_{XY}(n_1, n_2) = E[(X_{n_1} - m_{X_{n1}})(Y_{n_2} - m_{Y_{n2}})] = R_{XY}(n_1, n_2) - m_{X_{n1}} m_{Y_{n2}} \tag{1.111}$$

When $m_{X_{n1}} = m_{Y_{n2}} = 0$, we have

$$K_{XY}(n_1, n_2) = R_{XY}(n_1, n_2) \tag{1.112}$$

For two random joint stationary processes that are both stationary, the cross-correlation function is defined as

$$R_{XY}(m) = R_{XY}(n_1, n_2) = R_{XY}(n_1, n_1 + m) \tag{1.113}$$

The covariance function is defined as

$$K_{XY}(m) = K_{XY}(n_1, n_2) = E[(X_{n_1} - m_X)(Y_{n_2} - m_Y)] \tag{1.114}$$

Equation (1.114) can be written as:

$$K_{XY}(m) = R_{XY}(m) - m_X m_Y \tag{1.115}$$

For all m, we have

$$R_{XY}(m) = 0$$

We say that the random sequences $X(n)$, $Y(n)$ are orthogonal to each other.

If we have

$$K_{XY}(m) = 0 \quad \text{or} \quad R_{XY}(m) = m_X m_Y$$

we say that the random sequences $X(n)$ and $Y(n)$ are uncorrelated. Two statistically independent random processes must be uncorrelated, but two uncorrelated random processes may not be statistically independent.

Example 1.4.2. We study a Bernoulli random process with a coin toss $\{X_n\}$. A coin is thrown at every unit time to observe the results. Throw a coin at the moment, and if it comes up heads, we denote the result by $X(n) = +1$. If it comes up tails, we denote the result by $X(n) = -1$. An infinite sequence of coins can be obtained by throwing them all the time: $X_1, X_2, X_3, \ldots$. Because each time the result $X_n = X(n)$ is a random variable (its value is $+1$ or -1), the results of endless tosses form a sequence of random variables, and we call this a discrete random sequence by definition. The result of each toss is independent from the previous or next result, and the probability of $+1$ or -1 of the X_n is uncorrelated with toss time n.

Set the probability of heads in the n-th throw $P\{X_n = +1\} = p$.

The probability of tails in the n-th throw $P\{X_n = -1\} = 1-p$, where $P\{X_n = +1\} = p$ is correlated with n, and X_i is an independent random variable from $X_j (i \neq j)$. The Bernoulli process above is a stationary random process. Try to solve the mean value, mean square value, variance and autocorrelation function of the process.

Solution: The mean value of $X(n)$ is

$$m_X = (+1)P\{X_n = +1\} + (-1)P\{X_n = +1\} = 1 \cdot p + (-1) \cdot (1-p) = 2p - 1$$

The mean square value is

$$\psi_X^2 = (+1)^2 P\{X_n = +1\} + (-1)^2 P\{X_n = -1\} = 1 \cdot p + 1 \cdot (1-p) = 1$$

So, the variance is

$$\sigma_X^2 = \psi_X^2 - m_X^2 = 1 - (2p-1)^2 = 4p(1-p)$$

Because we assumed the random variable X_i to be independent of X_j $(i \neq j)$, the autocorrelation function of $X(n)$ is

$$R_X(m) = E[X(n)X(n+m)] = \begin{cases} E[X_n^2] = 1\,, & m = 0 \\ E[X_n] \cdot E[X_{n+m}] = m_X^2 = (2p-1)^2\,, & m \neq 0 \end{cases}$$

Especially when $p = 1/2$ (the probability of heads versus tails is equal)

$$m_X = 0\,, \quad \psi_X^2 = 1\,, \quad \sigma_X^2 = 1\,, \quad R_X(m) = \delta(m)$$

In general, whenever the random variables of a stationary random sequence are uncorrelated with each other (i.e., linearly independent), the autocorrelation function will always have the form of a δ function. Such random processes (so-called white noise processes) play an important role in dealing with many signal processing problems.

1.4.3.4 Ergodicity and time average

For discrete-time random processes, similarly to the continuous-time random process discussed above, we hope that through the study of one random sequence we can replace multiple random sequences of observations after much research.

For a random sequence $X(n)$, if its various time average (long enough) converges to the corresponding set average in probability 1, we say that the sequence $X(n)$ has strict (or narrow)-sense ergodicity and abbreviate this sequence as a strict ergodic sequence.

The time average of real discrete-time random process is defined as:

$$A\langle X(n)\rangle = \overline{X(n)} = \lim_{N\to\infty} \frac{1}{2N+1} \sum_{n=-N}^{N} X_n \tag{1.116}$$

Similarly, the time autocorrelation function of the random sequences $X(n)$ is defined as

$$R_X(n, n+m) = \overline{X(n)X(n+m)} = \lim_{N\to\infty} \frac{1}{2N+1} \sum_{n=-N}^{N} X_n X_{n+m} \tag{1.117}$$

Generally, the two time average values defined by the last two formulas are random variables.

Again, we want to introduce the concept of a wide-sense ergodic sequence. Setting $X(n)$ as a stationary random sequence, if both

$$A\langle X(n)\rangle = \overline{X(n)} = E[X(n)] = m_X \tag{1.118}$$

$$R_X(m) = \overline{X(n)X(n+m)} = E[X(n)X(n+m)] = R_X(m) \tag{1.119}$$

are established in probability 1, we call $X(n)$ a wide (or generalized)-sense ergodic sequence, or an ergodic sequence for short.

In the future, when we refer to the terms "stationary random sequence" or "ergodic sequence," they mean the wide-sense stationary random sequence or the wide-sense ergodic sequence unless specified otherwise.

For ergodic sequences, the time average obtained from the definition of the time value and time autocorrelation function converge to a certain nonrandom value. This implies that the time average values of almost all possible sample sequences are the same. So, the time average of ergodic sequences can be expressed by the time average of any sample sequence. In this way, we can use the time average of any sample sequence of traversal sequences instead of the statistical average of the whole sequence, namely:

$$E[X(n)] = \lim_{N\to\infty} \frac{1}{2N+1} \sum_{n=-N}^{N} x(n) \tag{1.120}$$

$$R_X(m) = \lim_{N\to\infty} \frac{1}{2N+1} \sum_{n=-N}^{N} x(n)x(n+m) \tag{1.121}$$

In fact, it is often assumed that the known sequence is a sample sequence of ergodic random sequences. So, we can work out the mean value and the correlation functions of the whole sequence. However, in the actual problem, the limits of these two types are difficult to obtain, so it is generally estimated with limited number n, namely:

$$\hat{m}_X = \overline{[x(n)]_N} = \frac{1}{2N+1} \sum_{n=-N}^{N} x(n) \tag{1.122}$$

$$\hat{R}_X(m) = \overline{[x(n)x(n+m)]_N} = \frac{1}{2N+1} \sum_{n=-N}^{N} x(n)x(n+m) \tag{1.123}$$

In practice, we tend to use these values as estimations (or estimate values) of the mean value and autocorrelation functions of ergodic random sequences.

In addition, when we consider two random sequences $X(n)$ and $Y(n)$, when they are joint stationary, their time cross-correlation functions can be defined as

$$R_{XY}(m) = \overline{X(n)Y(n+m)} = \lim_{N\to\infty} \frac{1}{2N+1} \sum_{n=-N}^{N} X(n)Y(n+m) \tag{1.124}$$

If it converges to the set cross-correlation functions $R_{XY}(m)$ in probability 1, namely:

$$R_{XY}(m) = \overline{X(n)Y(n+m)} = E[X(n)Y(n+m)] = R_{XY}(m) \tag{1.125}$$

We say that the sequences $X(n)$ and $Y(n)$ have joint wide-sense ergodicity.

1.4.4 Properties of correlation functions of stationary discrete-time random processes

For two real stationary random sequences $X(n)$ and $Y(n)$ the autocorrelation, autocovariance, cross-correlation and cross-covariance functions are, respectively,

$$R_X(m) = E[X(n)X(n+m)] = E[X_n X_{n+m}]$$
$$K_X(m) = E[(X(n) - m_X)(X(n+m) - m_X)] = E[(X_n - m_X)(X_{n+m} - m_X)]$$
$$R_{XY}(m) = E[X(n)Y(n+m)] = E[X_n Y_{n+m}]$$
$$K_{XY}(m) = E[(X(n) - m_X)(Y(n+m) - m_Y)] = E[(X_n - m_X)(Y_{n+m} - m_Y)]$$

where m_X and m_Y are the mean of the two sequences, respectively. According to the definition, it is easy to derive the following properties after some simple operations:

(1)
$$K_X(m) = R_X(m) - m_X^2$$
$$K_{XY}(m) = R_{XY}(m) - m_X m_Y$$

(2)
$$R_X(0) = E[X_n^2] = \psi_X^2 \geq 0$$
$$K_X(0) = E[(X_n - m_X)^2] = \sigma_X^2 \geq 0$$

(3)
$$R_X(m) = R_X(-m)$$
$$K_X(m) = K_X(-m)$$
$$R_{XY}(m) = R_{YX}(-m)$$
$$K_{XY}(m) = K_{YX}(-m)$$

(4)
$$R_X(0) \geq |R_X(m)|$$
$$K_X(0) \geq |K_X(m)|$$
$$R_X(0)R_Y(0) \geq |R_{XY}(m)|^2$$
$$K_X(0)K_Y(0) \geq |K_{XY}(m)|^2$$

(5) If $Y(n) = X(n - n_0)$, where n_0 is a fixed discrete moment, we have

$$R_Y(m) = R_X(m)\,, \quad K_Y(m) = K_X(m)$$

(6) If the stationary random sequence does not contain any periodic components, then, we have

$$\lim_{|m|\to\infty} R_X(m) = R_X(\infty) = m_X^2\,, \quad \lim_{|m|\to\infty} K_X(m) = K_X(\infty) = 0$$

Similarly, in order to exactly express the linear correlation degrees of two fluctuation values at different moments of the real stationary random sequences $X(n)$ and $Y(n)$, we introduced autocorrelation and cross-correlation coefficients. We define the autocorrelation coefficient as:

$$r_X(m) = \frac{K_X(m)}{K_X(0)} = \frac{R_X(m) - m_X^2}{\sigma_X^2} \tag{1.126}$$

$r_X(m)$ has the same property as $K_X(m)$, and we have $r_X(0) = 1$ and $|r_X(m)| < 1$. We define the cross-correlation coefficient as:

$$r_{XY}(m) = \frac{K_{XY}(m)}{\sqrt{K_X(0)K_Y(0)}} = \frac{R_{XY}(m) - m_X m_Y}{\sigma_X \sigma_Y} \tag{1.127}$$

And we have $|r_{XY}(m)| \leq 1$. When $r_{XY}(m) = 0$, the two stationary sequences $X(n)$ and $Y(n)$ are uncorrelated.

1.5 Normal random processes

The normal distribution is one of the most frequently encountered distributions in actual work, because the central limit theorem has proved that the limit distribution of the sum of a large number of independent random variables obeys the normal distribution. Similarly, in electronic information technology, the most commonly encountered one is the normal random process.

The normal random process has some characteristics, such as that its arbitrary n-dimensional distribution depends only on the first and second moment functions, and the normal random process is still normal after linear transformation. These features make it convenient as a random process for mathematical analysis.

1.5.1 General normal random processes

We know that the random process $X(t)$ consists of a group of random variables $X(t_1), X(t_2), \ldots, X(t_n)$ and we can write them for short as $(X_1, X_2, \ldots, X_n)$. Obviously, $(X_1, X_2, \ldots, X_n)$ is an n-dimensional random variable with its n-dimensional probability distribution.

If any n-dimension of the joint probability distribution of random process $X(t)$ is normal distribution, then we call $X(t)$ a normal random process or a Gaussian random process, or a Gaussian process for short. The n-dimensional joint probability density of the normal process $X(t)$ should satisfy the following formula:

$$\begin{aligned} p_n(x_1, x_2, \ldots, x_n; t_1, t_2, \ldots, t_n) \\ = \frac{1}{\sqrt{(2\pi)^n D} \cdot \sigma_1 \sigma_2 \ldots \sigma_n} \exp\left[-\frac{1}{2D} \sum_{i=1}^{n} \sum_{j=1}^{n} D_{ij} \frac{x_i - m_i}{\sigma_i} \cdot \frac{x_j - m_j}{\sigma_j} \right] \end{aligned} \quad (1.128)$$

where $m_i = E[X_i]$ is the mean value of the random variable X_i, $\sigma_i^2 = D[X_i]$ is the variance of the random variable X_i ; D is a determinant consisting of the correlation coefficient r_{ij} as follows:

$$D = \begin{vmatrix} r_{11} & r_{12} & \ldots & r_{1n} \\ r_{21} & r_{22} & \ldots & r_{2n} \\ \vdots & \vdots & & \vdots \\ r_{n1} & r_{n2} & \ldots & r_{nn} \end{vmatrix}$$

where $r_{ij} = \frac{E[(X_i - m_i)(X_j - m_j)]}{\sigma_i \sigma_j}$, and we have $r_{ii} = 1$, $r_{ij} = r_{ji}$. D_{ij} as the algebraic cofactor of the determinant D corresponding to element r_{ij}.

The above equation shows that the n-dimensional probability distribution of normal processes depends only on the first and second moment functions, which are only determined by the mean m_i, the variance σ_i^2 and the correlation coefficient r_{ij}. Therefore, for the normal process, the problem can be solved by applying the relevant theories. Moreover, the n-dimensional probability density corresponding to the n-dimensional characteristic function is

$$\Phi_X(\lambda_1, \lambda_2, \ldots, \lambda_n; t_1, t_2, \ldots, t_n) = \exp\left[\mathrm{j} \sum_{i=1}^{n} m_i \lambda_i - \frac{1}{2} \sum_{i=1}^{n} \sum_{j=1}^{n} C_X(t_i, t_j) \lambda_i \lambda_j \right] \quad (1.129)$$

where $C_X(t_i, t_j) = \sigma_i \sigma_j r_{ij}$ is the cross-covariance of the random variables X_i and X_j.

Example 1.5.1. Try to use definition equation of the normal process $X(t)$ to solve its two-dimensional probability density.

Solution: $n = 2$. Substituting the two-dimensional probability density in eq. (1.144), we have

$$p_2(x_1, x_2; t_1, t_2) = \frac{1}{2\pi\sqrt{D}\sigma_1\sigma_2} \exp\left\{-\frac{1}{2D} \cdot \left[D_{11}\frac{(x_1 - m_1)^2}{\sigma_1^2}\right.\right.$$
$$\left.\left. + D_{12}\frac{(x_1 - m_1)(x_2 - m_2)}{\sigma_1\sigma_2} + D_{21}\frac{(x_2 - m_2)(x_1 - m_1)}{\sigma_2\sigma_1} + D_{22}\frac{(x_2 - m_2)^2}{\sigma_2^2}\right]\right\}$$

We have $r_{11} = r_{22} = 1$, $r_{12} = r_{21} = r$, so $D = \left|\begin{smallmatrix} 1 & r \\ r & 1 \end{smallmatrix}\right| = 1 - r^2$

Then we can obtain $D_{11} = D_{22} = 1$, $D_{12} = D_{21} = -r$ and, finally, the expression equation is

$$p_2(x_1, x_2; t_1, t_2) = \frac{1}{2\pi\sigma_1\sigma_2\sqrt{1 - r^2}} \exp\left\{-\frac{1}{2(1 - r^2)} \cdot \left[\frac{(x_1 - m_1)^2}{\sigma_1^2}\right.\right.$$
$$\left.\left. -2r\frac{(x_1 - m_1)(x_2 - m_2)}{\sigma_1\sigma_2} + \frac{(x_2 - m_2)^2}{\sigma_2^2}\right]\right\}$$

Example 1.5.2. For the normal process $X(t)$, try to obtain its one-dimensional probability density and the two-dimensional probability density when uncorrelated by using the characteristic function method.

Solution: $n = 1$. Substituting the one-dimensional characteristic function in eq. (1.145), we have

$$\Phi_X(\lambda_1, t_1) = \exp\left[jm_1\lambda_1 - \frac{1}{2}\sigma_1^2\lambda_1^2\right]$$

Using eq. (1.20), we can obtain the one-dimensional probability density as follows:

$$p_1(x_1, t_1) = \frac{1}{2\pi}\int_{-\infty}^{\infty} \exp\left[-j(x_1 - m_1)\lambda_1 - \frac{1}{2}\sigma_1^2\lambda_1^2\right] d\lambda_1$$
$$= \frac{1}{\sqrt{2\pi}\sigma_1} \exp\left[-\frac{(x_1 - m_1)^2}{2\sigma_1^2}\right] \quad (1.130)$$

Substituting $n = 2$ in eq. (1.145), because $r_{11} = r_{22} = 1$, $r_{12} = r_{21} = 0$ when uncorrelated, the two-dimensional characteristic function is

$$\Phi_X(\lambda_1, \lambda_2; t_1, t_2) = \exp\left[j(m_1\lambda_1 + m_2\lambda_2) - \frac{1}{2}\left(\sigma_1^2\lambda_1^2 + \sigma_2^2\lambda_2^2\right)\right]$$

Using eq. (1.24), the two-dimensional probability density can be obtained as follows:

$$
\begin{aligned}
p_2(x_1, x_2; t_1, t_2) &= \frac{1}{2\pi}\int_{-\infty}^{\infty} \exp\left[-\mathrm{j}(x_1 - m_1)\lambda_1 - \frac{1}{2}\sigma_1^2\lambda_1^2\right] \mathrm{d}\lambda_1 \\
&\quad \times \frac{1}{2\pi}\int_{-\infty}^{\infty} \exp\left[-\mathrm{j}(x_2 - m_2)\lambda_1 - \frac{1}{2}\sigma_2^2\lambda_2^2\right] \mathrm{d}\lambda_2 \\
&= \frac{1}{\sqrt{2\pi}\sigma_1}\exp\left[-\frac{(x_1 - m_1)^2}{2\sigma_1^2}\right] \cdot \frac{1}{\sqrt{2\pi}\sigma_1}\exp\left[-\frac{(x_1 - m_1)^2}{2\sigma_1^2}\right] \\
&= p_1(x_1, t_1) \cdot p_1(x_2, t_2)
\end{aligned}
$$

Note: A special integral is used here: $\int_{-\infty}^{\infty} \mathrm{e}^{-x^2}\,\mathrm{d}x = \sqrt{\pi}(\int_{0}^{\infty} \mathrm{e}^{-x^2}\,\mathrm{d}x = \frac{\sqrt{\pi}}{2})$.

For general random processes, uncorrelated is not equal to statistically independent. However, it can be seen from this case that uncorrelated is equivalent to statistical independence for normal processes. This is a good property of Gaussian random processes.

1.5.2 Stationary normal random processes

The mathematical expectation of the normal process $X(t)$ is a constant that is independent of time, and the correlation function depends only on the time interval, namely:

$$m_i = m\,, \quad \sigma_i^2 = \sigma^2\,, \quad R(t_i, t_j) = R(\tau_{j-i})\,,$$

where $\tau_{j-i} = t_j - t_i$, $i, j = 1, 2, \dots, n$. According to the definition of wide-sense stationarity, this normal process is wide-sense and stationary, which is called a stationary normal process.

If the above conditions are substituted into eq. (1.144), the n-dimensional probability density function of the normal process $X(t)$ can be obtained

$$
\begin{aligned}
&p_n(x_1, x_2, \dots, x_n; \tau_1, \tau_2, \dots, \tau_{n-1}) \\
&= \frac{1}{\sqrt{(2\pi)^n D}\cdot\sigma^n}\exp\left[-\frac{1}{2D\sigma^n}\sum_{i=1}^{n}\sum_{j=1}^{n} D_{ij}(x_i - m_i)\cdot(x_j - m_j)\right]
\end{aligned}
\tag{1.131}
$$

This equation indicates that, the n-dimensional probability density depends only on the time interval $(\tau_1, \tau_2, \dots, \tau_{n-1})$ and it is uncorrelated to the selection of the starting point. Obviously, the process $X(t)$ meets strict-sense stationary conditions. Therefore, wide-sense stationarity and strict stationarity are equivalent to the normal process.

The n-dimensional characteristic function corresponding to the n-dimensional probability density function n is

$$\Phi_X(\lambda_1,\lambda_2,\ldots,\lambda_n;\tau_1,\tau_2,\ldots,\tau_{n-1}) = \exp\left[jm\sum_{i=1}^{n}\lambda_i - \frac{1}{2}\sum_{i=1}^{n}\sum_{j=1}^{n} C_X(\tau_{j-i})\lambda_i\lambda_j\right] \quad (1.132)$$

where $C_X(\tau_{j-i}) = \sigma^2 r(\tau_{j-i})$ is the covariance for random variables X_i with X_j.

Using the results of Example 1.5.1, the two-dimensional probability density of the stationary normal process $X(t)$ is

$$p_2(x_1,x_2;\tau) = \frac{1}{2\pi\sigma^2\sqrt{1-r^2}} \exp\left\{-\frac{(x_1-m)^2 - 2r(x_1-m)(x_2-m) + (x_2-m)^2}{2\sigma^2\left(1-r^2\right)}\right\} \quad (1.133)$$

where $r = r(\tau)$.

From eq. (1.23), we can obtain $X(t)$ and the two-dimensional characteristic function of the stationary normal process is

$$\Phi_X(\lambda_1,\lambda_2;\tau) = \exp\left[jm(\lambda_1+\lambda_2) - \frac{1}{2}\sigma^2\left(\lambda_1^2+\lambda_2^2+2\lambda_1\lambda_2 r\right)\right] \quad (1.134)$$

It is necessary to point out that, when detecting useful signals from background noise, we generally assume that the noise obeys the normal distribution and is stationary, namely that this noise is a stationary normal process. However, when useful signals are certain, their synthetic random process is generally not stationary. This is shown in the following example.

Example 1.5.3. The output random process of the frequency amplifier in the receiver is set as $X(t) = s(t) + n(t)$, where the mean value of $n(t)$ is zero, the variance is σ^2 and $n(t)$ is a stationary normal noise, while $s(t) = \cos(\omega_0 t + \varphi_0)$ is a certain signal. Try to solve the one-dimensional probability density of the random process $X(t)$ at any time t_1, and discriminate if $X(t)$ is stationary.

Solution: The value of the random process $X(t)$ at any time t_1 is a one-dimensional random variable $X(t_1) = s(t_1) + n(t_1)$, where $s(t_1) = \cos(\omega_0 t_1 + \varphi_0)$ is a certain value. Therefore, $X(t_1)$ is still a normal variable, and its mean and variance are:

$$E[X(t_1)] = E[s(t_1) + n(t_1)] = s(t_1)$$
$$D[X(t_1)] = D[s(t_1) + n(t_1)] = \sigma^2$$

Then, from eq. (1.130), the one-dimensional probability density for $X(t)$ is

$$p_1(x_1,t) = \frac{1}{\sqrt{2\pi}\sigma} \exp\left\{-\frac{[x_1 - s(t_1)]^2}{2\sigma^2}\right\}$$

This equation shows that the one-dimensional probability density is correlated with the time t_1, and we know that the synthetic random process $X(t)$ is no longer stationary.

1.5.3 Vector matrix representation of normal stochastic processes

The multidimensional probability distribution expression of the normal process described above is quite complex. If expressed in vector matrix form, it can be simplified.

Now, we write the random values $(x_1, x_2, \ldots, x_n)$ of the n-dimensional random variable $\boldsymbol{X} = (X_1, X_2, \ldots, X_n)$ and its mean value $(m_1, m_2, \ldots, m_n)$ into the column matrix as:

$$\boldsymbol{X} = \begin{pmatrix} x_1 \\ x_2 \\ \vdots \\ x_n \end{pmatrix}, \quad \boldsymbol{m} = \begin{pmatrix} m_1 \\ m_2 \\ \vdots \\ m_n \end{pmatrix}$$

Writing all covariance into a covariance matrix:

$$\boldsymbol{C} = \begin{pmatrix} C_{11} & C_{12} & \ldots & C_{1n} \\ C_{21} & C_{22} & \ldots & C_{21} \\ \vdots & \vdots & & \vdots \\ C_{n1} & C_{n2} & \ldots & C_{nn} \end{pmatrix} = \begin{pmatrix} \sigma_1^2 & r_{12}\sigma_1\sigma_2 & \ldots & r_{1n}\sigma_1\sigma_n \\ r_{21}\sigma_2\sigma_1 & \sigma_2^2 & \ldots & r_{21}\sigma_2\sigma_n \\ \vdots & \vdots & & \vdots \\ r_{n1}\sigma_n\sigma_1 & r_{n2}\sigma_n\sigma_2 & \ldots & \sigma_n^2 \end{pmatrix}$$

It is not difficult to change the n-dimensional joint probability density in eq. (1.128) into:

$$p_n(\boldsymbol{X}) = \frac{1}{\sqrt{(2\pi)^n |\boldsymbol{C}|}} \exp\left\{-\frac{1}{2|\boldsymbol{C}|} \sum_{i=1}^{n} \sum_{j=1}^{n} |\boldsymbol{C}_{ij}| (x_i - m_i) \cdot (x_j - m_j)\right\} \tag{1.135}$$

or:

$$p_n(\boldsymbol{X}) = \frac{1}{\sqrt{(2\pi)^n |\boldsymbol{C}|}} \exp\left\{-\frac{1}{2} [\boldsymbol{X} - \boldsymbol{m}]^{\mathrm{T}} \boldsymbol{C}^{-1} [\boldsymbol{X} - \boldsymbol{m}]\right\} \tag{1.136}$$

where $[\boldsymbol{X} - \boldsymbol{m}]^{\mathrm{T}}$ represents the transpose matrix of the matrix $[\boldsymbol{X} - \boldsymbol{m}]$; $\boldsymbol{C}^{-1}$ is the inverse matrix of the covariance matrix $\boldsymbol{C}$; $|\boldsymbol{C}|$ represents the determinant determined by the covariance matrix $\boldsymbol{C}$, while $|\boldsymbol{C}_{ij}|$ represents its algebraic cofactor.

Similarly, we can change the n-dimensional characteristic function in eq. (1.145) into

$$\Phi_X(\boldsymbol{\lambda}) = \exp\left\{\mathrm{j}\boldsymbol{m}^{\mathrm{T}}\boldsymbol{\lambda} - \frac{1}{2}\boldsymbol{\lambda}^{\mathrm{T}}\boldsymbol{C}\boldsymbol{\lambda}\right\} \tag{1.137}$$

where

$$\boldsymbol{m}^{\mathrm{T}} = [m_1, m_2, \ldots, m_n], \quad \boldsymbol{\lambda}^{\mathrm{T}} = [\lambda_1, \lambda_2, \ldots, \lambda_n], \quad \boldsymbol{\lambda} = \begin{pmatrix} \lambda_1 \\ \lambda_2 \\ \vdots \\ \lambda_n \end{pmatrix}$$

If the values of the random process $X(t)$ are uncorrelated at different moments, namely $r_{ij} = 0$ when $i \neq j$, the covariance matrix can be simplified as

$$\boldsymbol{C} = \begin{pmatrix} \sigma_1^2 & 0 & \dots & 0 \\ 0 & \sigma_2^2 & \dots & 0 \\ \vdots & \vdots & & \vdots \\ 0 & 0 & \dots & \sigma_n^2 \end{pmatrix}$$

The value of the determinant can be obtained as:

$$|\boldsymbol{C}| = \prod_{i=1}^{n} \sigma_i^2$$

Therefore,

$$\exp\left\{-\frac{1}{2}[\boldsymbol{X}-\boldsymbol{m}]^{\mathrm{T}}\boldsymbol{C}^{-1}[\boldsymbol{X}-\boldsymbol{m}]\right\} = \exp\left\{-\sum_{i=1}^{n}\frac{(x_i-m_i)^2}{2\sigma_i^2}\right\} = \prod_{i=1}^{n}\exp\left\{-\frac{(x_i-m_i)^2}{2\sigma_i^2}\right\}$$

Therefore, its n-dimensional probability density in this case is

$$p_n(\boldsymbol{X}) = \prod_{i=1}^{n}\frac{1}{\sqrt{(2\pi)}\cdot\sigma_i}\exp\left\{-\frac{(x_i-m_i)^2}{2\sigma_i^2}\right\} \tag{1.138}$$

The n-dimensional probability density of the stationary normal process $X(t)$ can also be expressed as

$$p_n(\boldsymbol{X}) = \frac{1}{\sqrt{(2\pi\sigma^2)^n\,|\boldsymbol{r}|}}\exp\left\{-\frac{1}{2\sigma^2}[\boldsymbol{X}-\boldsymbol{m}]^{\mathrm{T}}\boldsymbol{r}^{-1}[\boldsymbol{X}-\boldsymbol{m}]\right\} \tag{1.139}$$

where

$$\boldsymbol{r} = \boldsymbol{r}(\tau) = \begin{pmatrix} 1 & r_{12}(\tau_1) & \dots & r_{1n}(\tau_{n-1}) \\ r_{21}(-\tau_1) & 1 & \dots & r_{21}(\tau_{n-2}) \\ \vdots & \vdots & & \vdots \\ r_{n1}(-\tau_{n-1}) & r_{n2}(-\tau_{n-2}) & \dots & 1 \end{pmatrix} = \frac{\boldsymbol{C}}{\sigma^{2n}}$$

is the standardized covariance matrix; $\boldsymbol{r}^{-1}$ is inverse matrix of matrix $\boldsymbol{r}$. In the same way, the n-dimensional characteristic function can be represented as

$$\Phi_X(\boldsymbol{\lambda}) = \exp\left\{\mathrm{j}\boldsymbol{m}^{\mathrm{T}}\boldsymbol{\lambda} - \frac{1}{2}\sigma^2\boldsymbol{\lambda}^{\mathrm{T}}\boldsymbol{r}\boldsymbol{\lambda}\right\} \tag{1.140}$$

Example 1.5.4. Given are normal real random variables with zero mean value, X_1, X_2, X_3, X_4. Try to solve $\overline{X_1X_2X_3X_4}$.

Solution: The mean value is zero, according to eq. (1.153); making $n = 4$, the four-dimensional characteristic function is

$$\Phi_X(\boldsymbol{\lambda}) = \exp\left\{-\frac{1}{2}\boldsymbol{\lambda}^{\mathrm{T}}\boldsymbol{C}\boldsymbol{\lambda}\right\} = \exp\left\{-\frac{1}{2}\sum_{i=1}^{4}\sum_{j=1}^{4}\lambda_i\lambda_i C_{ij}\right\}$$

where $C_{ij} = R_{ij} - m_i m_j = R_{ij}$, that is,

$$\Phi_X(\lambda_1, \lambda_2, \lambda_3, \lambda_4) = \exp\left\{-\frac{1}{2}\sum_{i=1}^{4}\sum_{j=1}^{4}\lambda_i\lambda_i R_{ij}\right\}$$

The fourth-order mixing moment is:

$$\overline{X_1X_2X_3X_4} = \mathrm{j}^{-4}\frac{\partial^4}{\partial\lambda_1\partial\lambda_2\partial\lambda_3\partial\lambda_4}\Phi_X(\lambda_1, \lambda_2, \lambda_3, \lambda_4)\bigg|_{\text{all } \lambda_k=0}, \quad k = 1, 2, 3, 4 \tag{1.141}$$

Because

$$e^{-x} = 1 - x + \frac{x^2}{2!} - \frac{x^3}{3!} + \cdots = \sum_{L=0}^{\infty}(-1)^L\frac{x^L}{L!}$$

then

$$\Phi_X(\lambda_1, \lambda_2, \lambda_3, \lambda_4) = \sum_{L=0}^{\infty}\frac{(-1)^L}{L!}\frac{1}{2^L}\left[\sum_{i=1}^{4}\sum_{j=1}^{4}\lambda_i\lambda_i R_{ij}\right]^L \tag{1.142}$$

There are 16 items in this type of square brackets, but it can be seen from eq. (1.141) that only those whose power is $1(\lambda_1\lambda_2\lambda_3\lambda_4)$ will contribute, while from eq. (1.142) it can be seen that only if $L = 2$, will term $\lambda_1\lambda_2\lambda_3\lambda_4$ come up, so eq. (1.141) can be rewritten as

$$\overline{X_1X_2X_3X_4} = \frac{\partial^4}{\partial\lambda_1\partial\lambda_2\partial\lambda_3\partial\lambda_4}\left\{\frac{1}{8}\left[\sum_{i=1}^{4}\sum_{j=1}^{4}\lambda_i\lambda_i R_{ij}\right]^2\right\}\bigg|_{\text{all } \lambda_k=0}, \quad k = 1, 2, 3, 4$$

There are 256 items in this type of curly brackets with the binomial formula, but for the same reason, only the following items can contribute:

$$2\lambda_1\lambda_2\lambda_3\lambda_4\big|_{i\neq j\neq k\neq l}, \quad i, j, k, l = 1, 2, 3, 4$$

Obviously, there are only 12 terms that will make contributions: $R_{12}R_{34}$, $R_{12}R_{43}$, $R_{21}R_{34}$, $R_{21}R_{43}$; $R_{13}R_{24}$, $R_{13}R_{42}$, $R_{31}R_{24}$, $R_{31}R_{42}$; $R_{14}R_{23}$, $R_{14}R_{32}$, $R_{41}R_{23}$ and $R_{41}R_{32}$. Because $R_{ij} = R_{ji}$, these 12 items can be co-written as $4[R_{12}R_{34} + R_{13}R_{24} + R_{14}R_{23}]$. Thus, we obtain:

$$\begin{aligned}\overline{X_1X_2X_3X_4} &= \frac{1}{8} \times 2 \times 4[R_{12}R_{34} + R_{13}R_{24} + R_{14}R_{23}] \\ &= \overline{X_1X_2}\cdot\overline{X_3X_4} + \overline{X_1X_3}\cdot\overline{X_2X_4} + \overline{X_1X_4}\cdot\overline{X_2X_3}\end{aligned} \tag{1.143}$$

1.6 Spectral analysis of stationary random processes

1.6.1 Concept of spectral density

We know that both time-domain analysis and frequency-domain analysis can be used for certain signals, and there is a definite relationship between the two methods. Periodic signals can be expanded into Fourier series, and nonperiodic signals can be expressed as Fourier integrals. For example, for nonperiodic signals $s(t)$, the Fourier transform is

$$S(\omega) = \int_{-\infty}^{\infty} s(t)\mathrm{e}^{-\mathrm{j}\omega t}\,\mathrm{d}t \tag{1.144}$$

where $S(\omega)$ is the spectral density (spectrum for short) of the signal $s(t)$, which represents the distribution of each complex amplitude component of the signal in the frequency domain; $S(\omega) = |S(\omega)| \cdot \mathrm{e}^{\mathrm{j}\varphi_S(\omega)}$, where $|S(\omega)|$ is called the amplitude spectrum density, and $\varphi_S(\omega)$ is called the phase spectrum density.

Given the spectral density $S(\omega)$, its inverse Fourier transform is the time function of this signal, i.e.,

$$s(t) = \frac{1}{2\pi} \int_{-\infty}^{\infty} S(\omega)\mathrm{e}^{\mathrm{j}\omega t}\,\mathrm{d}\omega \tag{1.145}$$

where $s(t)$ with $S(\omega)$ is a pair of Fourier transforms, which can be abbreviated as $s(t) \to S(\omega)$.

The condition for spectral density to exist is

$$\int_{-\infty}^{\infty} |s(t)|\,\mathrm{d}t < \infty \tag{1.146}$$

The duration time of the signal $s(t)$ must be limited.

The Perceval theorem shows that the total energy of the signal is equal to the sum of the energy of each spectrum component:

$$E = \int_{-\infty}^{\infty} |s(t)|^2\,\mathrm{d}t = \frac{1}{2\pi} \int_{-\infty}^{\infty} |S(\omega)|^2\,\mathrm{d}\omega = \int_{-\infty}^{\infty} |S(2\pi f)|^2\,\mathrm{d}f \tag{1.147}$$

where $|S(\omega)|^2 = S(\omega) \cdot S^*(\omega)$, and the symbol "$*$" denotes a conjugate.

According to the above equation, the energy of each spectrum component is

$$\mathrm{d}E = \frac{1}{2\pi} |S(\omega)|^2\,\mathrm{d}\omega = |S(2\pi f)|^2\,\mathrm{d}f \tag{1.148}$$

or

$$|S(\omega)|^2 = 2\pi \frac{\mathrm{d}E}{\mathrm{d}\omega} = \frac{\mathrm{d}E}{\mathrm{d}f} = |S(2\pi f)|^2 (\mathrm{J/Hz}) \tag{1.149}$$

$|S(\omega)|^2$ or $|S(2\pi f)|^2$ are known as the signal $s(t)$; the energy spectrum density, or spectrum density for short, is the signal component energy in the unit band, indicating the distribution of each component of the signal in the frequency domain.

The existence condition of the spectral density is as follows:

$$\int_{-\infty}^{\infty} |s(t)|^2\,\mathrm{d}t < \infty \tag{1.150}$$

The total energy of the signal $s(t)$ must be limited.

Because of the general random process $X(t)$ with infinite time duration and total energy, the same is true for any sample $x_i(t)$. As a result, the process $X(t)$ does not satisfy the eqs. (1.146) and (1.150), so the spectral density and the energy density are nonexistent. However, the average power of a random process that can actually be produced is always limited, i.e.,

$$P = \lim_{T\to\infty} \frac{1}{2T} \int_{-T}^{T} |x(t)|^2 \, \mathrm{d}t < \infty \tag{1.151}$$

Therefore, the concept of power spectral density can be derived by using the extended spectrum analysis method.

1.6.2 Definition of power spectral density

A sample curve $x_i(t)$ of the random process $X(t)$ is shown in Fig. 1.13. A section of $2T$ length of the sample curve is called a truncated function, whose expression is:

$$x_{Ti}(t) = \begin{cases} x_i(t)\,, & |t| < T \\ 0\,, & |t| \geq T \end{cases} \tag{1.152}$$

Due to the limited duration of $x_{Ti}(t)$, its Fourier transform $X_{Ti}(\omega)$ exists, i.e.,

$$X_{Ti}(\omega) = \int_{-\infty}^{\infty} x_{Ti}(t)^{-\mathrm{j}\omega t} \, \mathrm{d}t = \int_{-T}^{T} x_{Ti}(t)\mathrm{e}^{-\mathrm{j}\omega t} \, \mathrm{d}t \tag{1.153}$$

$$x_{Ti}(t) = \frac{1}{2\pi} \int_{-\infty}^{\infty} X_{Ti}(\omega)\mathrm{e}^{\mathrm{j}\omega t} \, \mathrm{d}\omega \tag{1.154}$$

Assuming that $x_{Ti}(t)$ is a real time function, we have is $x^*_{Ti}(t) = x_{Ti}(t)$. Therefore, we carry out a conjugate operation for eq. (1.153) and obtain

$$X^*_{Ti}(\omega) = \int_{-T}^{T} x_{Ti}(t)\mathrm{e}^{\mathrm{j}\omega t} \, \mathrm{d}t \tag{1.155}$$

We have $x_{Ti}(t) = x_i(t)$ for $|t| < T$, so the sample average power:

$$P_i = \lim_{T\to\infty} \frac{1}{2\pi} \int_{-T}^{T} |x_i(t)|^2 \, \mathrm{d}t = \lim_{T\to\infty} \frac{1}{2\pi} \int_{-T}^{T} |x_{Ti}(t)|^2 \, \mathrm{d}t \tag{1.156}$$

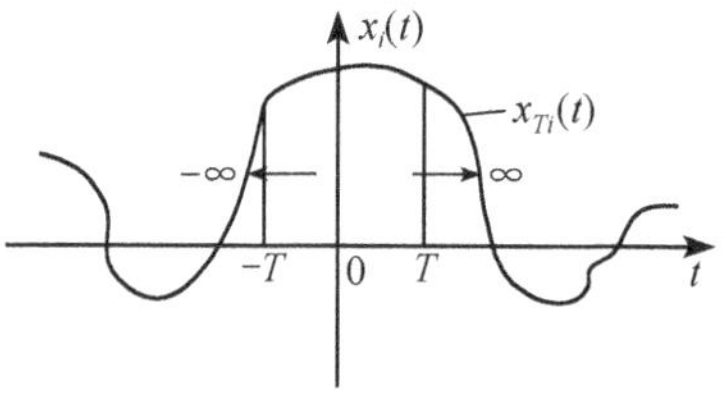

Fig. 1.13: Sample $x_i(t)$ and the truncated function $x_{Ti}(t)$.

where

$$
\begin{aligned}
\int_{-T}^{T} |x_{Ti}(t)|^2 \, dt &= \int_{-T}^{T} x_{Ti}^2(t) \, dt \\
&= \int_{-T}^{T} x_{Ti}(t) \left[\frac{1}{2\pi} \int_{-\infty}^{\infty} X_{Ti}(\omega) e^{j\omega t} \, d\omega \right] dt \\
&= \frac{1}{2\pi} \int_{-\infty}^{\infty} X_{Ti}(\omega) \left[\int_{-T}^{T} x_{Ti}(t) e^{j\omega t} \, dt \right] d\omega \\
&= \frac{1}{2\pi} \int_{-\infty}^{\infty} |X_{Ti}(\omega)|^2 \, d\omega
\end{aligned} \tag{1.157}
$$

Thus, eq. (1.156) can be rewritten as:

$$
P_i = \lim_{T\to\infty} \frac{1}{2T} \int_{-\infty}^{\infty} \frac{|X_{Ti}(\omega)|^2}{2\pi} \, d\omega = \frac{1}{2\pi} \int_{-\infty}^{\infty} \left[\lim_{T\to\infty} \frac{|X_{Ti}(\omega)|^2}{2T} \right] d\omega \tag{1.158}
$$

or

$$
P_i = \int_{-\infty}^{\infty} \left[\lim_{T\to\infty} \frac{|X_{Ti}(2\pi f)|^2}{2T} \right] df \tag{1.159}
$$

If $X(t)$ represents the voltage (V) or current (A), the amount of the two square brackets is divided by the unit resistance (1 Ω), with power dimension (W/Hz). So,

$$
G_i(\omega) = \lim_{T\to\infty} \frac{|X_{Ti}(\omega)|^2}{2T} \tag{1.160}
$$

or

$$
P_i = \int_{-\infty}^{\infty} \left[\lim_{T\to\infty} \frac{|X_{Ti}(2\pi f)|^2}{2T} \right] df \tag{1.161}
$$

This is called the power spectral density of a sample $x_i(t)$(sample power spectrum for short), which expresses that the average power of an ohm resister in the spectral component of the unit band is consumed. Integration in the whole frequency domain $(-\infty, \infty)$ leads to the $x_i(t)$ average power; $x_i(t)$ is a sample of a stationary random process $X(t)$, and a different sample $x_i(t)$ will have different power spectral density $G_i(\omega)$, which is a random function that varies with the sample. Carrying out a statistical average operation, the function with a certain frequency will be unrelated to the sample number i. So, it is defined as

$$
G_X(\omega) = E\left[\lim_{T\to\infty} \frac{|X_{Ti}(\omega)|^2}{2T} \right], \quad -\infty < \omega < \infty \tag{1.162}
$$

It is called the average power spectral density (power spectral density for short) of the stationary random process $X(t)$. The power spectral density definition of nonstationary random processes is defined in Appendix 1.

Power spectral density $G_X(\omega)$ is the statistical average of the process $X(t)$, which is the average power consumed by an ohm resistor of the spectral components of each sample in the unit band. Besides, $G_X(\omega)$ can also be regarded as the average statistical parameter of the process $X(t)$ in the aspect of the frequency domain description, which expresses the distribution of the average power in the frequency domain. It should be noted that, like normal power values, it only reflects the amplitude information of the random process and does not reflect the phase information.

1.6.3 Relation between the power spectral density and correlation functions

We have already talked about the certain time function $s(t)$ and its spectral density $S(\omega)$, which are a pair of Fourier transforms $s(t) \to S(\omega)$. In fact, for the random-time function $X(t)$ in the stationary condition, the correlation function $R_X(\tau)$ and power spectral density $G_X(\omega)$ are also a pair of Fourier transforms $R_X(\tau) \to G_X(\omega)$. Respectively, they describe the average statistical characteristic of the random process $X(t)$ from the aspect of the time domain and the frequency domain, which are the two main statistical parameters of the random process. As long as one of them is known, the other can be completely determined. Let us prove this important relationship.

Assuming that $X(t)$ is the real stationary process, substitute eqs. (1.153) and (1.155) into eq. (1.162) to obtain

$$\begin{aligned} G_X(\omega) &= E\left[\lim_{T\to\infty}\frac{1}{2T}\int_{-T}^{T} x_{Ti}(t_2)\mathrm{e}^{-\mathrm{j}\omega t_2}\,\mathrm{d}t_2\int_{-T}^{T} x_{Ti}(t_1)\mathrm{e}^{\mathrm{j}\omega t_1}\,\mathrm{d}t_1\right] \\ &= \lim_{T\to\infty}\frac{1}{2T}\int_{-T}^{T}\int_{-T}^{T} E\left[x_{Ti}(t_2)x_{Ti}(t_1)\right]\mathrm{e}^{-\mathrm{j}\omega(t_2-t_1)}\,\mathrm{d}t_1\,\mathrm{d}t_2 \end{aligned} \tag{1.163}$$

where $x_{Ti}(t_2)$ and $x_{Ti}(t_1)$ are the random variables that vary from sample to sample,

$$E\left[x_{Ti}(t_2)x_{Ti}(t_1)\right] = R_{XT}(t_2, t_1) \tag{1.164}$$

If $X(t)$ is a stationary process, we have

$$R_X(t_2, t_1) = R_X(-\tau) = R_X(\tau) \tag{1.165}$$

where $\tau = t_2 - t_1$. So, we have

$$G_X(\omega) = \lim_{T\to\infty}\frac{1}{2T}\int_{-T}^{T}\int_{-T}^{T} R_{XT}(\tau)\mathrm{e}^{-\mathrm{j}\omega\tau}\,\mathrm{d}t_1\,\mathrm{d}t_2 \tag{1.166}$$

Using the variable t_1, t_2 substitution method as shown in Fig. 1.5, substitute the variable into the upper formula. Make $t = -t_1$, $\tau = t_2 - t_1$ and then you can find:

$$\begin{aligned}\int_{-T}^{T}\int_{-T}^{T} R_{XT}(\tau)\mathrm{e}^{-\mathrm{j}\omega\tau}\,\mathrm{d}t_1\,\mathrm{d}t_2 &= \iint_{\diamond} R_{XT}(\tau)\mathrm{e}^{-\mathrm{j}\omega\tau}\,\mathrm{d}\tau\,\mathrm{d}t\\ &= \int_{0}^{2T} R_{XT}(\tau)\mathrm{e}^{-\mathrm{j}\omega\tau}\,\mathrm{d}\tau\int_{\tau-T}^{T}\mathrm{d}t + \int_{-2T}^{0} R_{XT}(\tau)\mathrm{e}^{-\mathrm{j}\omega\tau}\,\mathrm{d}\tau\int_{-T}^{\tau+T}\mathrm{d}t\\ &= \int_{0}^{2T}(2T-\tau)R_{XT}(\tau)\mathrm{e}^{-\mathrm{j}\omega\tau}\,\mathrm{d}\tau + \int_{-2T}^{0}(2T+\tau)R_{XT}(\tau)\mathrm{e}^{-\mathrm{j}\omega\tau}\,\mathrm{d}\tau\\ &= 2T\int_{-2T}^{2T}\left(1-\frac{|\tau|}{2T}\right)R_{XT}(\tau)\mathrm{e}^{-\mathrm{j}\omega\tau}\,\mathrm{d}\tau\end{aligned} \tag{1.167}$$

Substituting it into eq. (1.166), we obtain:

$$G_X(\omega) = \int_{-\infty}^{\infty} R_X(\tau)\mathrm{e}^{-\mathrm{j}\omega\tau}\,\mathrm{d}\tau \tag{1.168}$$

The above expression indicates that the power spectral density of the stationary process is the Fourier transform of the correlation function. Therefore, the correlation function is the inverse Fourier transform of the power spectral density,

$$R_X(\tau) = \frac{1}{2\pi}\int_{-\infty}^{\infty} G_X(\omega)\mathrm{e}^{\mathrm{j}\omega\tau}\,\mathrm{d}\omega \tag{1.169}$$

The above two formulas give the important relationship between the power spectral density and correlation function of stationary random process: $R_X(\tau) \to G_X(\omega)$, which shows the equivalent relationship between statistical properties of stationary random process from the frequency domain or the time domain. This conclusion is widely used in engineering technology. This relationship was put forward by Wiener and Khintchine, so it is also known as the Weiner–Khintchine theorem.

According to the existence conditions of Fourier transforms, the conditions of establishment are that $G_X(\omega)$ and $R_X(\tau)$ must be absolutely integral, i.e.,

$$\int_{-\infty}^{\infty} G_X(\omega)\,\mathrm{d}\omega < \infty \tag{1.170}$$

$$\int_{-\infty}^{\infty} |\tau R_X(\tau)|\,\mathrm{d}\tau < \infty \tag{1.171}$$

Conditional eq. (1.170) description: the total average power of the random process $X(t)$ must be limited, which can be satisfied for general random process. However, in order to satisfy the conditional eq. (1.171), the mathematical expectation of the random process $X(t)$ must be zero [namely, there is no DC component and actually, $R_X(\tau)$ is $C_X(\tau)$] and the random process does not contain periodic components. As the Fourier transform of certain signals, if we introduce the δ-function, (wide-sense function), the Fourier transforms of the random process even including certain DC component and periodic component still exist. For the common correlation function and power spectral density, the corresponding relation is shown in Tab. 1.1.

1.6.4 Properties of power spectral density

If $X(t)$ is the stationary real random process, we have the following properties for $G_X(\omega)$:

(1)
$$G_X(\omega) = G_X(-\omega) \geq 0\,. \tag{1.172}$$

That is, the power spectral density is a non-negative, real even function. This can be seen from the definition (1.6.19), because of $|X_{Ti}(\omega)|^2 \geq 0$, and:

$$|X_{Ti}(\omega)|^2 = X_{Ti}(\omega)X_{Xi}^*(\omega) = X_{Ti}(\omega)X_{Ti}(-\omega) = X_{Ti}(-\omega)X_{Ti}(\omega) \tag{1.173}$$

Because the correlation function of the stationary real random process $R_X(\tau)$ is a real even function, from eq. (1.168) we know that $G_X(\omega)$ will be a real even function.

(2)
$$G_X(\omega) \to R_X(\tau)\,.$$

They are a pair of Fourier transforms, and there is a pair of relationships:

$$\begin{cases} G_X(\omega) = \int_{-\infty}^{\infty} R_X(\tau)\mathrm{e}^{-\mathrm{j}\omega\tau}\,\mathrm{d}\tau \\ R_X(\tau) = \frac{1}{2\pi}\int_{-\infty}^{\infty} G_X(\omega)\mathrm{e}^{\mathrm{j}\omega\tau}\,\mathrm{d}\omega \end{cases} \tag{1.174}$$

For the most common stationary real random process because the correlation function is the real even function for τ, we have $R_X(\tau) = R_X(-\tau)$. The power spectral density is the real even function for Ω, and $G_X(\omega) = G_X(-\omega)$, while

$$\mathrm{e}^{\pm\mathrm{j}\omega\tau} = \cos\omega\tau \pm \mathrm{j}\sin\omega\tau \tag{1.175}$$

So, the Weiner–Khintchine theorem has the following forms:

$$\begin{cases} G_X(\omega) = 2\int_{0}^{\infty} R_X(\tau)\cos\omega\tau\,\mathrm{d}\tau \\ R_X(\tau) = \frac{1}{\pi}\int_{0}^{\infty} G_X(\omega)\cos\omega\tau\,\mathrm{d}\omega \end{cases} \tag{1.176}$$

$G_X(\omega)$ is defined in the $-\infty < \omega < \infty$. Namely, the power distribution is from $-\infty$ to $+\infty$ in the entire frequency domain, but, in fact, the negative frequency does not exist, and the actual power is only generated by the positive frequency components.

Tab. 1.1: The corresponding relationship of common correlation functions and power spectral density.

Correlation function		**Power spectrum density**	
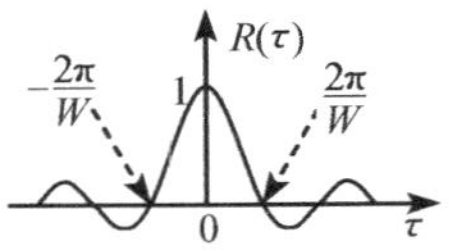	$R(\tau) = \mathrm{Sa}\left(\frac{W\tau}{2}\right)$	$G(\omega) = \begin{cases} \frac{2\pi}{W}, & \lvert\omega\rvert \le \frac{W}{2} \\ 0, & \lvert\omega\rvert > \frac{W}{2} \end{cases}$	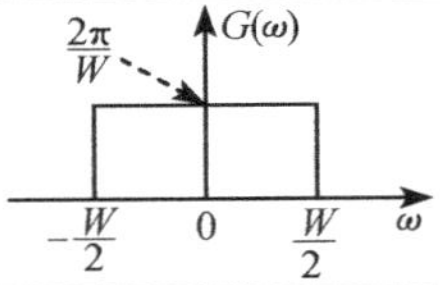
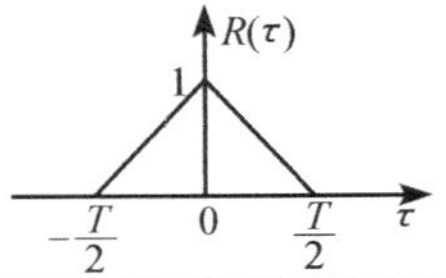	$R(\tau) = \begin{cases} 1 - \frac{2\lvert\tau\rvert}{T}, & \lvert\tau\rvert < \frac{T}{2} \\ 0, & \lvert\tau\rvert \ge \frac{T}{2} \end{cases}$	$G(\omega) = \frac{T}{2}\mathrm{Sa}^2\left(\frac{\omega T}{4}\right)$	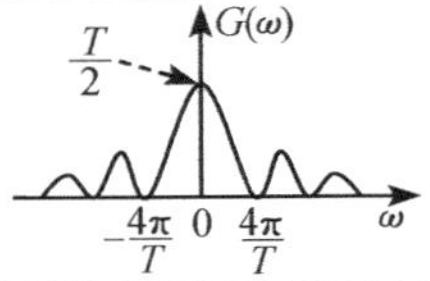
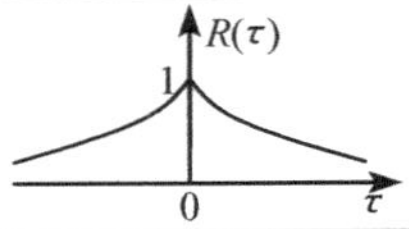	$R(\tau) = \mathrm{e}^{-\alpha\lvert\tau\rvert}$	$G(\omega) = \frac{2\alpha}{\alpha^2 + \omega^2}$	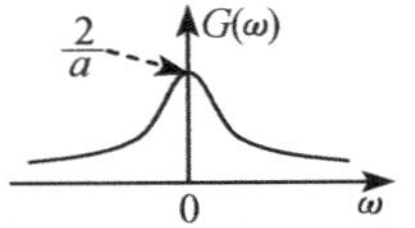
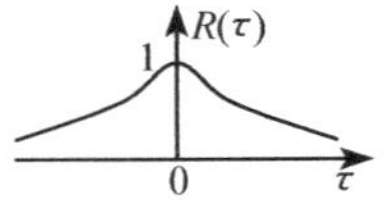	$R(\tau) = \mathrm{e}^{-\frac{\tau^2}{2\sigma^2}}$	$G(\omega) = \sqrt{2\pi}\sigma \cdot \mathrm{e}^{-\frac{(\omega\sigma)^2}{2}}$	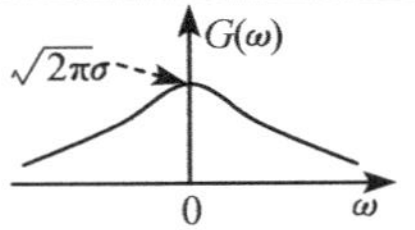
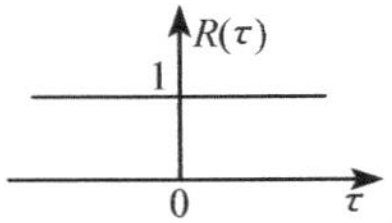	$R(\tau) = 1\,, \quad -\infty < \tau < \infty$	$G(\omega) = 2\pi\delta(\omega)$	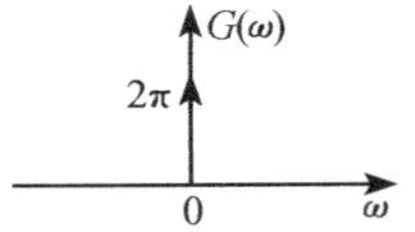
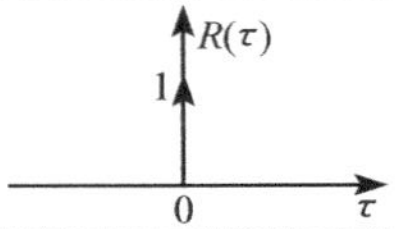	$R(\tau) = \delta(\tau)$	$G(\omega) = 1\,, \quad -\infty < \omega < \infty$	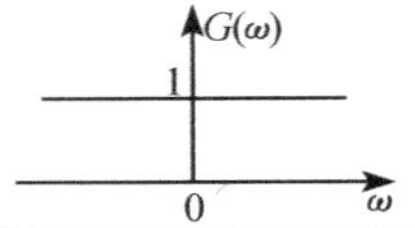

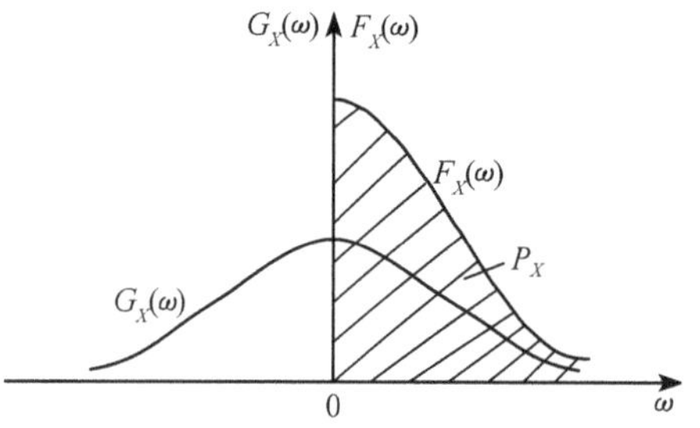

Fig. 1.14: Curve of the power spectral density $G_X(\omega)$ and $F_X(\omega)$.

In this book, the concept of negative frequency is adopted, and the negative frequency component and the normal frequency component both similarly generate power. For the sake of mathematical analysis, $G_X(\omega)$ is called the mathematical power spectral density or the double-side power spectral density. The actual measured power spectral density will only be distributed from 0 to $+\infty$ in the positive frequency domain, which is called the physical power spectral density or the unilateral power spectral density, i.e., $F_X(\omega)$.

Because $G_X(\omega)$ is the real even function for Ω, we have is following relationship:

$$F_X(\omega) = 2G_X(\omega), \quad \omega > 0 \tag{1.177}$$

as is shown in Fig. 1.14.

Therefore, a new expression of the Weiner–Khintchine theorem is obtained:

$$\begin{cases} F_X(\omega) = 4\int_0^\infty R_X(\tau)\cos\omega\tau\,\mathrm{d}\tau \\ R_X(\tau) = \frac{1}{2\pi}\int_0^\infty F_X(\omega)\cos\omega\tau\,\mathrm{d}\omega \end{cases} \tag{1.178}$$

or

$$\begin{cases} F_X(f) = 4\int_0^\infty R_X(\tau)\cos 2\pi f\tau\,\mathrm{d}\tau \\ R_X(\tau) = \int_0^\infty F_X(f)\cos 2\pi f\tau\,\mathrm{d}f \end{cases} \tag{1.179}$$

Power spectral density $G_X(\omega)$ or $F_X(\omega)$ is an important parameter to represent statistical characteristics from the aspect of the frequency domain. Once it is known, it can be worked out that the total average power of the stationary process $X(t)$ consumed on an ohm resistor is

$$P_X = R_X(0) = \frac{1}{2\pi}\int_{-\infty}^{\infty} G_X(\omega)\,\mathrm{d}\omega = \frac{1}{2\pi}\int_0^{\infty} F_X(\omega)\,\mathrm{d}\omega \tag{1.180}$$

(the shaded area shown in Fig. 1.14). In the same way, the interval at any given frequency (ω_1, ω_2) can be worked out. The average power is

$$P_X = \frac{1}{2\pi}\int_{\omega_1}^{\omega_2} F_X(\omega)\,\mathrm{d}\omega \tag{1.181}$$

Example 1.6.1. A random signal $X(t)$ with initial phase is shown in Example 1.2.1. Find the power spectral density $G_X(\omega)$, $F_X(\omega)$.

Solution: The correlation function is obtained by above example:

$$R_X(\tau) = \frac{A^2}{2}\cos\omega_0\tau$$

This can be rewritten as

$$R_X(\tau) = \frac{A^2}{4}[\mathrm{e}^{\mathrm{j}\omega_0\tau} + \mathrm{e}^{-\mathrm{j}\omega_0\tau}]$$

The above equation is substituted into eq. (1.174):

$$\begin{aligned}G_X(\omega) &= \int_{-\infty}^{\infty}\frac{A^2}{4}[\mathrm{e}^{\mathrm{j}(\omega-\omega_0)\tau} + \mathrm{e}^{-\mathrm{j}(\omega+\omega_0)\tau}]\,\mathrm{d}\tau\\ &= \frac{A^2}{4}[2\pi\delta(\omega-\omega_0) + 2\pi\delta(\omega+\omega_0)]\\ &= \frac{\pi A^2}{2}[\delta(\omega-\omega_0) + \delta(\omega+\omega_0)]\\ F_X(\omega) &= 2G_X(\omega)\big|_{\omega>0} = \pi A^2\delta(\omega-\omega_0)\end{aligned}$$

Example 1.6.2. The random telegraph signal is shown in Example 1.2.2 as $X(t)$. Find its power spectral density $G_X(\Omega)$.

Solution: The correlation function is obtained by the above example:

$$R_X(\tau) = \frac{1}{4} + \frac{1}{4}\mathrm{e}^{-2\lambda|\tau|}$$

The above equation is substituted into eq. (1.176):

$$G_X(\omega) = \frac{1}{2}\int_{-\infty}^{\infty}\cos\omega\tau\,\mathrm{d}\tau + \frac{1}{2}\int_{-\infty}^{\infty}\mathrm{e}^{-2\lambda\tau}\cos\omega\tau\,\mathrm{d}\tau = \frac{\pi}{2}\delta(\omega) + \frac{\lambda}{4\lambda^2+\omega^2}$$

As is shown in Fig. 1.15, the first part of the above equation represents the DC component, which is a discrete line spectrum $\delta(\omega)$ with the strength $\pi/2$. The second term is the undulating component, which is a continuous spectrum.

1.6.5 Mutual spectral density of joint stationary random processes

Like in the discussion of the power spectral density of a single random process, we assume that there are two joint stationary random processes $X(t)$ and $Y(t)$, and the truncated functions of any sample are $x_{Ti}(t)$ and $y_{Ti}(t)$. Their spectral density varies from sample number to random variables $X_{Ti}(\omega)$ and $Y_{Ti}(\omega)$. Like with eq. (1.162), define:

$$G_{XY}(\omega) = E\left[\lim_{T\to\infty}\frac{X_T^*(\omega)Y_T(\omega)}{2T}\right] \tag{1.182}$$

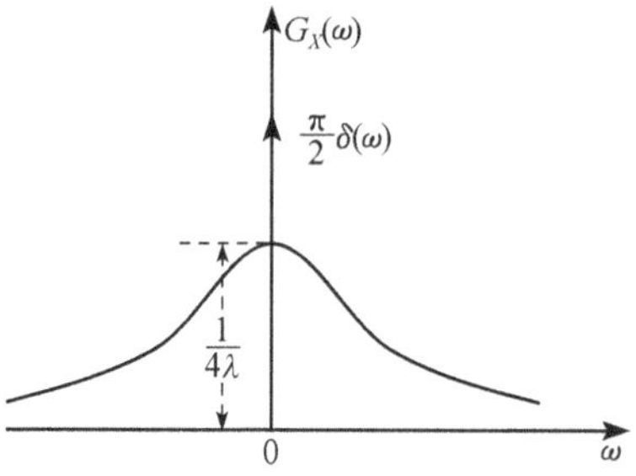

Fig. 1.15: Power spectrum of random telegraph signals.

or

$$G_{YX}(\omega) = E\left[\lim_{T\to\infty} \frac{Y_T^*(\omega)X_T(\omega)}{2T}\right] \tag{1.183}$$

where $X_T^*(\omega)$ and $Y_T^*(\omega)$, respectively, are the complex conjugates of $X_T(\omega)$ and $Y_T(\omega)$, and $G_{XY}(\omega)$ or $G_{YX}(\omega)$ are called the cross-spectral density of the random processes $X(t)$ and $Y(t)$.

The above equation shows that the cross-spectral density is different from the power spectral density of a single stationary random process. It may not be a non-negative, real even function.

A similar method to that discussed in Section 1.6.2 can prove that the cross-spectral density and cross-correlation function are also a pair of Fourier transforms, which are

$$\begin{cases} G_{XY}(\omega) = \int_{-\infty}^{\infty} R_{XY}(\tau)\mathrm{e}^{-\mathrm{j}\omega\tau}\,\mathrm{d}\tau \\ G_{YX}(\omega) = \int_{-\infty}^{\infty} R_{YX}(\tau)\mathrm{e}^{-\mathrm{j}\omega\tau}\,\mathrm{d}\tau \end{cases} \tag{1.184}$$

and

$$\begin{cases} R_{XY}(\tau) = \frac{1}{2\pi}\int_{-\infty}^{\infty} G_{XY}(\omega)\mathrm{e}^{\mathrm{j}\omega\tau}\,\mathrm{d}\tau \\ R_{YX}(\tau) = \frac{1}{2\pi}\int_{-\infty}^{\infty} G_{YX}(\omega)\mathrm{e}^{\mathrm{j}\omega\tau}\,\mathrm{d}\tau \end{cases} \tag{1.185}$$

The two Fourier transforms with the relation between the autocorrelation function and the spectral density is very similar, but it should be noted that the cross-correlation function is not an even function and the cross-spectral density may also not be an even function. So, the cross-correlation function and the cross-spectral density of the Fourier transforms cannot be written in a similar form to eqs. (1.176) and (1.178)

The cross-spectral density has the following properties:

(1)
$$G_{XY}(\omega) = G_{YX}^*(\omega) \tag{1.186}$$

Namely, $G_{XY}(\omega)$ with $G_{YX}(\omega)$ is defined by the mutual conjugate function.

(2) If $G_{XY}(\omega) = a(\omega) + \mathrm{j}b(\omega)$, then the real part is the even function for ω, namely, $a(\omega) = a(-\omega)$ and the imaginary part is the odd function for ω, namely, $b(\omega) = -b(-\omega)$.

(3) If the process $X(t)$ with $Y(t)$ is orthogonal, we have

$$G_{XY}(\omega) = G_{YX}(\omega) = 0\,. \tag{1.187}$$

Example 1.6.3. The known real random processes $X(t)$ with $Y(t)$ are joint stationary and orthogonal. Find the autocorrelation function and autospectral density expression of $Z(t) = X(t) + Y(t)$.

Solution: The autocorrelation function of $Z(t)$ is

$$\begin{aligned} R_Z(t, t+\tau) &= E\,[Z(t)Z(t+\tau)] \\ &= E\,\{[X(t) + Y(t)]\,[X(t+\tau) + Y(t+\tau)]\} \\ &= R_X(t, t+\tau) + R_Y(t, t+\tau) + R_{XY}(t, t+\tau) + R_{YX}(t, t+\tau) \end{aligned}$$

Because the process $X(t)$ with $Y(t)$ is joint stationary, we have

$$R_Z(t, t+\tau) = R_X(\tau) + R_Y(\tau) + R_{XY}(\tau) + R_{YX}(\tau) = R_Z(\tau)$$

Because the process $X(t)$ with $Y(t)$ is orthogonal, $R_{XY}(\tau) = R_{YX}(\tau) = 0$. Therefore, we have

$$R_Z(\tau) = R_X(\tau) + R_Y(\tau)$$

Because $R(\tau) \to G(\omega)$, the auto self-spectral density of $Z(t)$ is

$$G_Z(\omega) = G_X(\omega) + G_Y(\omega)$$

1.6.6 Power spectral density of discrete-time random processes

In the preceding sections, we discussed the power spectral density and its properties of the continuous-time random process and deduced a very important relationship, the Weiner–Khintchine theorem. With the rapid development of digital technology, the analysis of discrete-time random processes is becoming an increasingly important in the field of electronic engineering. In this section, we will extend the concept of power spectral density to discrete-time random processes.

Assuming that $X(n)$ is a wide-sense stationary discrete-time random process, or a wide-sense stationary random sequence for short, with zero mean value, its autocorrelation function is

$$R_X(m) = E[X(nT)X(nT + mT)] \tag{1.188}$$

Or for short

$$R_X(m) = E[X(n)X(n + m)] \tag{1.189}$$

When $R_X(m)$ satisfies the condition $\sum_{m=-\infty}^{\infty} |R_X(m)| < \infty$, we define the power spectral density of $X(n)$ as the discrete Fourier transform of the autocorrelation function $R_X(m)$

and write it as $G_X(\omega)$

$$G_X(\omega) = \sum_{m=-\infty}^{\infty} R_X(m)e^{-jm\omega T} \tag{1.190}$$

where T is the interval between the values of the random sequence. From the above formula, it can be seen that $G_X(\omega)$ is a periodic continuous function with frequency ω and periodic $2\omega_q$, and the coefficient of the Fourier series of $G_X(\omega)$ is exactly $R_X(m)$:

$$\omega_q = \pi/T \tag{1.191}$$

This is the Nyquist frequency that we are familiar with.

According to eqs. (1.190) and (1.191), we obtain

$$R_X(m) = \frac{1}{2\omega_q} \int_{-\omega_q}^{\omega_q} G_X(\omega)e^{jm\omega T}\,d\omega \tag{1.192}$$

When $m = 0$, we have

$$E[|X(n)|^2] = R_X(0) = \frac{1}{2\omega_q} \int_{-\omega_q}^{\omega_q} G_X(\omega)\,d\omega \tag{1.193}$$

In discrete-time system analysis, sometimes Z-transformation is more convenient, so the power spectral density of the wide-sense stationary discrete-time random process is also defined as the Z-transform of $R_X(m)$ and denoted by $G'_X(z)$, i.e.,

$$G'_X(z) = \sum_{m=-\infty}^{\infty} R_X(m)z^{-m} \tag{1.194}$$

where, $z = e^{j\omega T}$ and

$$G'_X(e^{j\omega T}) = G_X(\omega) \tag{1.195}$$

$R_X(m)$ is the inverse Z-transformation of $G'_X(z)$, i.e.,

$$R_X(m) = \frac{1}{2\pi j} \oint_D G'_X(z)z^{m-1}\,dz \tag{1.196}$$

where D is a closed contour in the convergence domain of $G'_X(z)$, which surrounds the origin of the Z-plane in counterclockwise direction.

According to the symmetry of the autocorrelation functions of the stationary random process, i.e., $R_X(m) = R_X(-m)$, we can easily obtain the property of power spectral density:

$$G'_X(z) = G'_X\left(\frac{1}{z}\right) \tag{1.197}$$

Example 1.6.4. Assuming $R_X(m) = a^{|m|}$ for $|a| < 1$, find $G'_X(z)$ and $G_X(\omega)$.

Solution: Substitute $R_X(m)$ into the definition to obtain:

$$G'_X(z) = \sum_{m=-\infty}^{-1} a^{-m}z^{-m} + \sum_{m=0}^{\infty} a^m z^{-m} = \frac{az}{1-az} + \frac{z}{z-a} = \frac{(1-a^2)z}{(z-a)(1-az)}$$

After finishing, we have

$$G'_X(z) = \frac{(1-a^2)}{(1-az^{-1})(1-az)} = \frac{a^{-1}-a}{(a^{-1}+a)-(z^{-1}+z)}$$

Substituting $z = \mathrm{e}^{\mathrm{j}\omega T}$ into the above formula, we obtain

$$G_X(\omega) = \frac{a^{-1}-a}{a^{-1}+a-2\cos\omega T}$$

1.7 White noise

Noise is also a random process. According to its probability distribution, it can be divided into normal noise, Rayleigh noise and so on. Besides, it can also be divided by the power spectrum characteristic into white noise or colored noise, and colored noise can be divided into broadband noise or narrowband noise. If the noise power spectrum distribution is within the broad-frequency band, it is broadband noise, while if the distribution is within a narrow-frequency band, it is narrowband noise.

If the random process $n(t)$ is stationary with zero mean, and its power spectral density is a nonzero constant in an infinitely wide-frequency domain, i.e.,

$$G(\omega) = \frac{N_0}{2}, \quad -\infty < \omega < \infty \tag{1.198}$$

or

$$F(\omega) = N_0, \quad 0 < \omega < \infty \tag{1.199}$$

we call the process $n(t)$ white (colored) noise. The "white" character is borrowed from the concept of a uniform spectrum of white light (referred as Fig. 1.16). Nonwhite noise is called colored noise.

The relation between the power spectrum and the correlation function can be used to obtain the correlation function (referred as Fig. 1.17) of white noise as follows:

$$R(\tau) = \frac{1}{2\pi}\int_{-\infty}^{\infty} G(\omega)\mathrm{e}^{\mathrm{j}\omega\tau}\,\mathrm{d}\omega = \frac{N_0}{2}\left[\frac{1}{2\pi}\int_{-\infty}^{\infty}\mathrm{e}^{\mathrm{j}\omega\tau}\,\mathrm{d}\omega\right] = \frac{N_0}{2}\delta(\tau) \tag{1.200}$$

Because white noise does not contain a determination component like DC, it conforms to property (4) of the correlation function: $R_X(\infty) = m_X^2$. Therefore, when $\tau = \infty$,

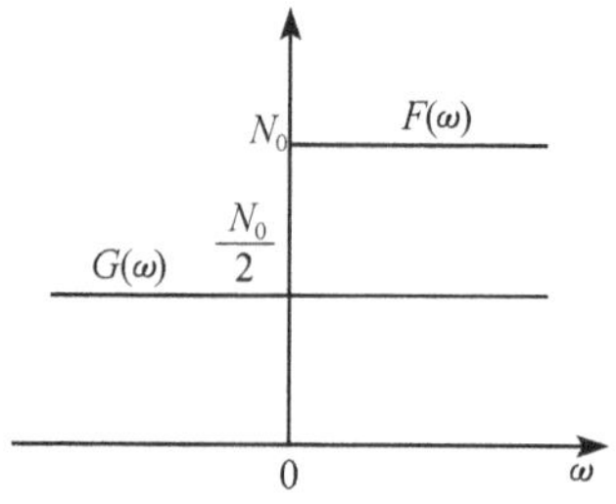

Fig. 1.16: Power spectral density of white noise.

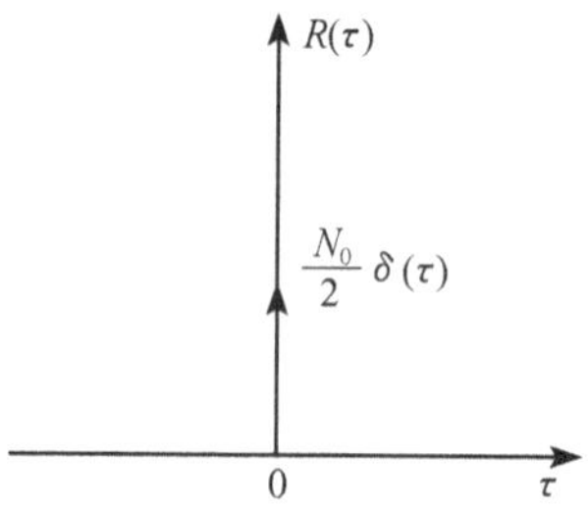

Fig. 1.17: Correlation function of white noise.

we have

$$m^2 = R(\infty) = \frac{N_0}{2}\delta(\infty) = 0 \tag{1.201}$$

$$C(\tau) = R(\tau) - m^2 = \frac{N_0}{2}\delta(\tau) \tag{1.202}$$

$$\sigma^2 = C(0) = \frac{N_0}{2}\delta(0) \tag{1.203}$$

Therefore, the correlation coefficient of white noise is

$$r(\tau) = \frac{C(\tau)}{\sigma^2} = \begin{cases} 1\,, & \tau = 0 \\ 0\,, & \tau \neq 0 \end{cases} \tag{1.204}$$

The correlation time is

$$\tau_0 = \int_0^{\infty} r(\tau)\,\mathrm{d}\tau = 0 \tag{1.205}$$

The above two expressions indicate that white noise is correlated only for $\tau = 0$. When the interval is not zero, the two values are uncorrelated. Therefore, the white noise waveform with infinitely narrow pulse width is a series of random pulses with extremely fast fluctuation. The waveform and power spectrum of a white noise sample are shown by solid lines in Fig. 1.18.

White noise is only an idealized model, so it does not exist. This is because if the power spectrum is constant in the infinite-frequency domain, the average power will be infinite, i.e.,

$$R(0) = \frac{1}{2\pi}\int_0^{\infty} F(\omega)\,\mathrm{d}\omega = \frac{1}{2\pi}\int_0^{\infty} N_0\,\mathrm{d}\omega = \infty \tag{1.206}$$

However, for any actual noise, its average power must be limited, and the actual power spectrum in a finite width of the frequency domain is almost a constant. Moreover, the high-end spectrum usually declines significantly, as is shown by dotted lines in Fig. 1.18 (a). The actual noise pulse waveform, even if it is very narrow, must have a certain width, as shown by dotted lines in Fig. 1.18 (b). Therefore, when the interval is

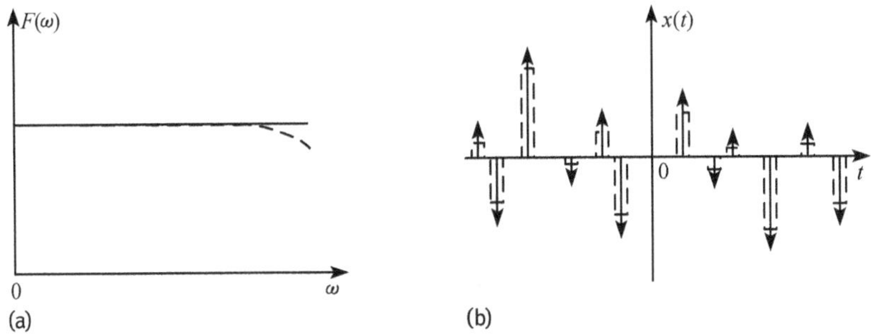

Fig. 1.18: Sample waveform and power spectrum of white noise.

extremely small, the two values of the interval are still correlated and not completely uncorrelated.

In fact, the bandwidth of the radio system must be limited. As long as the noise power spectrum width Δf_n is wider than the system power spectrum width Δf, and the system power spectral density in the system band is relatively uniform, broadband noise can be treated as white noise. This kind of noise, which has uniform power spectral density in a limited band range, is called limited white noise. According to the high–low frequency, it can also be subdivided into intermediate frequency limited white noise and video frequency limited white noise.

Here, we should point out that white noise is classified according to the spectrum property, but based on the probability distribution, noise can be divided into normal noise, Rayleigh noise, etc. This should be the main classification method for noise.

Exercises

1.1 Using trial of repeated tosses of a coin, a random process is defined as

$$X(t) = \begin{cases} \cos \pi t\,, & \text{head} \\ 2t\,, & \text{tail} \end{cases}$$

Suppose the probability of heads or tails is 1/2.

(1) Determine the one-dimensional distribution function $F_1(x, 1/2)$, $F1(x, 1)$ of $X(t)$;

(2) Determine the two-dimensional distribution function $F_2(x_1, x_2; 1/2, 1)$ of $X(t)$.

1.2 Given random process $X(t)$, x is any real number; now another random process is defined as

$$Y(t) = \begin{cases} 1\,, & X(t) \le x \\ 0\,, & X(t) > x\,. \end{cases}$$

Proof. The mean value and the autocorrelation function of $Y(t)$ are, respectively, the one-dimensional and two-dimensional distribution function of $X(t)$. □

1.3 Given a random signal $X(t) = X\cos\omega t$, where ω is a constant and X is a random variable of uniform distribution in the interval (0, 1), try to solve the one-dimensional probability density of $X(t)$ for the moment $t = 0, \pi/4\omega, \pi/2\omega$.

1.4 The mean value $m_X(t)$ and the covariance function $C_X(t_1, t_2)$ of random processes $X(t)$ are known functions, and $s(t)$ is a certain function. Try to solve the mean value and the covariance function of the random process $Y(t) = X(t) + s(t)$.

1.5 Given a random process $X(t) = A\cos\omega t + B\sin\omega t$, where ω is a constant and A and B are standard normal random variables and they are statistically independent, try to find the mean value, variance, correlation function, covariance function, mean square value and standard deviation of $X(t)$.

1.6 Try to find the characteristic function of random amplitude signal $X(t)$ shown in the test example 1.1.2 when $t_2 = \pi/3\omega_0$ and solve its mathematical expectation, variance and mean square value in this case.

1.7 There is a random process $X(t) = X\cos(\omega t + \theta)$, where X is a random variable obeying Rayleigh distribution and its probability density is

$$p(x) = \begin{cases} \frac{x}{\sigma^2} e^{-\frac{x^2}{2\sigma^2}}, & x \leq 0 \\ 0, & x > 0 \end{cases}$$

θ is a random variable of uniform distribution in $(0, 2\pi)$ and it is statistically independent with X, and ω is a constant. Is $X(t)$ a stationary random process?

1.8 Set $Z(t) = A\cos\omega t + n(t)$, where A and ω are constants; $n(t)$ is the stationary normal random process whose mean value is zero and whose variance is σ^2.
   (1) Try to solve the one-dimensional probability density of $Z(t)$.
   (2) Is $Z(t)$ stationary?

1.9 Given a random process $X(t) = X\cos(\omega t + \theta)$, where ω is a constant, X obeys the normal distribution $N(0, \sigma^2)$, θ obeys uniform distribution in $(0, 2\pi)$ and X and θ are statistically independent.
   (1) Try to solve mathematical expectation, variance and correlation function of $X(t)$.
   (2) Is $Z(t)$ stationary?

1.10 Set $s(t)$ as a function with a period T; θ is a random variable of the uniform distribution in $(0, T)$. Try to prove that the periodic random initial phase process $X(t) = s(t + q)$ is an ergodic process.

1.11 It is known that the correlation function of the stationary random process $X(t)$ is $R_X(\tau) = \sigma^2 e^{-\alpha|\tau|}\cos\beta\tau$.
   (1) Determine its power spectral density $G_X(\omega)$, $F_X(\omega)$ and $F_X(f)$.
   (2) Determine its correlation coefficient $r_X(\tau)$ and related time τ_0.

1.12 Find the autocorrelation function and power spectral density $G(f)$ of the following stationary random process and draw their graphs:
   (1) $u_1(t) = A\cos(\omega_0 t + \theta)$;
   (2) $u_2(t) = [A\cos(\omega_0 t + \theta)]^2$.

In the last two formulas, A and ω_0 are both constants, θ is evenly distributed in $(0, 2\pi)$.

1.13 Which of the following functions are the correct expressions for power spectral density? Why?

(1) $G_1(\omega) = \dfrac{\omega^2 + 9}{(\omega^2 + 4)(\omega + 1)^2}$;

(2) $G_2(\omega) = \dfrac{\omega^2 + 1}{\omega^4 + 5\omega^2 + 6}$;

(3) $G_3(\omega) = \dfrac{\omega^2 + 4}{\omega^4 - 4\omega^2 + 3}$;

(4) $G_4(\omega) = \dfrac{e^{-j\omega^2}}{\omega^2 + 2}$.

1.14 A random process $X(t) = A\cos(\omega t + \theta)$, where A is a constant and ω and θ are statistically independent random variables; θ is evenly distributed in $(0, 2\pi)$ and ω has symmetric probability density $p(\omega) = p(-\omega)$.

(1) Is $X(t)$ stationary?

(2) Try to solve the variance and power spectral density of $X(t)$.

1.15 Set $X(t)$ and $Y(t)$ as the statistically independent stationary process, neither of the mean values m_X and m_Y are zero. Now, defining $Z(t) = X(t) + Y(t)$, try to solve cross-spectral densities $G_{XY}(\omega)$ and $G_{XZ}(\omega)$.

1.16 Set $X(t)$ and $Y(t)$ as the joint stationary process. Try to prove

$$\mathrm{Re}[G_{XY}(\omega)] = \mathrm{Re}[G_{YX}(\omega)] : \mathrm{Im}[G_{XY}(\omega)] = -\mathrm{Im}[G_{YX}(\omega)]\ .$$

1.17 $X(t)$ and $Y(t)$ are statistically independent stationary processes, their mean values are m_X and m_Y, respectively; the covariance functions are $C_X(\tau) = e^{-\alpha|\tau|}$ and $C_Y(\tau) = e^{\beta|\tau|}$. Try to solve the correlation function and power spectral density of $Z(t) = X(t)Y(t)$.

1.18 Random processes $X(t)$ and $Y(t)$ are joint stationary.

(1) Solve the autocorrelation function of $Z(t) = X(t) + Y(t)$.

(2) If $X(t)$ and $Y(t)$ are statistically independent, try to solve the autocorrelation function of $Z(t)$.

(3) If $X(t)$ and $Y(t)$ are statistically independent and $E[X] = 0$, try to solve the autocorrelation function of $Z(t)$.

1.19 It is known that $Z(t) = X\cos\omega t + Y\sin\omega t$, where X and Y are random variables.

(1) What conditions do the moment functions of X, Y satisfy, is $Z(t)$ the stationary process?

(2) What distribution do X, Y obey, if $Z(t)$ obeys the normal distribution? Try to solve the one-dimensional probability density of $Z(t)$ under the condition of (1).

1.20 It is known that the random process $Y(t) = X(t) + s(t)$, where $s(t) = \cos(\omega_0 t + \theta)$, ω_0 is a constant, $X(t)$ is a stationary normal process with zero value, and it is statistically independent from $s(t)$. Find when in the following two cases of θ, the random process $Y(t)$ is the normal distribution or not and stationary.

(1) θ is a random variable and it obeys uniform distribution in $(0, 2\pi)$;

(2) $\theta = 0$.

1.21 We know that the mean value and the covariance function of normal random process $X(t)$ are respectively $m_X(t) = 3$ and $C_X(t_1, t_2) = 4e^{-0.2|t_2-t_1|}$. Try to find the probability of $X(5) \le 2$.

1.22 Suppose there are $2n$ random variables A_i, B_i,and none of them are correlated to each other. Their mean values are zero and $E[A_i^2] = E[B_i^2] = \sigma_i^2$. Try to find the mean and the autocorrelation function of the random process

$$X(t) = \sum_{i=1}^{N}(A_i \cos \omega_i t + B_i \sin \omega_i t)$$

and try to solve whether or not $X(t)$ is a stationary process.

1.23 Programming problem: There is a random initial phase signal $X(t) = 5\cos(t+\varphi)$, in which phase φ is a random variable uniform distributed in the range $(0, 2\pi)$. Try to generate three sample functions by using MATLAB programming.

2 Linear transformation of random processes

2.1 Linear transformation and linear system overview

2.1.1 Basic concepts of linear systems

2.1.1.1 Definition and characteristics of the system

Radio systems are usually divided into linear systems and nonlinear systems. Systems with superposition and proportionality are called linear systems. Instead, it is called a nonlinear system.

Assuming a linear system, with the ascertained signal $x(t)$ as the input and the known signal $y(t)$ as the output, the $y(t)$ could be viewed as the result undergoing some certain mathematical operations by the linear system. These mathematical operations are some kind of linear operation, such as addition, multiplication, differentiation, integration, etc. If the linear operator notation $L[\bullet]$ is used to express the linear transformation, it can generally be as shown in Fig. 2.1 or eq. (2.1):

$$y(t) = L[x(t)] \tag{2.1}$$

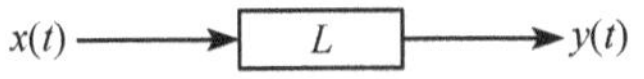

Fig. 2.1: Linear operation of the known signal.

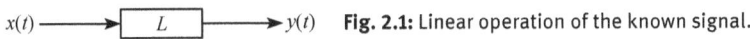

If you take any constant a_1, a_2,

$$y(t) = L[a_1 x_1(t) + a_2 x_2(t)] = a_1 L[x_1(t)] + a_2 L[x_2(t)] \tag{2.2}$$

This is called linear transformation and has the following two basic characteristics:

(1) Superposition: The linear transformation of the sum of functions is equal to the sum of the linear transformations of each function. That is,

$$L\,[x_1(t) + x_2(t)] = L[x_1(t)] + L[x_2(t)] \tag{2.3}$$

(2) Proportionality: The linear transformation of the multiplication between any function and a constant is equal to the multiplication between the linear transformation of the function and this constant. That is,

$$L[ax(t)] = aL[x_1(t)] \tag{2.4}$$

In addition, the radio system can also be divided into time-variant systems and time-invariant systems, according to whether they have time-invariant properties. If there is causality, it is divided into causal systems. Otherwise, it is divided into noncausal systems. If at any time t_0, we have:

$$y(t + t_0) = L\,[x(t + t_0)] \tag{2.5}$$

https://doi.org/10.1515/9783110593808-003

it is said that the system is a time-invariant linear system, and the time-invariant characteristic is that the system response is not dependent on the selection of the timing starting point, which means that if the time of input signal is earlier (or later) in t_0, the time of output signal is the same as that of the input signal, and the waveform remains the same.

Select the timing starting point $t_0 = 0$. While the input signal $x(t)$ of the system contains $x(t) = 0$ when $t < 0$, there is a corresponding output signal $y(t) = 0$. This system is called a causal system. The idea is that when stimulus is added, the response output follows. The stimulus is the cause of the response, and the response is the result of the excitation, which is called causality. Systems that can be physically implemented are all causal systems, so the causal system is called the physical realizable system. On the contrary, the noncausal system without causality is called a physically unrealized system.

2.1.1.2 Analysis of linear systems

The response feature of a linear system is available, namely, the n-th-order linear differential equation is expressed as follows:

$$a_n \frac{\mathrm{d}^n y(t)}{\mathrm{d}t_n} + a_{n-1} \frac{\mathrm{d}^{n-1} y(t)}{\mathrm{d}t_{n-1}} + \cdots + a_1 \frac{\mathrm{d}y(t)}{\mathrm{d}t} + a_0 y(t) = b_m \frac{\mathrm{d}^m x(t)}{\mathrm{d}t_m} + b_{m-1} \frac{\mathrm{d}^{m-1} x(t)}{\mathrm{d}t_{m-1}} + \cdots + b_1 \frac{\mathrm{d}x(t)}{\mathrm{d}t} + b_0 x(t) \tag{2.6}$$

In the formula, if all the coefficients a, b are constants, it is a time-invariant system. If it is a time function, it is a time-variant system.

For time-invariant linear systems, eq. (2.6) is a linear differential equation with constant coefficients and can be used to directly solve the output signal $y(t)$ with the classic differential equation method, but its calculation is generally complicated. So, the common method of time domain analysis with the is convolution integral, with which, calculation is usually easier.

When the unit impulsive $\delta(t)$ is regarded as the input, its system response expresses $h(t)$. Therefore, when the input of a linear system is a known signal $x(t)$, the response output $y(t)$ corresponds to $x(t)$. This transformation is shown in Fig. 2.2.

Enter the input signal $x(t)$ as shown in Fig. 2.3(a). Using the superposition principle, it is broken down into many adjacent narrow pulses (as shown in Fig. 2.3), where the pulse width is $\Delta\tau$ and the pulse amplitude is $x(\tau)$. The smaller $\Delta\tau$ is, the closer the pulse amplitude is to the function value. When $\Delta\tau \to 0$, the narrow pulse becomes the impulse function, and the impulse intensity equals the area of the narrow pulse $x(\tau)\Delta\tau$. According to time invariance, the corresponding output is $x(\tau)\Delta\tau h(t-\tau)$, as shown in Fig. 2.3(b).

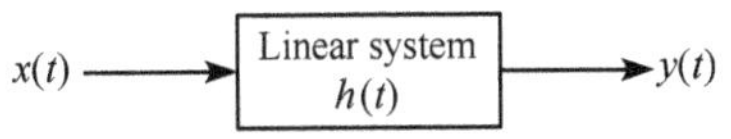

Fig. 2.2: Linear transformation of the known signal.

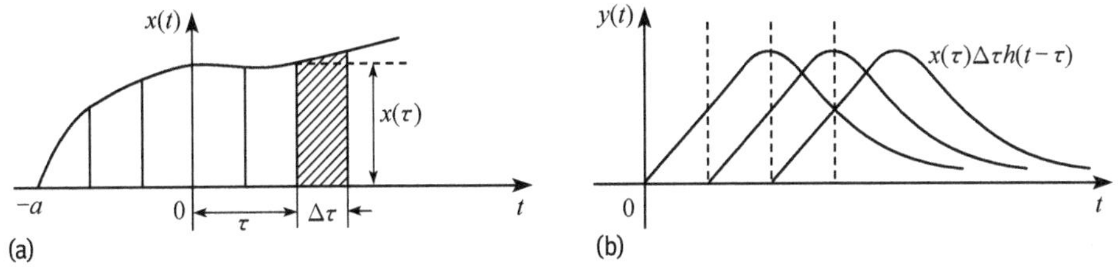

Fig. 2.3: Impulse response method with a known signal.

For the causal system, the response output $y(t_1)$ at any time t_1, is the sum of the response outputs of the previous narrow pulse, which can be written as:

$$y(t_1) = \lim_{\Delta\tau \to 0} \sum_{t=-a}^{t_1} x(\tau)h(t-\tau)\Delta\tau \tag{2.7}$$

or

$$y(t_1) = \int_{-a}^{t_1} x(\tau)h(t-\tau)\,\mathrm{d}\tau \tag{2.8}$$

Assuming that the linear system is a physically stable system, because of causality, if $t < 0$, we have $h(t) < 0$ and because of stability, the output signal with the bounded input signal should also be bounded, so as to guarantee the stability of the system. So, $h(t)$, the impulse response of the system, must be absolutely integral, which means:

$$\int_{-\infty}^{+\infty} |h(t)|\,\mathrm{d}t < \infty \tag{2.9}$$

As a result,

$$\lim_{t \to \infty} h(t) = 0 \tag{2.10}$$

This is when the upper limit of eq. (2.8) is extended $+\infty$ and the lower bound is extended to $-\infty$, but there is no impact on the results. So, we have:

$$y(t) = \int_{-\infty}^{\infty} x(\tau)h(t-\tau)\,\mathrm{d}\tau \tag{2.11}$$

This is called the convolution integral, which is represented by the notation $\otimes$, so this is the shorthand form:

$$y(t) = x(t) \otimes h(t) \tag{2.12}$$

With the variable substitution $t' = t - \tau$, the above integral can be rewritten as:

$$y(t) = \int_{-\infty}^{\infty} x(t-\tau)h(\tau)\,\mathrm{d}\tau \tag{2.13}$$

Or:

$$y(t) = h(t) \otimes x(t) \tag{2.14}$$

For time-invariant linear systems, frequency domain analysis can be used. The superposition principle is superimposed on the frequency domain, which is the Fourier transform. In the complex frequency domain, that is the Laplace transform.

Using the Laplace transform with differential properties to solve the n-th-order linear differential functions of the linear system, we can obtain:

$$\left(a_n s^n + a_{n-1} s^{n-1} + \cdots + a_1 s + a_0\right) Y(s) = \left(b_m s^m + b_{m-1} s^{m-1} + \cdots + b_1 s + b_0\right) X(s) \tag{2.15}$$

where $X(s) = \int_{-\infty}^{\infty} x(t) \mathrm{e}^{-st}\,\mathrm{d}t$, $Y(s) = \int_{-\infty}^{\infty} y(t) \mathrm{e}^{-st}\,\mathrm{d}t$, $s = \sigma + \mathrm{j}\omega$. The above formula can be also written as:

$$Y(s) = H(s)X(s) \tag{2.16}$$

where

$$H(s) = \frac{Y(s)}{X(s)} = \frac{b_m s^m + b_{m-1} s^{m-1} + \cdots + b_1 s + b_0}{a_n s^n + a_{n-1} s^{n-1} + \cdots + a_1 s + a_0} \tag{2.17}$$

It is only related to the characteristic of the system, which is called the transfer function of the linear system. The transfer function $H(s)$ is the Laplace transform of an impulse response $h(t)$, as follows:

$$H(s) = \int_{-\infty}^{\infty} h(t) \mathrm{e}^{-st}\,\mathrm{d}t \tag{2.18}$$

Therefore, when solving the linear system output with the Laplace transform, we can take the following steps. First, $H(s)$, the transfer function, can be obtained according to the specific composition of the linear system. Second, $X(s)$, the Laplace transform of the known input signal $x(t)$, can also be obtained. Finally, find out $Y(s) = H(s)X(s)$, take the inverse Laplace transform, and obtain the output signal $y(t)$.

The Fourier transform is just a special case of the Laplace transform, where $s = \mathrm{j}\omega$. So, by eq. (2.18):

$$H(\omega) = \int_{-\infty}^{\infty} h(t) \mathrm{e}^{-\mathrm{j}\omega t}\,\mathrm{d}t \tag{2.19}$$

$H(\omega)$ is called as a transmission function or frequency response characteristic of a linear system. In general, it is a complex number and can be written as:

$$H(\omega) = |H(\omega)|\, \mathrm{e}^{\mathrm{j}\varphi(\omega)} \tag{2.20}$$

where, the absolute value $|H(\omega)|$ is called the transfer function or the amplitude-frequency characteristic of a linear system, and the phase angle $\varphi(\omega)$ is called the phase-frequency characteristic.

If you know the transfer function $H(\omega)$, then the inverse Fourier transform of the impulse response is made:

$$h(t) = \frac{1}{2\pi} \int_{-\infty}^{\infty} H(\omega) e^{j\omega t} \, d\omega \tag{2.21}$$

It is a pair of Fourier transforms, which is related:

$$h(t) \leftrightarrow H(\omega) \tag{2.22}$$

For the causal system, we know $h(t) = 0$ for $t < 0$ and as a result:

$$H(\omega) = \int_{0}^{\infty} h(t) e^{-j\omega t} \, dt \tag{2.23}$$

When using Fourier transforms to solve the output signal, if the input signal is $x(t)$ and the transfer function $H(\omega)$ is known, then the output signal is:

$$y(t) = \frac{1}{2\pi} \int_{-\infty}^{\infty} X(\omega) H(\omega) e^{j\omega t} \, d\omega \tag{2.24}$$

where $X(\omega)$ is the spectrum of the input signal $x(t)$, and $X(\omega)/H(\omega)$ is the spectrum of the output signal $y(t)$.

Example 2.1.1. The RC integral circuit is shown in Fig. 2.4, so we can find its transfer function and the impulse response.

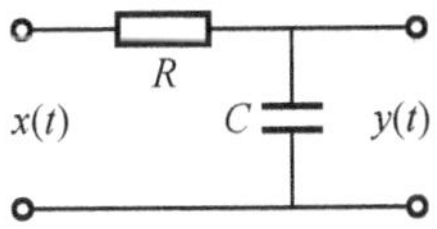

Fig. 2.4: RC integrator circuit.

Solution: Figure 2.4 shows the relation between output voltage $y(t)$ and input voltage $x(t)$:

$$RC \frac{dy(t)}{dt} + y(t) = x(t) \tag{2.25}$$

make $a = 1/RC$, and the Laplace transform of both sides of the upper equation:

$$(s + a)Y(s) = aX(s) \tag{2.26}$$

Therefore, the transfer function is:

$$H(s) = \frac{Y(s)}{X(s)} = \frac{a}{s + a} \tag{2.27}$$

For the inverse Laplace transform, the impulse response is:

$$h(t) = \begin{cases} ae^{-at}, & t \geq 0 \\ 0, & t < 0 \end{cases} \tag{2.28}$$

Of course, the Fourier transform method can be used to obtain the transfer function directly with the principle of partial pressure:

$$H(\omega) = \frac{1/\mathrm{j}\omega C}{R + 1/\mathrm{j}\omega C} = \frac{1}{1 + \mathrm{j}\omega RC} \tag{2.29}$$

2.1.2 Research topic of linear transformation of random processes

When the input of the linear system is random process $X(t)$, then its response output $Y(t)$ is still a random process (Fig. 2.5). For a sample $x(t)$ of the random processes $X(t)$, because it is a definite time function, the linear transformation method can be used directly. For the whole random process $X(t)$, in this chapter, the method of differential equations, impulse response and the spectral method can be used to solve the output random process $Y(t)$. However, when studying a random process, one does not need to obtain the specific detailed output processes, but only needs to solve the statistical properties of the output random process $Y(t)$ with the statistical properties of the known input random process $X(t)$. Therefore, the research topic of random process linear transformation is mainly the following two types of problems:

(1) The moment function (or power spectral density) of the input process $X(t)$ is known, then solve the moment function (or power spectral density) of the output process $Y(t)$.

For the linear transformation of the moment function, the main research is the relationships between the statistical mean $m_Y(t)$ and $m_X(t)$, also autocorrelation $R_Y(t_1, t_2)$ and $R_X(t_1, t_2)$. If the process is stationary, we have:

$$R_X(\tau) \leftrightarrow G_X(\omega), R_Y(\tau) \leftrightarrow G_Y(\omega) \tag{2.30}$$

Therefore, the relationship between the power spectral densities $G_Y(\omega)$ and $G_X(\omega)$ can be obtained.

(2) The input process $X(t)$ probability distribution is known, then solve the output process $Y(t)$ probability distribution.

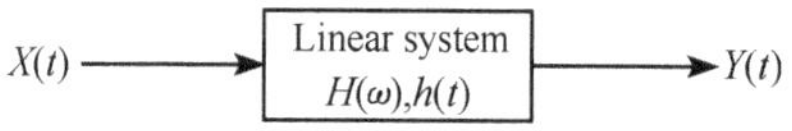

Fig. 2.5: Linear transformation of random processes.

2.2 Differentiation and integration in stochastic processes

2.2.1 Limit of the random process

2.2.1.1 Limit of the sequence of random variables

The definition of the limit of random variables is different from that of ordinary variables. There are two common definitions:

(1) A sequence of random variables $\{X_n\}$, $n = 1, 2, \ldots$, for any small positive integer ε, constantly has:

$$\lim_{n\to\infty} P\{|X_n - X| > \varepsilon\} = 0 \tag{2.31}$$

This is called a sequence of random variables $\{X_n\}$, which converges to the random variable X in the meaning of convergence in probability, or the variable X is the limit of the sequence $\{ X_n\}$ in the meaning of convergence in probability, written as:

$$\lim_{n\to\infty} X_n \overset{P}{=} X \tag{2.32}$$

The explanation is that for any small positive number ε and η, there are $N = N(\varepsilon, \eta)$, which makes it constantly satisfy the following for $n > N$:

$$P\{|X_n - X| > \varepsilon\} < \eta \tag{2.33}$$

This means that the event of the deviation between X_n with X larger than ε is almost impossible to come by.

(2) We assume that both random variables X and X_n $(n = 1, 2, \ldots)$ have second moments, if we have:

$$\lim_{n\to\infty} E\{|X_n - X|^2\} = 0 \tag{2.34}$$

This is called a sequence of random variables $\{X_n\}$, which converges to the random variable X in the meaning of convergence in the mean square, or the variable X is the limit of the sequence $\{ X_n\}$ with the meaning of convergence in the mean square, written as:

$$\text{l.i.m}\, X_n = X \tag{2.35}$$

where, l.i.m. is the abbreviation for limit in the mean square, which means the limit of the convergence in the mean square.

By the Chebyshev inequality:

$$P\{|X_n - X| > \varepsilon\} \le \frac{E\{|X_n - X|^2\}}{\varepsilon^2} \tag{2.36}$$

We know that, if the random variable sequence $\{X_n\}$ converges to the random variable X with the meaning of the mean square, it is bound to be the same as converging to X with the meaning of probability. However, the opposite is not true. The following mainly uses the concept of convergence in the mean square, while the concept of convergence in probability is used only as a statistical explanation.

2.2.1.2 Limits of random processes

Random processes are a group of random variables that change over time. The sequence of random variables is the same as a continuous variable that is generalized to a continuous variable by the limit definition of a sequence $\{X_n\}$. The limit definition can be generalized to random processes (random functions).

(1) If the random process $X(t)$ for any small positive number ε, for $t \to t_0$, we always have:

$$\lim_{t \to t_0} P\{|X(t) - X| > \varepsilon\} = 0 \tag{2.37}$$

which is called the random process $X(t)$, for $t \to t_0$, which converges to the random variable X in the meaning of probability. Or variable X is the limit of the process $X(t)$, for $t \to t_0$, in the meaning of convergence in probability, written as:

$$\lim_{t \to t_0} X(t) \overset{P}{=} X \tag{2.38}$$

(2) Assuming that the random process $X(t)$ and random variables X both have second moments, for $t \to t_0$, we have:

$$\lim_{t \to t_0} E\{|X(t) - X|^2\} = 0 \tag{2.39}$$

Which is called the random process $X(t)$, which, for $t \to t_0$, converges to the random variable X in the meaning of convergence in the mean square. Or variable X is the limit of the process $X(t)$, for $t \to t_0$, in the meaning of convergence in the mean square:

$$\underset{t \to t_0}{\text{l.i.m}}\, X(t) = X \tag{2.40}$$

2.2.2 Continuity of stochastic processes

The so-called continuous process of the random process does not require each sample to be continuous, but only in the continuous case of convergence in the mean square.

2.2.2.1 Definition of the continuous mean square

In the random process $X(t)$ in t nearby, we have:

$$\lim_{\Delta t \to 0} E\{|X(t + \Delta t) - X(t)|^2\} = 0 \tag{2.41}$$

or

$$\underset{\Delta t \to 0}{\text{l.i.m}}\, X(t + \Delta t) = X(t) \tag{2.42}$$

which is called the random process $X(t)$, which is convergence in the mean square at point t and must be continuous, and also $X(t)$ is mean square continuous at point t for short.

Convergence in the mean square must lead to convergence in probability, so we have:

$$\lim_{\Delta t \to 0} X(t + \Delta t) \overset{P}{=} X(t) \tag{2.43}$$

That is, when $\Delta t \to 0$, we have:

$$P\{|X(t + \Delta t) - X(t)| > \varepsilon\} < \eta \tag{2.44}$$

where, ε and η can be any small positive number. The above formula (2.44) shows that when time t changes a little, it is impossible that the deviation of $X(t + \Delta t)$ and $X(t)$ be greater than ε.

2.2.2.2 Conditions of the continuous mean square

The correlation function $R_X(t_1, t_2)$ at $t_1 = t_2 = t$ is continuous, which is the necessary and sufficient condition in the case that real random process $X(t)$ at point t is continuous with mean square. This is because:

$$\begin{aligned} &E\{|X(t + \Delta t) - X(t)|^2\} \\ &\quad = E[X(t + \Delta t)X(t + \Delta t)] - E[X(t + \Delta t)X(t)] - E[X(t)X(t + \Delta t)] + E[X(t)X(t)] \\ &\quad = R_X(t + \Delta t, t + \Delta t) - R_X(t + \Delta t, t) - R_X(t, t + \Delta t) + R_X(t, t) \end{aligned} \tag{2.45}$$

When $\Delta t \to 0$, the right-hand side of the above equation approaches zero, and we have:

$$E\{|X(t + \Delta t) - X(t)|^2\} \to 0 \tag{2.46}$$

So, we know that process $X(t)$ at point t is continuous with mean square.

If $X(t)$ is the stationary process, its correlation function is:

$$R_X(\tau) = E\left[X(t)X(t + \tau)\right] = E\left[X(t - \tau)X(t)\right] \tag{2.47}$$

At this point, we have:

$$E\{|X(t + \Delta t) - X(t)|^2\} = 2[R_X(0) - R_X(\Delta t)] \tag{2.48}$$

Knowing the stationary process $X(t)$, the necessary and continuous requirements of the continuous mean square are:

$$\lim_{\Delta t \to 0} R_X(\Delta t) = R_X(0) \tag{2.49}$$

or:

$$\lim_{\tau \to 0} R_X(\tau) = R_X(0) \tag{2.50}$$

That is, as long as $R_X(\tau)$ for $\tau = 0$ is continuous, the stationary process $X(t)$ for any time t is continuous.

2.2.2.3 Continuity of the mean

If the random process $X(t)$ is continuous in the mean square, the mean $E[X(t)]$ must be continuous, namely:

$$\lim_{\Delta t\to 0} E\left[X(t+\Delta t)\right] = E[X(t)] \tag{2.51}$$

The proof is as follows:

Assuming random variable $Z = \overset{\circ}{Z} + E[Z]$, we have:

$$E[Z^2] = \sigma_Z^2 + [E(Z)]^2 \geq [E(Z)]^2 \tag{2.52}$$

Similarly, if you have a random variable $Z(t) = X(t+\Delta t) - X(t)$, then:

$$E\{[X(t+\Delta t) - X(t)]^2\} \geq \{E[X(t+\Delta t) - X(t)]\}^2 \tag{2.53}$$

If it is known that process $X(t)$ is continuous in the mean square, then:

$$\lim_{\Delta t\to 0} E\{|X(t+\Delta t) - X(t)|^2\} = 0 \tag{2.54}$$

It can be seen from eq. (2.54) that the right-hand side of this inequality must approach zero, i.e.,

$$\lim_{\Delta t\to 0} E\left[X(t+\Delta t) - X(t)\right] = 0 \tag{2.55}$$

or

$$\lim_{\Delta t\to 0} E\left[X(t+\Delta t)\right] = E[X(t)] \tag{2.56}$$

which is the proof.

If eq. (2.42) is substituted into eq. (2.51), then we have:

$$\lim_{\Delta t\to 0} E\left[X(t+\Delta t)\right] = E\left[\underset{\Delta t\to 0}{\mathrm{l.i.m}}\, X(t+\Delta t)\right] \tag{2.57}$$

The above formula shows that the order of operations can be exchanged when we solve the mean and the limit of the random process under the condition of mean square continuity. However, it should be noted that the meaning of the two limits in the upper form is different, the left-hand side is the limit of the ordinary function, but the right-hand side is the limit of the convergence in the mean square.

2.2.3 Differential of stochastic processes (derivatives)

2.2.3.1 Definition of the mean square derivative

There is a random process $X(t)$. If the increment ratio $[X(t+\Delta t) - X(t)]/\Delta t$, for $\Delta t \to 0$, converges to a random variable (and related t) in the mean square, then the random variable is called the mean square derivative of the random process $X(t)$ at the point t. Remember it as $\mathrm{d}X(t)/\mathrm{d}t$, or $\dot{X}(t)$ for short, i.e.,

$$\dot{X}(t) = \frac{\mathrm{d}X(t)}{\mathrm{d}t} = \underset{\Delta t\to 0}{\mathrm{l.i.m}}\, \frac{X(t+\Delta t) - X(t)}{\Delta t} \tag{2.58}$$

or

$$\lim_{\Delta t\to 0} E\left\{\left|\frac{X(t+\Delta t)-X(t)}{\Delta t}-\dot{X}(t)\right|^2\right\}=0 \tag{2.59}$$

Like above, we can also define the second-order mean square of a random process $X(t)$ as follows:

$$\ddot{X}(t)=\frac{\mathrm{d}^2X(t)}{\mathrm{d}t^2}=\underset{\Delta t\to 0}{\mathrm{l.i.m}}\frac{\dot{X}(t+\Delta t)-\dot{X}(t)}{\Delta t} \tag{2.60}$$

2.2.3.2 Conditions of the mean square differential (derivative)

If the random process $X(t)$ is mean square differential at point t, it should be mean square continuous at point t; at the same time, the right or left limit of the mean square is equal, i.e.,

$$\lim_{\Delta t_1,\Delta t_2\to 0} E\left\{\left[\frac{X(t+\Delta t_1)-X(t)}{\Delta t_1}-\frac{X(t+\Delta t_2)-X(t)}{\Delta t_2}\right]^2\right\}=0 \tag{2.61}$$

Where:

$$\begin{aligned}
&E\left\{\left[\frac{X(t+\Delta t_1)-X(t)}{\Delta t_1}-\frac{X(t+\Delta t_2)-X(t)}{\Delta t_2}\right]^2\right\}\\
&=\frac{1}{\Delta t_1^2}\{[R_X(t+\Delta t_1,t+\Delta t_1)-R_X(t,t+\Delta t_1)]-[R_X(t+\Delta t_1,t)-R_X(t,t)]\}\\
&\quad+\frac{1}{\Delta t_2^2}\{[R_X(t+\Delta t_2,t+\Delta t_2)-R_X(t,t+\Delta t_2)]-[R_X(t+\Delta t_2,t)-R_X(t,t)]\}\\
&\quad-2\frac{1}{\Delta t_1\Delta t_2}\{[R_X(t+\Delta t_1,t+\Delta t_2)-R_X(t,t+\Delta t_2)]-[R_X(t+\Delta t_1,t)-R_X(t,t)]\}
\end{aligned} \tag{2.62}$$

If there is a second-order mixed partial derivative $\frac{\partial^2 R_X(t_1,t_2)}{\partial t_1\partial t_2}$ for $t_1=t_2=t$, then the limit of the third item of the upper right-hand side of the above equation is as follows:

$$\begin{aligned}
&\lim_{\Delta t_1,\Delta t_2\to 0}\frac{1}{\Delta t_1\Delta t_2}\{[R_X(t+\Delta t_1,t+\Delta t_2)-R_X(t,t+\Delta t_2)]-[R_X(t+\Delta t_1,t)-R_X(t,t)]\}\\
&\qquad=\frac{\partial^2 R_X(t_1,t_2)}{\partial t_1\partial t_2}
\end{aligned} \tag{2.63}$$

The limit of the first and second items on the right-hand side is:

$$\begin{aligned}
&\lim_{\Delta t_1=\Delta t_2\to 0}\frac{1}{\Delta t_1\Delta t_2}\{[R_X(t+\Delta t_1,t+\Delta t_2)-R_X(t,t+\Delta t_2)]-[R_X(t+\Delta t_1,t)-R_X(t,t)]\}\\
&\qquad=\frac{\partial^2 R_X(t_1,t_2)}{\partial t_1\partial t_2}
\end{aligned} \tag{2.64}$$

and we have:

$$\lim_{\Delta t_1, \Delta t_2 \to 0} E\left\{\left[\frac{X(t+\Delta t_1) - X(t)}{\Delta t_1} - \frac{X(t+\Delta t_2) - X(t)}{\Delta t_2}\right]^2\right\}$$
$$= \frac{\partial^2 R_X(t_1, t_2)}{\partial t_1 \partial t_2} + \frac{\partial^2 R_X(t_1, t_2)}{\partial t_1 \partial t_2} - 2\frac{\partial^2 R_X(t_1, t_2)}{\partial t_1 \partial t_2} = 0 \quad (2.65)$$

It can be seen for random process $X(t)$ that the necessary and sufficient condition of the mean square differential is that the correlation function $R_X(t_1, t_2)$ have a second-order mixed partial derivative for the independent variable $t_1 = t_2 = t$.

If $X(t)$ is a stationary process, because of

$$R_X(t_1, t_2) = R_X(t_2 - t_1) = R_X(\tau) \quad (2.66)$$

We have $\tau = t_2 - t_1$, and then:

$$\left.\frac{\partial^2 R_X(t_1, t_2)}{\partial t_1 \partial t_2}\right|_{t_1=t_2=t} = -\left.\frac{d^2 R_X(\tau)}{d\tau^2}\right|_{\tau=0} = -R''_X(0) \quad (2.67)$$

It can be seen for random process $X(t)$ that the necessary and sufficient condition of the mean square differential is that the correlation function $R_X(\tau)$ have a second derivative $R''_X(0)$ for the independent variable $\tau = 0$. Obviously, to make $R''_X(0)$ exist, $R'_X(\tau)$ must be continuous for $\tau = 0$.

Example 2.2.1. Given the correlation function $R_X(\tau) = e^{-a\tau^2}$ of a random process $X(t)$, the process $X(t)$ is the mean square continuous or not?

Solution: $R_X(\tau) = e^{-a\tau^2}$ belongs to the elementary function, the correlation function $R_X(\tau)$ is continuous for $\tau = 0$, so the process $X(t)$ is mean square continuous for any time t:

$$R'_X(\tau) = \frac{dR_X(\tau)}{d\tau} = -2a\tau e^{-a\tau^2} \quad (2.68)$$

$$R''_X(0) = \frac{d^2 R_X(\tau)}{d\tau^2}|_{\tau=0} = -2a \quad (2.69)$$

Because of the $R''_X(0)$, we know that the process $X(t)$ is mean square differential.

Example 2.2.2. Given the correlation function $R_X(\tau) = \sigma^2 e^{-a|\tau|}$ of a random process $X(t)$, try to determine its continuity and differentiability.

Solution:

$$R_X(\tau) = \sigma^2 e^{-a|\tau|} = \begin{cases} \sigma^2 e^{-a\tau}, & \tau \geq 0 \\ \sigma^2 e^{a\tau}, & \tau < 0 \end{cases} \quad (2.70)$$

It belongs to the elementary function. The correlation function $R_X(\tau)$ is continuous for $\tau = 0$, as is shown in Fig. 2.6 (a). So, we know that process $X(t)$ is mean square

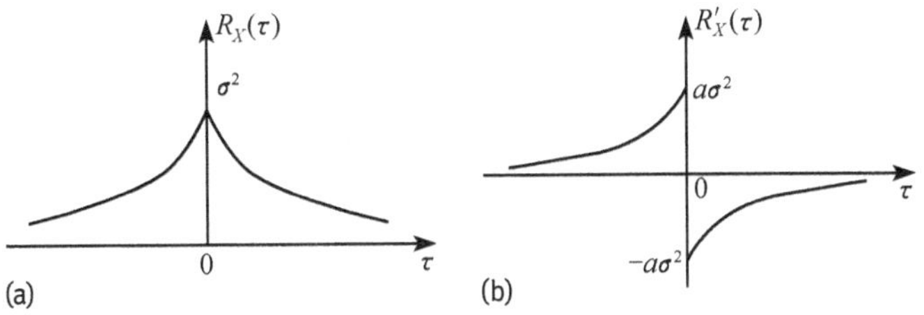

Fig. 2.6: Correlation function graph of Example 2.2.2.

continuous for any time t:

$$R'_X(\tau) = \begin{cases} -a\sigma^2 e^{-a\tau}, & \tau \geq 0 \\ a\sigma^2 e^{a\tau}, & \tau < 0 \end{cases} \tag{2.71}$$

As is shown in Fig. 2.6 (b), because the first derivative $R'_X(\tau)$ is discontinuous, the second derivative $R''_X(0)$ does not exist for $\tau = 0$. Therefore, the mean square differential of this random process does not exist either.

2.2.4 Differential transformation of stochastic processes

If the random process $X(t)$ satisfies the mean square differentiable, then the mean square derivative $\frac{dX(t)}{dt}$ is still a random process, written as $\dot{X}(t)$ for short, i.e.,

$$Y(t) = \frac{dX(t)}{dt} = \dot{X}(t) \tag{2.72}$$

It is the output of the input random process $X(t)$ with the differential transformation of the ideal differentiator.

2.2.4.1 Mean (or mathematical expectation) of the mean square derivatives of random processes

Because:

$$\dot{X}(t) = \frac{dX(t)}{dt} = \lim_{\Delta t \to 0} \frac{X(t+\Delta t) - X(t)}{\Delta t} \tag{2.73}$$

so,

$$\begin{aligned} m_{\dot{X}}(t) = E[\dot{X}(t)] &= E\left[\underset{\Delta t \to 0}{\mathrm{l.i.m}} \frac{X(t+\Delta t) - X(t)}{\Delta t}\right] = \lim_{\Delta t \to 0} E\left[\frac{X(t+\Delta t) - X(t)}{\Delta t}\right] \\ &= \lim_{\Delta t \to 0} \frac{m_X(t+\Delta t) - m_X(t)}{\Delta t} = \frac{dm_X(t)}{dt} \end{aligned} \tag{2.74}$$

The mean of the derivative of the random process is equal to the derivative of the mean of the process. That is,

$$E\left[\frac{dX(t)}{dt}\right] = \frac{d}{dt}E[X(t)] \tag{2.75}$$

So. we can see that the derivative and the mean operation of the random process can be reversed.

If $X(t)$ is the stationary process, because its mean $m_X(t)$ is independent of the time t, we know that the mean of the random process $\dot{X}(t)$ is $m_{\dot{X}}(t) = 0$. In other words, for the changing rate of all the samples of process $X(t)$ at the same time t, the statistical average is zero.

2.2.4.2 Correlation function between random processes and the mean square derivative

Based on the definition of the mean square derivative, we obtain:

$$\dot{X}(t_2) = \lim_{\Delta t_2 \to 0} \frac{X(t_2 + \Delta t_2) - X(t_2)}{\Delta t_2} \tag{2.76}$$

while

$$\begin{aligned} R_{X\dot{X}}(t_1, t_2) = E[X(t_1)\dot{X}(t_2)] &= E\left[X(t_1) \lim_{\Delta t_2 \to 0} \frac{X(t_2 + \Delta t_2) - X(t_2)}{\Delta t_2}\right] \\ &= \lim_{\Delta t_2 \to 0} \frac{1}{\Delta t_2} E\left[X(t_1)X(t_2 + \Delta t_2) - X(t_1)X(t_2)\right] \\ &= \lim_{\Delta t_2 \to 0} \frac{1}{\Delta t_2} \left[R_X(t_1, t_2 + \Delta t_2) - R_X(t_1, t_2)\right] \\ &= \frac{\partial}{\partial t_2} R_X(t_1, t_2) \end{aligned} \tag{2.77}$$

Similarly:

$$R_{\dot{X}X}(t_1, t_2) = \frac{\partial}{\partial t_1} R_X(t_1, t_2) \tag{2.78}$$

If $X(t)$ is a stationary process and $\tau = t_2 - t_1$, we have:

$$R_{\dot{X}X}(\tau) = -\frac{\mathrm{d}}{\mathrm{d}\tau} R_X(\tau) \tag{2.79}$$

$$R_{X\dot{X}}(\tau) = \frac{\mathrm{d}}{\mathrm{d}\tau} R_X(\tau) \tag{2.80}$$

For $t_1 = t_2 = t$, we have

$$R_{X\dot{X}}(\tau)\big|_{\tau=0} = -\, R_{\dot{X}X}(\tau)\big|_{\tau=0} = 0 \tag{2.81}$$

This shows that the value $X(t_1)$ and the derivative e $\dot{X}(t_1)$ of the stationary process $X(t)$ for any fixed time t_1 are not related. If $X(t)$ is the normal process, then $\dot{X}(t)$ is still a normal process, which leads to $X(t_1)$ and $\dot{X}(t_1)$ being unrelated. Namely, it is statistically independent. So, we can see that the value $X(t_1)$ and the slope at that point $\dot{X}(t_1)$ of a stationary normal process, for any time t_1, are statistically independent.

2.2.4.3 Autocorrelation function of the mean square derivative of random processes

$$\begin{aligned}R_{\dot{X}}(t_1,t_2) &= E[\dot{X}(t_1)\dot{X}(t_2)] = E\left[\lim_{\Delta t_1\to 0}\frac{X(t_1+\Delta t_1)-X(t_1)}{\Delta t_1}\cdot \dot{X}(t_2)\right]\\ &= \lim_{\Delta t_1\to 0}\frac{1}{\Delta t_1}E[X(t_1+\Delta t_1)\dot{X}(t_2)-X(t_1)\dot{X}(t_2)]\\ &= \lim_{\Delta t_1\to 0}\frac{1}{\Delta t_1}[R_{X\dot{X}}(t_1+\Delta t_1,t_2)-R_{\dot{X}X}(t_1,t_2)]\\ &= \frac{\partial}{\partial t_1}R_{X\dot{X}}(t_1,t_2)\end{aligned} \tag{2.82}$$

Similarly:

$$R_{\dot{X}}(t_1,t_2) = \frac{\partial}{\partial t_2}R_{\dot{X}X}(t_1,t_2) \tag{2.83}$$

by substituting eq. (2.79) or eq. (2.80) into eq. (2.82) and eq. (2.83), we have

$$R_{\dot{X}}(t_1,t_2) = \frac{\partial^2}{\partial t_1\partial t_2}R_X(t_1,t_2) \tag{2.84}$$

The eq. (2.84) shows that before the autocorrelation function of the process $X(t)$ is found, we need to do twice differential transformations of the autocorrelation function of the original process $X(t)$ (first do the once differential for a variable t_1 or t_2, then do the once differential for another variable t_2 or t_1).

If $X(t)$ is a stationary process and $\tau = t_2 - t_1$, then we have

$$R_{\dot{X}}(\tau) = -\frac{\mathrm{d}^2}{\mathrm{d}\tau^2}R_X(\tau) \tag{2.85}$$

This shows that the derivative of the stationary process is still a stationary process.

Example 2.2.3. Given the autocorrelation function $R_X(\tau) = \mathrm{e}^{-a\tau^2}$ of a stationary process $X(t)$, find the correlation coefficients $r_X(\tau)$, $r_{\dot{X}X}(\tau)$, $r_{\dot{X}}(\tau)$ and draw their graph.

Solution: The mean:

$$m_X = \pm\sqrt{R_X(\infty)} = 0$$

Then

$$C_X(\tau) = R_X(\tau) = \mathrm{e}^{-a\tau^2} \tag{2.86}$$

$$C_{\dot{X}X}(\tau) = -\frac{\mathrm{d}C_X(\tau)}{\mathrm{d}\tau} = 2a\tau\mathrm{e}^{-a\tau^2} \tag{2.87}$$

$$C_{\dot{X}}(\tau) = -\frac{\mathrm{d}^2C_X(\tau)}{\mathrm{d}\tau^2} = 2a\left(1-2a\tau^2\right)\mathrm{e}^{-a\tau^2} \tag{2.88}$$

so $\sigma_X^2 = C_X(0) = 1$ and $\sigma_{\dot{X}}^2 = C_{\dot{X}}(0) = 2a$.

Finally, the correlation coefficients:

$$r_X(\tau) = \frac{C_X(\tau)}{\sigma_X^2} = \mathrm{e}^{-a\tau^2} \tag{2.89}$$

$$r_{\dot{X}X}(\tau) = \frac{C_{\dot{X}X}(\tau)}{\sigma_{\dot{X}}\sigma_X} = \sqrt{2a}\tau\mathrm{e}^{-a\tau^2} \tag{2.90}$$

$$r_{\dot{X}}(\tau) = \frac{C_{\dot{X}}(\tau)}{\sigma_{\dot{X}}^2} = (1 - 2a\tau^2)\mathrm{e}^{-a\tau^2} \tag{2.91}$$

The graph of each correlation coefficient is shown in Fig. 2.7.

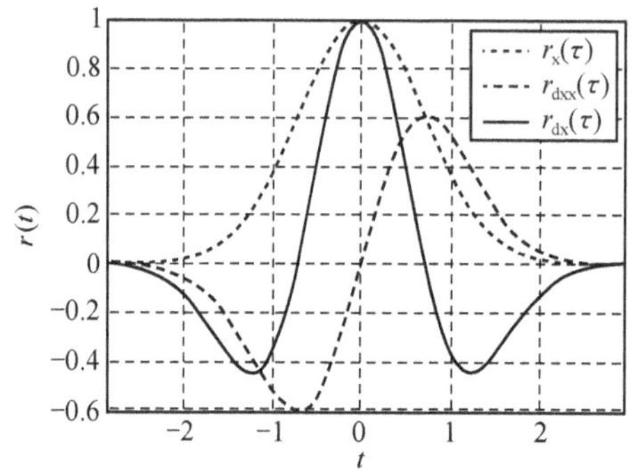

Fig. 2.7: Correlation function graph of Example 2.2.3.

2.2.4.4 Power spectral density of mean square derivatives of random processes

The power spectral density and correlation function of the stationary process are a pair of Fourier transforms:

$$R_X(\tau) = \frac{1}{2\pi}\int_{-\infty}^{\infty} G_X(\omega)\mathrm{e}^{\mathrm{j}\omega\tau}\,\mathrm{d}\omega \tag{2.92}$$

$$R_{\dot{X}}(\tau) = \frac{1}{2\pi}\int_{-\infty}^{\infty} G_{\dot{X}}(\omega)\mathrm{e}^{\mathrm{j}\omega\tau}\,\mathrm{d}\omega \tag{2.93}$$

The correlation functions of the process $\dot{X}(t)$ are:

$$R_{\dot{X}}(\tau) = -\frac{\mathrm{d}^2}{\mathrm{d}\tau^2}R_X(\tau) = \frac{1}{2\pi}\int_{-\infty}^{\infty} \omega^2 G_X(\omega)\mathrm{e}^{\mathrm{j}\omega\tau}\,\mathrm{d}\omega \tag{2.94}$$

So, we can obtain the relationship between the power spectral density of $\dot{X}(t)$ and $X(t)$ with eqs. (2.93) and (2.94), which is

$$G_{\dot{X}}(\omega) = \omega^2 G_X(\omega) \tag{2.95}$$

Now we find that the variance of the zero mean random process $\dot{X}(t)$ is

$$\sigma_{\dot{X}}^2 = C_{\dot{X}}(0) = R_{\dot{X}}(0) = \frac{1}{2\pi}\int_{-\infty}^{\infty} \omega^2 G_X(\omega)\,\mathrm{d}\omega \tag{2.96}$$

The formula above shows that to make the right integral, when ω is increasing, power spectral density $G_X(\omega)$ must decrease faster than ω^{-2} does.

2.2.5 Integrals of random processes

Random process $X(t)$ for any of the samples $x_i(t)$ is the determine function of time t. When it is integrated in the determined interval (a, b), the integral value y is equal to the limit of the sum of the area of all of its subintervals, i.e.,

$$y = \int_a^b x_i(t)\,\mathrm{d}t = \lim_{\Delta t_k \to 0} \sum_{k=1}^{n} x_i(t_k)\Delta t_k \tag{2.97}$$

where $\Delta t_k = t_k - t_{k-1}$, $k = 1, 2, \ldots$.

Obviously, for any sample $x_i(t)$, the integral value y is always fixed, but it is different when the sample is different. So, the integral value of the random process in the determined interval is a random variable. On the meaning of convergence in the mean square, the integral value of the process is a fixed value, and the mean square integral of the random process $X(t)$ can be seen as the following definition. If we have

$$\lim_{\Delta t_k \to 0} E\left\{\left|\sum_{k=1}^{n} X(t_k)\Delta t_k - \int_a^b X(t)\,\mathrm{d}t\right|^2\right\} = 0 \tag{2.98}$$

the random variables are called the mean square integral of the random process $X(t)$ in the determination interval (a, b)

$$Y = \int_a^b X(t)\,\mathrm{d}t = \lim_{\Delta t_k \to 0} \sum_{k=1}^{n} X(t_k)\Delta t_k \tag{2.99}$$

It should be pointed out that although the integral symbol without differentiation of the generally ascertained function is still used here, its meaning is different. In this case, it represents the integral in the meaning of the convergence in mean square.

The integral definition can be used for generalized integral promotion:

$$Y = \int_{-\infty}^{\infty} X(t)\,\mathrm{d}t = \lim_{\substack{a \to -\infty \\ b \to \infty}} \int_a^b X(t)\,\mathrm{d}t \tag{2.100}$$

The necessary and sufficient condition of the mean square integral of a random process $X(t)$ is that the double integral as shown in the following formula (integral under the general meaning) should exist:

$$\int_a^b \int_a^b R_X(t_1, t_2)\, \mathrm{d}t_1\, \mathrm{d}t_2 < \infty \tag{2.101}$$

When researching the linear transformation of a random process $X(t)$, it is often necessary to compute the weighted integral with a certain function $w(t)$, and then the integral value is still a random variable:

$$Y = \int_a^b w(t)X(t)\, \mathrm{d}t \tag{2.102}$$

However, in practical applications, such as the transient process of the random process, it is commonly used to integrate the random process with variable upper limits. Namely, the lower limit is zero and the upper limit is the time variable t. So, the integral value is a random variable that changes over time, and it becomes a random process, which is:

$$Y(t) = \int_0^t w(a)X(a)\, \mathrm{d}a \tag{2.103}$$

2.2.6 Integral transformation of random processes

If a random process $X(t)$ is a mean square integral, then the integral $Y(t) = \int_0^t X(a)\, \mathrm{d}a$ is still a random process. It is the output of the input random process $X(t)$ with the integral transformation of the ideal integrator.

2.2.6.1 Mean value (or mathematical expectation) of the mean square integral of random processes

$$\begin{aligned} m_Y(t) = E[Y(t)] &= E\left[\int_0^t X(a)\, \mathrm{d}a\right] \\ &= E\left[\lim_{\Delta t_k \to 0} \sum_{k=1}^{n} X(t_k)\Delta t_k\right] = \lim_{\Delta t_k \to 0} E\left[\sum_{k=1}^{n} X(t_k)\Delta t_k\right] \\ &= \lim_{\Delta t_k \to 0} \sum_{k=1}^{n} E[X(t_k)] \cdot \Delta t_k = \int_0^t E[X(a)]\, \mathrm{d}a \\ &= \int_0^t m_X(t)\, \mathrm{d}a \end{aligned} \tag{2.104}$$

The above formula shows that the mean of the random process integral is equal to the integral of the mean of the process. That is:

$$E\left[\int_0^t X(a)\,\mathrm{d}a\right] = \int_0^t E[X(a)]\,\mathrm{d}a \tag{2.105}$$

So, we can see that the integral and the mean operation of the random process can be reversed.

If $X(t)$ is the stationary process, its mean $m_X(t)$ is a constant (assuming C), which is unrelated to time t and then:

$$m_Y(t) = \int_0^t m_X(t)\,\mathrm{d}a = \int_0^t c\,\mathrm{d}a = ct \tag{2.106}$$

So, even if the input random process $X(t)$ is stationary, its integral $Y(t)$ is related to time, and as a result, the process is not stationary.

2.2.6.2 Correlation functions of the mean square integral of random processes

Based on the definition of the correlation function, we have

$$R_Y(t_1, t_2) = E[Y(t_1)Y(t_2)] \tag{2.107}$$

where $Y(t_1) = \int_0^{t_1} X(a_1)\,\mathrm{d}a_1$: $Y(t_2) = \int_0^{t_2} X(a_2)\,\mathrm{d}a_2$.

Exchanging the mean operation and the integral operation, we have

$$\begin{aligned} R_Y(t_1, t_2) &= E\left[\int_0^{t_1} X(a_1)\,\mathrm{d}a_1 \cdot \int_0^{t_2} X(a_2)\,\mathrm{d}a_2\right] = \int_0^{t_1}\int_0^{t_2} E[X(a_1)X(a_2)]\,\mathrm{d}a_1\,\mathrm{d}a_2 \\ &= \int_0^{t_1}\int_0^{t_2} R_X(a_1, a_2)\,\mathrm{d}a_1\,\mathrm{d}a_2 \end{aligned} \tag{2.108}$$

The above formula (2.108) shows that to obtain the autocorrelation function of the process $Y(t)$, the original process $X(t)$ of the autocorrelation function is required to do a twice integral transform (first integration with a variable t_1 or t_2, and then integration with another variable t_2 or t_1).

Similarly, the mutual correlation function of random process $X(t)$ and its integral $Y(t)$, is:

$$R_{XY}(t_1, t_2) = E[X(t_1)Y(t_2)] = E\left[X(t_1) \cdot \int_0^{t_2} X(a_2)\,\mathrm{d}a_2\right] = \int_0^{t_2} R_X(t_1, a_2)\,\mathrm{d}a_2 \tag{2.109}$$

$$R_{YX}(t_1, t_2) = \int_0^{t_1} R_X(a_1, t_2)\,\mathrm{d}a_1 \tag{2.110}$$

The above three formulas show that assuming $X(t)$ is a stationary process, although the function $R_X(t_1, t_2)$ is equal to $R_X(\tau)$, which is unrelated to time t, both the autocorrelation function and the cross-correlation function of the process $Y(t)$ are related to time t.

Example 2.2.4. The random signal $X(t) = A\cos(\omega_0 t + \varphi)$ with initial phase, where A and ω_0 are constants. Given $m_X(t) = 0$, $R_X(\tau) = (A^2/2)\cos\omega_0\tau$, $\tau = t_2 - t_1$, the integral value of signal $X(t)$ in the time T is $Y(T) = \int_0^T X(t)\,dt$, try to find the mean and the variance of $Y(T)$.

Solution: The mean is:

$$m_Y(T) = \int_0^T m_X(t)\,dt = 0 \tag{2.111}$$

The correlation functions are:

$$\begin{aligned} R_Y(T, T+\tau) &= \int_0^T \int_0^{T+\tau} R_X(t_1, t_2)\,dt_1\,dt_2 \\ &= \int_0^T \int_0^{T+\tau} \frac{A^2}{2}\cos\omega_0(t_2 - t_1)\,dt_1\,dt_2 \\ &= \frac{A^2}{2\omega_0^2}[\cos\omega_0\tau - \cos\omega_0(T+\tau) - \cos\omega_0 T + 1] \end{aligned} \tag{2.112}$$

So, the variance is

$$\sigma_Y^2(T) = C_Y(T, T) = R_Y(T, T) = \frac{A^2}{\omega_0^2}(1 - \cos\omega_0 T) \tag{2.113}$$

Summarizing the differential transformation and integral transformation of this random process $X(t)$, we know that the linear transformation can be expressed as

$$Y(t) = L[Y(t)] \tag{2.114}$$

Knowing the mathematical expectation $m_X(t)$ and the correlation function $R_X(t_1, t_2)$ of the process $X(t)$, in order to obtain the mathematical expectation of the process $Y(t)$, the mathematical expectation of the process $X(t)$ should take the same linear operation, i.e.,

$$m_Y(t) = L[m_X(t)] \tag{2.115}$$

In order to obtain the correlation functions of the process $Y(t)$, the mathematical expectation of the process $X(t)$ should take the same linear operation twice (first the operation with one variable, then the operation with another variable), i.e.,

$$R_Y(t_1, t_2) = L_{t_1} L_{t_2}[R_X(t_1, t_2)] \tag{2.116}$$

where L_{t_1} and L_{t_2} are the result of linear operations for the variables t_1 and t_2.

2.3 Analysis of random processes through continuous-time systems

2.3.1 Impulse response method

When a certain signal $x(t)$ is added to a linear system with the impulse response $h(t)$, the response output of the known signal is the convolution of $x(t)$ with $h(t)$, i.e.,

$$y(t) = x(t) \otimes h(t) = \int_{-\infty}^{\infty} x(\tau)h(t-\tau)\,\mathrm{d}\tau \tag{2.117}$$

or

$$y(t) = h(t) \otimes x(t) = \int_{-\infty}^{\infty} x(t-\tau)h(\tau)\,\mathrm{d}\tau \tag{2.118}$$

Assuming that the input of this system is a random process $X(t)$, because each sample $x(t)$ satisfies the upper relationship, the output $Y(t)$ is still a random process. It is the convolution of $X(t)$ with $h(t)$:

$$Y(t) = X(t) \otimes h(t) = \int_{-\infty}^{\infty} X(\tau)h(t-\tau)\,\mathrm{d}\tau \tag{2.119}$$

or

$$Y(t) = h(t) \otimes X(t) = \int_{-\infty}^{\infty} X(t-\tau)h(\tau)\,\mathrm{d}\tau \tag{2.120}$$

The bounds of $-\infty$ and ∞ in the two superscripts are a general representation. If given $X(t)$ and $h(t)$, the limit should be specific and the principle is that on the time axis, the overlaps of $X(t)$ with $h(t-\tau)$ or $X(t-\tau)$ with $h(\tau)$ correspond to a range of τ, which is regarded as the true integral interval.

2.3.1.1 Impulse response method of causal systems

Generally, the linear systems studied are physical, and they are causal systems. Namely, $h(t) = 0$ for $t < 0$. Input process $X(t)$ can be divided into two situations: (1) $X(t)$ is sent to the system for $t = -\infty$, and we called the input signal $X(t)$ a double side signal; (2) $X(t)$ is sent to the system for $t = 0$, and the input signal is $X(t)U(t)$ ($U(t)$ is a step function), which is called a right signal.

(1) Double-sided signal input

If the input signal is a double-sided signal, i.e., it is delivered to the system for $t = -\infty$, the output process is

$$Y(t) = \int_{-\infty}^{t} X(\tau)h(t-\tau)\,\mathrm{d}\tau \tag{2.121}$$

or

$$Y(t) = \int_0^\infty X(t-\tau)h(\tau)\,\mathrm{d}\tau \tag{2.122}$$

In this case, whether the output process $Y(t)$ is stationary depends on whether the input process was stationary before the moment of observation t. If the process $X(t)$ is stationary before the time t, but it is not stationary after the time t, then $Y(t)$ is stationary before the time t. This is because it is not possible for the nonstationary process of the input to affect the output. Similarly, $Y(t)$ is not stationary after the time t, because the input nonstationary process can already influence it.

(2) Right-sided signal input
If the input signal is a right-sided signal, i.e., it is not delivered to the system until $t = 0$, the output process is

$$Y(t) = \int_0^t X(\tau)h(t-\tau)\,\mathrm{d}\tau \tag{2.123}$$

or

$$Y(t) = \int_0^t X(t-\tau)h(\tau)\,\mathrm{d}\tau \tag{2.124}$$

If the input process $X(t)$ was stationary before the moment t, it was nonstationary after the moment t. Although the input process is stationary before time t, the impulsive response $h(\tau)$ or $h(t-\tau)$ was not to become zero, and the output process is still in the transient process period, so the output process $Y(t)$ is not stationary. Just the response time of the impulse response $h(\tau)$ is short, and the transient process is basically over at the moment t. In this situation, the process $Y(t)$ is almost a stationary random process.

From the above analysis, the following can be seen. For the causal system, the simple calculation formula of the integral limit can be chosen. If the input is a double-sided signal, the steady-state output of the demand system is formulated with the formula: $Y(t) = \int_0^\infty X(t-\tau)h(\tau)\,\mathrm{d}\tau$. If the input is a right-sided signal, the transient state output of the demand system at the moment t is formulated with the formula: $Y(t) = \int_0^t X(t-\tau)h(\tau)\,\mathrm{d}\tau$.

2.3.1.2 Mean and correlation functions of the output process

(1) Steady-state output
Carrying out the mean operation for both sides of eq. (2.122), we have

$$E[Y(t)] = E\left[\int_0^\infty X(t-\tau)h(\tau)\,\mathrm{d}\tau\right] \tag{2.125}$$

Exchanging the order of the mean and the integral, the mean of the process $Y(t)$ is

$$m_Y(t) = \int_0^{\infty} E\left[X(t-\tau)\right] h(\tau)\, \mathrm{d}\tau \tag{2.126}$$

If $X(t)$ is a stationary process, we have

$$E\left[X(t-\tau)\right] = E[X(t)] = m_X \tag{2.127}$$

so,

$$m_Y(t) = m_Y = m_X \int_0^{\infty} h(\tau)\, \mathrm{d}\tau \tag{2.128}$$

The correlation function of $Y(t)$ is:

$$R_Y(t_1, t_2) = E[Y(t_1)Y(t_2)] = E\left[\int_0^{\infty} X(t_1-\tau)h(\tau)\, \mathrm{d}\tau \int_0^{\infty} X(t_2-\tau)h(\tau)\, \mathrm{d}\tau\right] \tag{2.129}$$

Exchange the order of mean and integration:

$$\begin{aligned} R_Y(t_1, t_2) &= \int_0^{\infty}\int_0^{\infty} E\left[X(t_1-u)X(t_2-v)\right] h(u)h(v)\, \mathrm{d}u\, \mathrm{d}v \\ &= \int_0^{\infty}\int_0^{\infty} R_X(t_1-u, t_2-v)h(u)h(v)\, \mathrm{d}u\, \mathrm{d}v \end{aligned} \tag{2.130}$$

if $X(t)$ is a stationary process, there is $R_X(t_1-u, t_2-v) = R_X(\tau+u-v)$, where $\tau = t_2-t_1$, then:

$$R_Y(\tau) = \int_0^{\infty}\int_0^{\infty} R_X(\tau+u-v)h(u)h(v)\, \mathrm{d}u\, \mathrm{d}v \tag{2.131}$$

So, if the input $X(t)$ is a nonstationary processes, output $Y(t)$ is also the nonstationary process. If the input $X(t)$ is a stationary process, output $Y(t)$ it is also a stationary process.

(2) Transient state output.

The upper limit should be taken as the observation time t. Now just analyze the input stationary process $X(t)$. The result of the analysis can be:

$$m_Y(t) = m_X \int_0^{t} h(\tau)\, \mathrm{d}\tau \tag{2.132}$$

$$R_Y(t_1, t_2) = \int_0^{t_1}\int_0^{t_2} R_X(\tau+u-v)h(u)h(v)\, \mathrm{d}u\, \mathrm{d}v \tag{2.133}$$

Even though the input is stationary, because the system has a transient process, the output $Y(t)$ is no longer a stationary process. Just under the observed moment $t(t_1, t_2) \to \infty Y(t)$ is the stationary process when the transient state process is over.

2.3.2 Spectrum method

If the input $X(t)$ and the output $Y(t)$ of the linear system are both stationary processes, the spectrum method can be applied. So, for the sake of analysis and solution, first solve $R_{XY}(\tau)$ and then obtain $R_Y(\tau)$. In order to make the analytical results with generality, use eq. (2.120), i.e.,

$$Y(t) = \int_{-\infty}^{\infty} X(t-u)h(u)\,\mathrm{d}u \tag{2.134}$$

Multiply by $X(t-\tau)$ on both sides and carry out the mean operation:

$$E[X(t-\tau)Y(t)] = E\left[\int_{-\infty}^{\infty} X(t-\tau)X(t-u)h(u)\,\mathrm{d}u\right] \tag{2.135}$$

Exchanging the order of the mean and the integral, the cross-correlation function is:

$$R_{XY}(\tau) = \int_{-\infty}^{\infty} E[X(t-\tau)X(t-u)]h(u)\,\mathrm{d}u = \int_{-\infty}^{\infty} R_X(\tau-u)h(u)\,\mathrm{d}u \tag{2.136}$$

or

$$R_{XY}(\tau) = h(\tau) \otimes R_X(\tau) = R_X(\tau) \otimes h(\tau) \tag{2.137}$$

In the same way, multiply by $Y(t+\tau)$ on both sides of eq. (2.134) and carry out the mean operation

$$E[Y(t)Y(t+\tau)] = E\left[\int_{-\infty}^{\infty} X(t-\tau)Y(t+u)h(u)\,\mathrm{d}u\right] \tag{2.138}$$

Exchanging the order of the mean and the integral, the correlation function is:

$$R_Y(\tau) = \int_{-\infty}^{\infty} E\left[X(t-\tau)Y(t+u)\right] h(u)\,\mathrm{d}u = \int_{-\infty}^{\infty} R_{XY}(\tau+u)h(u)\,\mathrm{d}u \tag{2.139}$$

making $u = -u'$, we have

$$R_Y(\tau) = \int_{-\infty}^{\infty} R_{XY}\left(\tau - u'\right) h\left(-u'\right) du' \tag{2.140}$$

or

$$R_Y(\tau) = h(-\tau) \otimes R_{XY}(\tau) = R_{XY}(\tau) \otimes h(-\tau) \tag{2.141}$$

Substituting eq. (2.137) into the above equation yields:

$$R_Y(\tau) = R_X(\tau) \otimes h(\tau) \otimes h(-\tau) \tag{2.142}$$

Because the Fourier transform of the convolution of two functions is equal to the product of the Fourier transform of these two functions, we obtain from eq. (2.137):

$$G_{XY}(\omega) = G_X(\omega)H(\omega) \tag{2.143}$$

From eq. (2.142), we have

$$G_Y(\omega) = G_X(\omega)H(\omega)H^*(\omega) = G_X(\omega)\,|H(\omega)|^2 \tag{2.144}$$

where $H^*(\omega)$ is the complex conjugate of the system transfer function and $|H(\omega)|^2$ is the power transfer function of the system.

Given the transfer function $H(\omega)$ of the linear system and the power spectral density $G_X(\omega)$ of the given input stationary process $X(t)$, we can directly use the equation to solve the power spectral density $G_Y(\omega)$ of the output stationary process $Y(t)$. Then, according to $R_Y(\tau) \leftrightarrow G_Y(\omega)$, we solve the correlation functions of the process: $Y(t)$:

$$R_Y(\tau) = \frac{1}{2\pi}\int_{-\infty}^{\infty} G_Y(\omega)\mathrm{e}^{\mathrm{i}\omega\tau}\,\mathrm{d}\omega = \frac{1}{\pi}\int_{0}^{\infty} G_X(\omega)\,|H(\omega)|^2\cos\omega\tau\,\mathrm{d}\omega \tag{2.145}$$

The same way, we can also find the cross-spectral density $G_{XY}(\omega)$ and the cross-correlation functions $R_{XY}(\tau)$. This is called the spectrum method, and we can directly use transfer function $H(\omega)$ to calculate the power spectral density and correlation functions.

Although the spectrum method is easier to use than the impulse response method, we should note that the spectrum method only uses the amplitude-frequency characteristics and does not relate to the phase-frequency characteristics.

A stationary random process was discussed above, along with the nonstationary random process; we can also use the same method. Now, we have

$$Y(t_1) = \int_{-\infty}^{\infty} X(t_1 - u)h(u)\,\mathrm{d}u \tag{2.146}$$

$$Y(t_2) = \int_{-\infty}^{\infty} X(t_2 - v)h(v)\,\mathrm{d}v \tag{2.147}$$

Multiplying by $X(t_1)$ on both sides of the second equation (2.3.31) and carrying out the mean operation, we have:

$$E[X(t_1)Y(t_2)] = E\left[\int_{-\infty}^{\infty} X(t_1)X(t_2 - v)h(v)\,\mathrm{d}v\right] \tag{2.148}$$

Exchanging the order of convolution and integration, we then obtain the correlation function:

$$\begin{aligned} R_{XY}(t_1, t_2) &= \int_{-\infty}^{\infty} E[X(t_1)X(t_2 - v)]h(v)\,\mathrm{d}v = \int_{-\infty}^{\infty} R_X(t_1, t_2 - v)h(v)\,\mathrm{d}v \\ &= R_X(t_1, t_2) \otimes h(t_2) \end{aligned} \tag{2.149}$$

Multiplying by $Y(t_2)$ on both sides of eq. (2.146) and carrying out the mean operation, we have:

$$E[Y(t_1)Y(t_2)] = E\left[\int_{-\infty}^{\infty} X(t_1 - u)Y(t_2)h(u)\,\mathrm{d}u\right] \tag{2.150}$$

Exchanging the order of convolution and integration, we obtain the correlation functions:

$$\begin{aligned} R_Y(t_1, t_2) &= \int_{-\infty}^{\infty} E\left[X(t_1 - u)Y(t_2)\right] h(v)\,\mathrm{d}v = \int_{-\infty}^{\infty} R_{XY}(t_1 - u, t_2)h(u)\,\mathrm{d}u \\ &= R_{XY}(t_1, t_2) \otimes h(t_1) \end{aligned} \tag{2.151}$$

Substituting eq. (2.149) into the above formula, we obtain:

$$R_Y(t_1, t_2) = R_X(t_1, t_2) \otimes h(t_1) \otimes h(t_2) \tag{2.152}$$

or

$$R_Y(t_1, t_2) = \int_{-\infty}^{\infty}\int_{-\infty}^{\infty} R_X(t_1 - u, t_2 - v)h(u)h(v)\,\mathrm{d}u\,\mathrm{d}v \tag{2.153}$$

It can be seen from eq. (2.147) that the normal expression of the power spectral density cannot be written as in eq. (2.144) for the nonstationary random process. The difference between eqs. (2.147) and (2.130) is that the impulse response of eq. (2.147) can be noncausal. If it is a causal system, then the lower limit of the integration should be zero, i.e., eq. (2.130).

Example 2.3.1. Given the correlation function $R_X(\tau) = \sigma_X^2 \mathrm{e}^{-\beta|\tau|}$ of input stationary process $X(t)$, where $\tau = t_2 - t_1$, find the correlation function of the output random process $Y(t)$ of the RC integral circuit at steady state.

Solution 1. Impulse response method.

According to the previous example, the impulse response of the RC integral circuit is

$$h(t) = \begin{cases} a\mathrm{e}^{-at}, & t \geq 0 \\ 0, & t < 0 \end{cases} \tag{2.154}$$

where $a = 1/RC$.

Because this is just the steady-state solution of the output process $Y(t)$, we obtain:

$$\begin{aligned} R_Y(\tau) &= \int_0^{\infty}\int_0^{\infty} \sigma_X^2 \mathrm{e}^{-\beta|\tau+u-v|} a\mathrm{e}^{-au} a\mathrm{e}^{-av}\,\mathrm{d}u\,\mathrm{d}v \\ &= a^2\sigma_X^2 \int_0^{\infty}\int_0^{\infty} \mathrm{e}^{-a(u+v)}\mathrm{e}^{-\beta|\tau+u-v|}\,\mathrm{d}u\,\mathrm{d}v \end{aligned} \tag{2.155}$$

The above equation is divided into two parts: $\tau + u - v > 0$ and $\tau + u - v < 0$, namely $v < \tau + u$ and $v > \tau + u$.

Therefore, the above equation can be rewritten as

$$R_Y(\tau) = a^2\sigma_X^2 \int_0^{\infty}\left[\int_0^{\tau+u} e^{-a(u+v)}e^{-\beta(\tau+u-v)}\,dv\right]du + a^2\sigma_X^2 \int_0^{\infty}\left[\int_{\tau+u}^{\infty} e^{-a(u+v)}e^{-\beta(-\tau-u+v)}\,dv\right]du \tag{2.156}$$

The first integral of this is available as

$$\int_0^{\infty}\left[\int_0^{\tau+u} e^{-a(u+v)}e^{-\beta(\tau+u-v)}\,dv\right]du = \int_0^{\infty} e^{-(a+\beta)u}e^{-\beta\tau}\frac{1}{\beta-a}[e^{(\beta-a)(\tau+u)}-1]\,du = \frac{e^{-a\tau}}{2a(\beta-a)} - \frac{e^{-\beta\tau}}{\beta^2-a^2} \tag{2.157}$$

For the second integral, we can obtain:

$$\int_0^{\infty}\left[\int_{\tau+u}^{\infty} e^{-a(u+v)}e^{-\beta(-\tau-u+v)}\,dv\right]du = \int_0^{\infty} e^{-(a-\beta)u}e^{\beta\tau}\left[\frac{e^{-(a+\beta)(\tau+u)}}{\beta-a}\right]du = \frac{e^{-a\tau}}{2a(a+\beta)} \tag{2.158}$$

Combining the two integrals:

$$R_Y(\tau) = a^2\sigma_X^2\left[\frac{e^{-a\tau}}{2a(\beta-a)} - \frac{e^{-\beta\tau}}{\beta^2-a^2} + \frac{e^{-a\tau}}{2a(a+\beta)}\right] = \frac{a\sigma_X^2}{\beta^2-a^2}[ae^{-\beta\tau}-\beta e^{-a\tau}] \tag{2.159}$$

Because the output process $Y(t)$ only has the steady state, the autocorrelation function is an even function, so we obtain:

$$R_Y(\tau) = \frac{a\sigma_X^2}{\beta^2-a^2}[ae^{-\beta|\tau|}-\beta e^{-a|\tau|}] \tag{2.160}$$

Solution 2. The Spectral method.

The input and output processes are both stationary processes. Because of $G_X(\omega) \leftrightarrow R_X(\tau)$, the power spectral density of the input process $X(t)$ is:

$$G_X(\omega) = 2\int_0^{\infty} R_X(\tau)\cos\omega\tau\,d\tau = 2\int_0^{\infty}\sigma_X^2 e^{-\beta\tau}\cos\omega\tau\,d\tau = 2\sigma_X^2\frac{\beta}{\beta^2+\omega^2} \tag{2.161}$$

According to the structure of RC integral circuit, the transmission function of RC integral circuit is:

$$H(\omega) = \frac{1}{1+j\omega RC} \tag{2.162}$$

Then,

$$|H(\omega)|^2 = \frac{1}{1 + (\omega RC)^2} = \frac{a^2}{a^2 + \omega^2}, \quad a = \frac{1}{RC} \tag{2.163}$$

Therefore, we obtain

$$\begin{aligned} R_Y(\tau) &= \frac{1}{\pi} \int_0^\infty G_X(\omega)\,|H(\omega)|^2 \cos\omega\tau \,\mathrm{d}\omega \\ &= \frac{1}{\pi} \int_0^\infty 2\sigma_X^2 \frac{\beta}{\beta^2 + \omega^2} \cdot \frac{a^2}{a^2 + \omega^2} \cos\omega\tau \,\mathrm{d}\omega \\ &= \frac{a\sigma_X^2}{a^2 - \beta^2} [a\mathrm{e}^{-\beta\tau} - \beta \mathrm{e}^{-a\tau}] \end{aligned} \tag{2.164}$$

Because the output process $Y(t)$ is a stationary process, the autocorrelation function is an even function:

$$R_Y(\tau) = \frac{a\sigma_X^2}{a^2 - \beta^2} [a\mathrm{e}^{-\beta|\tau|} - \beta \mathrm{e}^{-a|\tau|}] \tag{2.165}$$

By comparison, the spectrum method is simpler than the impulse response method. Because spectrum method is simpler, it is widely applied. However, we need to note that the spectrum method has limitations, because it is only applicable to stationary random processes and can only be used for the steady-state solution but cannot be applied to the transient solution. The impulse response method is general and can be used for stationary or nonstationary random processes, which can be used for the steady-state solution and transient solution. When impulse response is simpler, in general, it is easier to calculate with the impulse response method.

Example 2.3.2. The circuit of the linear system is shown in Fig. 2.8 (a). Given the power spectral density of the input of stationary process $X(t)$, it is $G_X(\omega) = \frac{4\lambda}{4\lambda^2 + \omega^2}$. Try to find the power spectral density $G_Z(\omega)$, $G_Y(\omega)$, $G_{ZY}(\omega)$ of the random voltage $Z(t)$ and $Y(t)$ on the two resistors.

Solution: These two random voltages can be considered as the output of two single-end linear systems, which have a common input process $X(t)$, as shown in Fig. 2.8 (b). The transfer functions of two single-end linear systems can be obtained as follows:

$$H_1(\omega) = \frac{a + \mathrm{j}\omega}{2a + \mathrm{j}\omega}, \quad H_2(\omega) = \frac{a}{2a + \mathrm{j}\omega} \tag{2.166}$$

where $a = 1/RC$.

Therefore, spectral method can be used to obtain the autospectral density:

$$G_Z(\omega) = G_X(\omega)\,|H_1(\omega)|^2 = \frac{4\lambda}{4\lambda^2 + \omega^2} \cdot \frac{a^2 + \omega^2}{4a^2 + \omega^2} \tag{2.167}$$

$$G_Y(\omega) = G_X(\omega)\,|H_2(\omega)|^2 = \frac{4\lambda}{4\lambda^2 + \omega^2} \cdot \frac{a^2}{4a^2 + \omega^2} \tag{2.168}$$

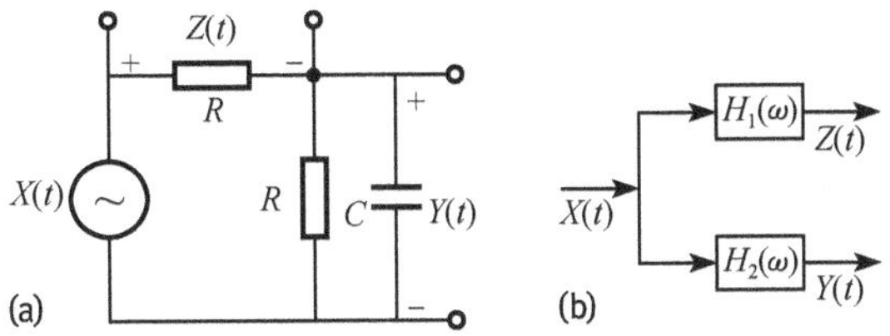

Fig. 2.8: Multiterminal linear systems.

The cross-spectral density is:

$$G_{ZY}(\omega) = G_X(\omega)H_1^*(\omega)H_2(\omega) = \frac{4\lambda}{4\lambda^2+\omega^2}\cdot\frac{a-\mathrm{j}\omega}{2a-\mathrm{j}\omega}\cdot\frac{a}{2a+\mathrm{j}\omega} \tag{2.169}$$

With the inverse Fourier transform for the above three, we can obtain the autocorrelation functions $R_Z(\tau)$, $R_Y(\tau)$ and the cross-correlation function $R_{ZY}(\tau)$.

2.4 White noise through linear systems

2.4.1 General relations

The impulse response and transfer function of a given linear system are $h(t)$ and $H(\omega)$. The input stationary white noise is $X(t)$, then its power spectral density is

$$G_N(\omega) = \frac{N_0}{2}, \quad -\infty < \omega < \infty \tag{2.170}$$

(1) Spectrum method
According to the previous analysis result of eq. (2.144), the power spectral density of the output process $Y(t)$ is:

$$G_Y(\omega) = G_X(\omega)\,|H(\omega)|^2 = \frac{N_0}{2}\,|H(\omega)|^2 \tag{2.171}$$

The above expression (2.171) indicates that due to the frequency response characteristics of the system, only the components corresponding to this frequency response can pass through the system, so the power spectrum of the output process is no longer uniform.

Then with the previous analysis result of eq. (2.145), the correlation function of $Y(t)$ can be obtained:

$$R_Y(\tau) = \frac{1}{\pi}\int_0^\infty G_X(\omega)\,|H(\omega)|^2\cos\omega\tau\,\mathrm{d}\omega = \frac{N_0}{2\pi}\int_0^\infty |H(\omega)|^2\cos\omega\tau\,\mathrm{d}\omega \tag{2.172}$$

Because the mean of the stationary white noise is zero, so, from the above eq. (2.172), the variance of $Y(t)$, for $\tau = 0$,is

$$\sigma_Y^2 = C_Y(0) = R_Y(0) = \frac{N_0}{2\pi}\int_0^\infty |H(\omega)|^2\,\mathrm{d}\omega \tag{2.173}$$

or

$$\sigma_Y^2 = N_0 \int_0^\infty |H(f)|^2 \,\mathrm{d}f \tag{2.174}$$

(2) Impulse response method
The correlation function of stationary white noise $X(t)$ is:

$$R_X(\tau) = \frac{N_0}{2}\delta(\tau) \tag{2.175}$$

According to the previous analysis result of eq. (2.142), the correlation function of the process $Y(t)$ is

$$R_Y(\tau) = R_X(\tau) \otimes h(\tau) \otimes h(-\tau) = \frac{N_0}{2}\delta(\tau) \otimes h(\tau) \otimes h(-\tau) \tag{2.176}$$

Using the convolution property of the DRT function: $\delta(\tau) \otimes h(\tau) = h(\tau)$, we obtain:

$$R_Y(\tau) = \frac{N_0}{2} h(\tau) \otimes h(-\tau) \tag{2.177}$$

i.e.,

$$R_Y(\tau) = \frac{N_0}{2} \int_{-\infty}^{\infty} h(u)h(\tau + u)\,\mathrm{d}u \tag{2.178}$$

When $\tau = 0$, the variance of $Y(t)$ can be obtained directly:

$$\sigma_Y^2 = R_Y(0) = \frac{N_0}{2} \int_{-\infty}^{\infty} h^2(u)\,\mathrm{d}u \tag{2.179}$$

The results of these two methods are different, but they are consistent. Obviously, the simple integral operation is easier than the convolution operation, so the following analyses all use the simpler spectrum method.

2.4.2 Noise equivalent passband

After white noise goes through a linear system, the output power spectrum $F_Y(\omega)$ is no longer uniform, as shown with the solid line in Fig. 2.9.

To facilitate the analysis and calculation, usually based on the principle of equivalent power, the power spectrum of nonuniform is equivalent to the uniform power spectrum within a certain band, as shown with the dotted line in Fig. 2.9. The width of this equivalent rectangular power spectrum is called the (output) equivalent spectrum width of noise, and the equivalent passband of noise is also called the noise passband, which is Δf_n.

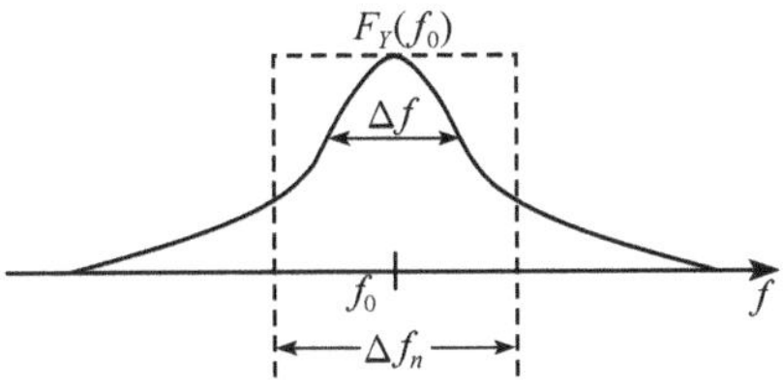

Fig. 2.9: Equivalent passband of noise.

The so-called equivalent refers to the equal power. That is, the rectangular area in the graph (its height is the maximum output power spectral density $F_Y(f_0)$, width is Δf_n (is equal to the area which is enclosed by $F_Y(f)$ and the f-axis, namely:

$$F_Y(f_0) \cdot \Delta f_n = \sigma_Y^2 \tag{2.180}$$

Substituting eqs. (2.171) and (2.174) into the above formula, we have (note: the lack of 2 is because this definition is the single-sided power spectrum):

$$N_0 \left|H(\omega_0)\right|^2 \cdot \Delta f_n = N_0 \int_0^{\infty} \left|H(\omega)\right|^2 \mathrm{d}f \tag{2.181}$$

Therefore, the noise passband is:

$$\Delta f_n = \frac{\int_0^{\infty} \left|H(\omega)\right|^2 \mathrm{d}f}{\left|H(\omega_0)\right|^2} \tag{2.182}$$

The above shows that the noise passband Δf_n and signal (half-power) passband Δf are the same and are determined only by the parameters of the circuit itself. When the form and series of the linear circuit are determined, they are fixed and mutually assured. In engineering, the signal passband Δf is often used to approximate the noise passband Δf_n. The signal passband is used to indicate the signal spectrum selectivity of the linear circuit, and the noise passband is used to show the noise power spectrum selectivity of the linear circuit. The closer the frequency response of the linear circuit is to a rectangle, the closer the noise passband is to a signal passband.

2.4.3 White noise through *RC* integral circuits

According to the previous analysis, the transmission function of an *RC* integral circuit is:

$$H(\omega) = \frac{a}{a + \mathrm{j}\omega} \tag{2.183}$$

where $a = 1/RC$, and therefore, we have

$$\left|H(\omega)\right|^2 = \frac{1}{1 + \left(\frac{\omega}{a}\right)^2} \tag{2.184}$$

Given the power spectral density of white noise $N_0/2$, with eq. (2.171), the power spectral density of the output process $Y(t)$ is

$$G_Y(\omega) = \frac{\frac{N_0}{2}}{1 + \left(\frac{\omega}{a}\right)^2} \tag{2.185}$$

According to eq. (2.172), the correlation function is:

$$R_Y(\tau) = \frac{N_0}{2\pi} \int_0^{\infty} \frac{\cos \omega\tau}{1 + \left(\frac{\omega}{a}\right)^2} \, d\omega = \frac{N_0 a}{4} e^{-a|\tau|} \tag{2.186}$$

Because

$$R_Y(\infty) = 0 \tag{2.187}$$

$$\sigma_Y^2 = C_Y(0) = R_Y(0) - R_Y(\infty) = \frac{N_0 a}{4} \tag{2.188}$$

Correlation coefficient:

$$r_Y(\tau) = \frac{C_Y(\tau)}{\sigma_Y^2} = e^{-a|\tau|} \tag{2.189}$$

The correlation time of process $Y(t)$ is:

$$\tau_0 = \int_0^{\infty} r_Y(\tau) \, d\tau = \int_0^{\infty} e^{-a\tau} \, d\tau = \frac{1}{a} \tag{2.190}$$

According to the expression of the transfer function, we have the maximum transfer function $H(\omega_0) = 1$ for $\omega = 0$. Therefore, with eq. (2.182), the noise-frequency band is:

$$\Delta f_n = \frac{1}{2\pi} \int_0^{\infty} |H(\omega)|^2 \, d\omega = \frac{1}{2\pi} \int_0^{\infty} \frac{1}{1 + \left(\frac{\omega}{a}\right)^2} \, d\omega = \frac{a}{4} \tag{2.191}$$

The following can be obtained from the above equation:

$$\tau_0 = \frac{1}{4\Delta f_n} \tag{2.192}$$

The above analysis results show that the value of an RC integral circuit is very large, i.e., a is small, so the noise passband Δf_n is very small (for low pass), and the correlation time τ_0 is very large. In this situation, only the low-frequency components can be passed by the circuit, thus the output noise fluctuation is very slow. In other words, after the rapidly changing white noise passes through the integral circuit, the correlation of the output is stronger due to the smoothing effect of the integrated circuit.

2.4.4 White noise through ideal lowpass linear systems

The amplitude-frequency characteristic of the ideal lowpass linear system is (as shown in Fig. 2.10):

$$|H(\omega)| = \begin{cases} K_0\,, & -\Delta\Omega < \omega < \Delta\Omega \\ 0\,, & \omega < -\Delta\Omega\,,\ \ \omega > \Delta\Omega \end{cases} \tag{2.193}$$

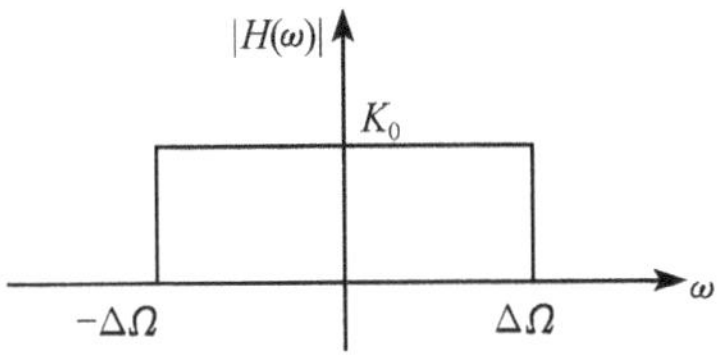

Fig. 2.10: Ideal lowpass amplitude-frequency characteristic.

Given the power spectral density of white noise $N_0/2$, the power spectral density of process $Y(t)$ is

$$G_Y(\omega) = G_X(\omega)\,|H(\omega)|^2 = \begin{cases} \frac{N_0 K_0^2}{2}\,, & -\Delta\Omega < \omega < \Delta\Omega \\ 0\,, & \omega < -\Delta\Omega\,,\ \ \omega > \Delta\Omega \end{cases} \tag{2.194}$$

The above expression indicates that the whole signal is filtered other than the components of the passband. The correlation function is:

$$R_Y(\tau) = \frac{1}{\pi}\int_0^{\infty} G_Y(\omega)\cos\omega\tau\,\mathrm{d}\omega = \frac{N_0 K_0^2}{2\pi}\int_0^{\Delta\Omega}\cos\omega\tau\,\mathrm{d}\omega = \frac{N_0 K_0^2 \Delta\Omega}{2\pi}\cdot\frac{\sin\Delta\Omega\tau}{\Delta\Omega\tau} \tag{2.195}$$

The variance is as follows:

$$\sigma_Y^2 = C_Y(0) = R_Y(0) = \frac{N_0 K_0^2 \Delta\Omega}{2\pi} \tag{2.196}$$

The correlation coefficient is

$$r_Y(\tau) = \frac{C_Y(\tau)}{\sigma_Y^2} = \frac{\sin\Delta\Omega\tau}{\Delta\Omega\tau} \tag{2.197}$$

The correlation time is:

$$\tau_0 = \int_0^{\infty}\frac{\sin\Delta\Omega\tau}{\Delta\Omega\tau}\,\mathrm{d}\tau = \frac{\pi}{2\Delta\Omega} = \frac{1}{4\Delta f} \tag{2.198}$$

where $\Delta\Omega = 2\pi\Delta f$.

Because the power spectrum of the output is close to rectangular, the noise passband of the system is equal to the signal passband:

$$\Delta f_n = \Delta f = \frac{\Delta\Omega}{2\pi} \tag{2.199}$$

So, there is a relationship:

$$\tau_0 \Delta f_n = \frac{1}{4} \tag{2.200}$$

The above formula shows that the correlation time τ_0 is inversely proportional to the passband Δf (or Δf_n) of the system. If $\Delta f \to \infty$, then $\tau_0 \to 0$ and in this situation, the output process is still white noise. Big Δf is equivalent to small τ_0, and in this situation, the fluctuation of the output process changes fast. On the other hand, small Δf is equivalent to big τ_0, and in this situation, the fluctuation of the output process changes slowly.

2.4.5 White noise through ideal bandpass linear systems

We have the following rectangular amplitude-frequency characteristics in the ideal bandpass linear system (as shown in Fig. 2.11):

$$|H(\omega)| = \begin{cases} K_0\,, & |\omega - \omega_0| \le \frac{\Delta\omega}{2}\,,\ |\omega + \omega_0| \le \frac{\Delta\omega}{2} \\ 0\,, & \text{others} \end{cases} \tag{2.201}$$

The multilevel cascaded tuning intermediate-frequency amplifier is similar to the above characteristics. As mentioned above, the noise passband of the system is equal to the signal passband.

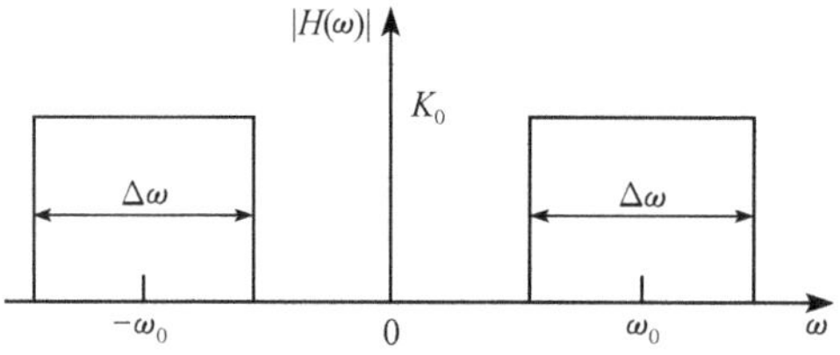

Fig. 2.11: Ideal passband amplitude-frequency characteristics.

Given the power spectral density $N_0/2$ of the input white noise, the power spectral density of the output process $Y(t)$ is

$$G_Y(\omega) = G_X(\omega)\,|H(\omega)|^2 = \begin{cases} \frac{N_0 K_0^2}{2}\,, & |\omega - \omega_0| \le \frac{\Delta\omega}{2}\,,\ |\omega + \omega_0| \le \frac{\Delta\omega}{2} \\ 0\,, & \text{others} \end{cases} \tag{2.202}$$

The correlation function is

$$\begin{aligned} R_Y(\tau) &= \frac{1}{\pi}\int_0^{\infty} G_Y(\omega)\cos\omega\tau\,\mathrm{d}\omega = \frac{N_0 K_0^2}{2\pi}\int_{\omega_0-\frac{\Delta\omega}{2}}^{\omega_0+\frac{\Delta\omega}{2}} \cos\omega\tau\,\mathrm{d}\omega \\ &= \frac{N_0 K_0^2 \Delta\omega}{2\pi}\cdot\frac{\sin\left(\frac{\Delta\omega\tau}{2}\right)}{\frac{\Delta\omega\tau}{2}}\cos\omega_0\tau \end{aligned} \tag{2.203}$$

Making

$$A(\tau) = \frac{N_0 K_0^2 \Delta\omega}{2\pi} \cdot \frac{\sin\left(\frac{\Delta\omega\tau}{2}\right)}{\frac{\Delta\omega\tau}{2}} \tag{2.204}$$

Then

$$R_Y(\tau) = A(\tau) \cos \omega_0 \tau \tag{2.205}$$

Like in the previous method, the correlation coefficient of process $Y(t)$ is

$$r'_Y(\tau) = \frac{C_Y(\tau)}{\sigma_Y^2} = \frac{\sin\left(\frac{\Delta\omega\tau}{2}\right)}{\frac{\Delta\omega\tau}{2}} \cos \omega_0 \tau \tag{2.206}$$

This can be rewritten as

$$r'_Y(\tau) = r_Y(\tau) \cos \omega_0 \tau \tag{2.207}$$

where: $r_Y(\tau) = \frac{\sin(\Delta\omega\tau/2)}{\Delta\omega\tau/2}$ is the correlation coefficient. Then, we can obtain the curve of the correlation coefficient. Since the correlation coefficient is an even function, the right half is generally only drawn.

By comparison, $r_Y(\tau)$ only depends on the low spectrum width $\Delta f = \Delta\Omega/2\pi$, while $r'_Y(\tau)$ is also related to the carrier angular frequency. So, the fluctuation of the instantaneous value of the output process $Y(t)$ is much faster than the fluctuation of the envelope value, as shown in Fig. 2.12.

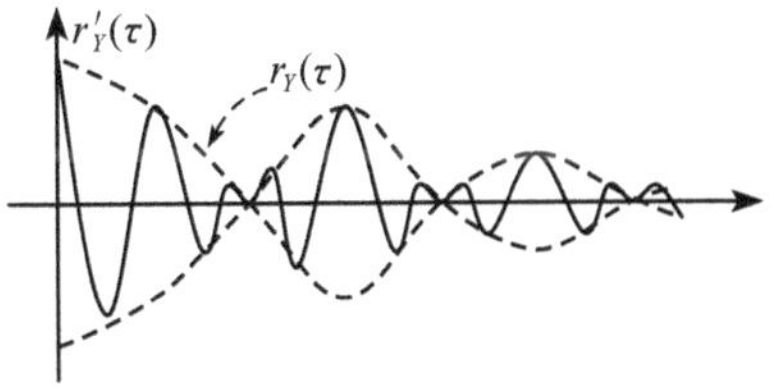

Fig. 2.12: Correlation coefficients of narrowband rectangular spectrum noise.

If a linear system satisfies $\Delta\omega \ll \omega_0$, it is a narrowband system and its output noise is narrowband noise. The expression of the correlation function and the correlation coefficient can be divided into two parts, $\cos \omega_0 \tau$ for the carrier, which is the fast changing part, while $A(\tau)$ or $r_Y(\tau)$ for the corresponding envelope is the slow changing part related to the carrier. Since the envelope contains information of amplitude modulation signal, generally, we do not use the correlation coefficient $r'_Y(\tau)$ to define the correlation time for the narrowband stochastic process, but rather use the slow change part $r_Y(\tau)$. Therefore, the correlation time of the output narrowband noise is:

$$\tau_0 = \int_0^\infty \frac{\sin\left(\frac{\Delta\omega\tau}{2}\right)}{\frac{\Delta\omega\tau}{2}} d\tau = \frac{\pi}{\Delta\omega} = \frac{1}{2\Delta f} \tag{2.208}$$

where Δf is the passband of the narrowband system.

The above expression indicates that the correlation time τ_0 of the narrowband noise is used to show how fast the envelope fluctuation changes. The smaller the passband Δf is, the larger τ_0. Namely, the envelope fluctuation changes more slowly.

2.4.6 White noise through a linear system with a Gaussian band

The amplitude-frequency characteristic of the passband linear system is a Gaussian shape (as shown in Fig. 2.13), which is:

$$|H(\omega)| = \begin{cases} K_0 \exp\left[-\dfrac{(\omega-\omega_0)^2}{2\beta^2}\right], & \omega > 0 \\ K_0 \exp\left[-\dfrac{(\omega+\omega_0)^2}{2\beta^2}\right], & \omega < 0 \end{cases} \tag{2.209}$$

where the parameter β is proportional to the system passband Δf. The total amplitude-frequency characteristic of a multilevel single-tuned intermediate frequency amplifier is similar to that of a Gaussian one.

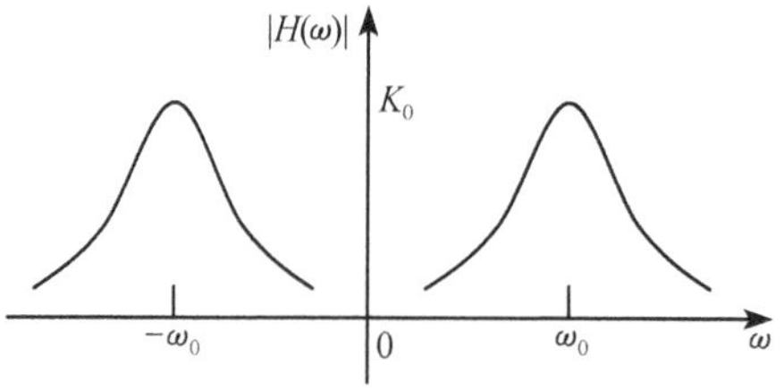

Fig. 2.13: Gaussian bandpass amplitude-frequency characteristic.

Given the power spectral density of the input white noise $N_0/2$, the power spectral density of output process $Y(t)$ is

$$G_Y(\omega) = G_X(\omega)\,|H(\omega)|^2 = \begin{cases} \dfrac{N_0K_0^2}{2} \exp\left[-\frac{(\omega-\omega_0)^2}{2\beta^2}\right], & \omega > 0 \\ \dfrac{N_0K_0^2}{2} \exp\left[-\frac{(\omega+\omega_0)^2}{2\beta^2}\right], & \omega < 0 \end{cases} \tag{2.210}$$

The correlation functions is:

$$R_Y(\tau) = \frac{1}{\pi}\int_0^\infty G_Y(\omega)\cos\omega\tau\,\mathrm{d}\omega = \frac{N_0K_0^2}{2\pi}\int_0^\infty \exp\left[-\frac{(\omega-\omega_0)^2}{2\beta^2}\right]\cos\omega\tau\,\mathrm{d}\omega \tag{2.211}$$

Making $\omega - \omega_0 = \Omega$, then $\cos\omega\tau\,\mathrm{d}\omega = \cos(\omega_0 + \Omega)\tau\,\mathrm{d}\Omega$

$$\begin{aligned} R_Y(\tau) &= \frac{N_0K_0^2}{2\pi}\int_{-\omega_0}^\infty \exp\left(-\frac{\Omega^2}{\beta^2}\right)[\cos\Omega\tau\cos\omega_0\tau - \sin\Omega\tau\sin\omega_0\tau]\,\mathrm{d}\Omega \\ &= \frac{N_0K_0^2}{2\pi}\cos\omega_0\tau\int_{-\omega_0}^\infty \exp\left(-\frac{\Omega^2}{\beta^2}\right)\cos\Omega\tau\,\mathrm{d}\Omega \\ &= \frac{N_0K_0^2\beta}{2\sqrt{\pi}}\exp\left(-\frac{\beta^2\tau^2}{4}\right)\cos\omega_0\tau \end{aligned} \tag{2.212}$$

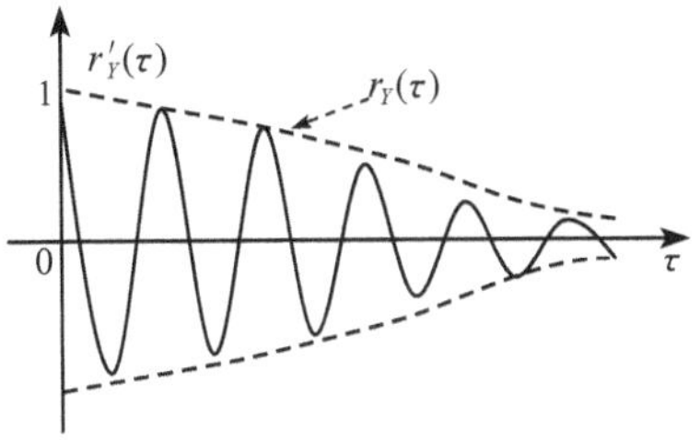

Fig. 2.14: Correlation coefficients of narrowband Gaussian spectral noise.

Making $A(\tau) = \frac{N_0 K_0^2 \beta}{2\sqrt{\pi}} \exp\left(-\frac{\beta^2\tau^2}{4}\right)$, the correlation coefficient is as follows:

$$r'_Y(\tau) = \exp\left(-\frac{\beta^2\tau^2}{4}\right)\cos\omega_0\tau \tag{2.213}$$

The curve is shown in Fig. 2.14.

The correlation time of this narrowband noise is:

$$\tau_0 = \int_0^{\infty} \exp\left(-\frac{\beta^2\tau^2}{4}\right) d\tau = \frac{\sqrt{\pi}}{\beta} \tag{2.214}$$

Due to the parametric $\beta \propto \Delta f$, therefore,

$$\tau_0 \propto \frac{1}{\Delta f} \tag{2.215}$$

This shows that the correlation time τ_0 is inversely proportional to the passband of the system.

2.5 Probability distribution of the linear transformation of random processes

We have discussed the linear transformation of the random process moment function. Although this has been able to solve many practical engineering problems, on a number of occasions, we also need to know the probability distribution of a random process after the linear transformation. The problem that needs to be studied is: given the probability distribution of the input random process and the transmission characteristics of the linear system, how can we obtain the probability distribution of the random process of the linear system's output terminal? The general solution to the random probability distribution rate is as follows.

We know that the linear transformation of the random process is actually the linear transformation of the multidimensional random variables (or random vector), so we can analyze it with the help of linear transformation of the multidimensional random variables.

Assuming that the output random vector is $Y = (\boldsymbol{Y})$ is the result of the input random vector $X = (\boldsymbol{X})$ with the linear transformation, the output can be expressed as the linear combination of these random variables $X_1, X_2, \ldots, X_n$. The values of these random variables can be expressed by the following n-th element linear equations:

$$\begin{cases} y_1 = l_{11}x_1 + l_{12}x_2 + \cdots + l_{1n}x_n \\ y_2 = l_{21}x_1 + l_{22}x_2 + \cdots + l_{2n}x_n \\ \vdots \\ y_n = l_{n1}x_1 + l_{n2}x_2 + \cdots + l_{nn}x_n \end{cases} \tag{2.216}$$

This equation can be written in simple matrix form:

$$\boldsymbol{Y} = \boldsymbol{L}\boldsymbol{X} \tag{2.217}$$

where

$$\boldsymbol{Y} = \begin{pmatrix} y_1 \\ y_2 \\ \vdots \\ y_n \end{pmatrix}, \quad \boldsymbol{X} = \begin{pmatrix} x_1 \\ x_2 \\ \vdots \\ x_n \end{pmatrix}, \quad \boldsymbol{L} = \begin{bmatrix} l_{11} & l_{12} & \ldots & l_{1n} \\ l_{21} & l_{22} & \ldots & l_{2n} \\ \vdots & \vdots & & \vdots \\ l_{n1} & l_{n2} & \ldots & l_{nn} \end{bmatrix}$$

Therefore, we have:

$$\boldsymbol{X} = \boldsymbol{L}^{-1}\boldsymbol{Y} \tag{2.218}$$

where $\boldsymbol{L}^{-1}$ is the inverse matrix of matrix $\boldsymbol{L}$.

From the probability density transformation of the random vector, the n-dimensional probability density of the random vector Y can be obtained as follows:

$$P_n(\boldsymbol{Y}) = |J|\, P_n(\boldsymbol{X}) = |J|\, P_n\left(\boldsymbol{L}^{-1}\boldsymbol{Y}\right) \tag{2.219}$$

where, the Jacobian determinant is

$$J = \frac{\partial \boldsymbol{X}}{\partial \boldsymbol{Y}} = \frac{\partial(x_1, x_2, \ldots, x_n)}{\partial(y_1, y_2, \ldots, y_n)} = \begin{vmatrix} \frac{\partial x_1}{\partial y_1} & \frac{\partial x_1}{\partial y_2} & \cdots & \frac{\partial x_1}{\partial y_n} \\ \frac{\partial x_2}{\partial y_1} & \frac{\partial x_2}{\partial y_2} & \cdots & \frac{\partial x_2}{\partial y_n} \\ \vdots & \vdots & & \vdots \\ \frac{\partial x_n}{\partial y_1} & \frac{\partial x_n}{\partial y_2} & \cdots & \frac{\partial x_n}{\partial y_n} \end{vmatrix} = \frac{1}{|\boldsymbol{L}|}, \quad |\boldsymbol{L}| \neq 0 \tag{2.220}$$

where, $|\boldsymbol{L}|$ is the determinant of matrix $\boldsymbol{L}$.

2.5.1 Input is normal and output is still normal

Given a normal random process $X(t)$ with zero mean, i.e., the mean of the random vector $\boldsymbol{X}$ satisfies $\boldsymbol{m} = \boldsymbol{0}$ and the covariance matrix satisfies $\boldsymbol{C} = \boldsymbol{R}$, we know that the n-dimensional probability density of the normal vector $\boldsymbol{X}$ is:

$$p_n(\boldsymbol{X}) = \frac{1}{\sqrt{(2\pi)^n\, |\boldsymbol{R}|}} \exp\left\{-\frac{1}{2}\boldsymbol{X}^{\mathrm{T}} R^{-1} \boldsymbol{X}\right\} \tag{2.221}$$

Substitute the above formula into eq. (2.219), then obtain the n-dimensional probability density of the random vector Y as follows:

$$p_n(\boldsymbol{Y}) = \frac{1}{\sqrt{(2\pi)^n |\boldsymbol{L}|^2 |\boldsymbol{R}|}} \exp\left\{-\frac{1}{2}(\boldsymbol{L}^{-1}\boldsymbol{Y})^{\mathrm{T}}\boldsymbol{R}^{-1}(\boldsymbol{L}^{-1}\boldsymbol{Y})\right\} \tag{2.222}$$

Using a matrix algorithm, obtain:

$$(\boldsymbol{L}^{-1}\boldsymbol{Y})^{\mathrm{T}} = \boldsymbol{Y}^{\mathrm{T}}(\boldsymbol{L}^{-1})^{\mathrm{T}} = \boldsymbol{Y}^{\mathrm{T}}(\boldsymbol{L}^{\mathrm{T}})^{-1} \tag{2.223}$$

then:

$$(\boldsymbol{L}^{-1}\boldsymbol{Y})^{\mathrm{T}}\boldsymbol{R}^{-1}(\boldsymbol{L}^{-1}\boldsymbol{Y}) = \boldsymbol{Y}^{\mathrm{T}}(\boldsymbol{L}^{\mathrm{T}})^{-1}\boldsymbol{R}^{-1}\boldsymbol{L}^{-1}\boldsymbol{Y} = \boldsymbol{Y}^{\mathrm{T}}(\boldsymbol{L}\boldsymbol{R}\boldsymbol{L}^{\mathrm{T}})^{-1}\boldsymbol{Y} = \boldsymbol{Y}^{\mathrm{T}}\boldsymbol{Q}^{-1}\boldsymbol{Y} \tag{2.224}$$

where $\boldsymbol{Q} = \boldsymbol{L}\boldsymbol{R}\boldsymbol{L}^{\mathrm{T}}$. The determinant is:

$$|\boldsymbol{Q}| = |\boldsymbol{L}\boldsymbol{R}\boldsymbol{L}^{\mathrm{T}}| = |\boldsymbol{L}| \cdot |\boldsymbol{R}| \cdot |\boldsymbol{L}^{\mathrm{T}}| = |\boldsymbol{L}|^2 \cdot |\boldsymbol{R}| \tag{2.225}$$

Substituting the above two formulas into eq. (2.222), we obtain:

$$p_n(\boldsymbol{Y}) = \frac{1}{\sqrt{(2\pi)^n |\boldsymbol{Q}|}} \exp\left\{-\frac{1}{2}\boldsymbol{Y}^{\mathrm{T}}\boldsymbol{Q}^{-1}\boldsymbol{Y}\right\} \tag{2.226}$$

When comparing the n-dimensional probability density expressions of $\boldsymbol{Y}$ and $\boldsymbol{X}$, it can be seen that the two forms are the same, so the vector $\boldsymbol{Y}$, the result of normal vector $\boldsymbol{X}$ through the linear transformation, is still a normal distribution. Its mean is still zero, and its covariance changes. This proves the important conclusion that the normal process is still in the normal state through the linear system. Although the zero mean assumption is used in the above derivation, the conclusion is still valid when the mean is nonzero.

The above conclusion can be explained with the superposition principle. We have talked about the relationship between the output process $Y(t)$ and the input process $X(t)$ of the linear system:

$$Y(t) = \int_{-\infty}^{\infty} X(\tau)h(t-\tau)\,\mathrm{d}\tau \tag{2.227}$$

The integral in this equation can be regarded as the limit of the sum of the last equation (refer to Fig. 2.15):

$$Y(t) = \lim_{\substack{\Delta\tau\to 0\\ n\to\infty}} \sum_{i=1}^{n} X(\tau_i)h(t-\tau_i)\Delta\tau \tag{2.228}$$

where $X(\tau_i)$ is the random amplitude of the narrow pulse for the moment, τ_i is a normal random variable, $X(\tau_i)\Delta\tau$ is the area of the narrow pulse, i.e., the intensity of the impulse function for $\Delta\tau \to 0$; $h(t-\tau_i)$, as a nonrandom variable, is the impulse response of the linear system for the moment τ_i.

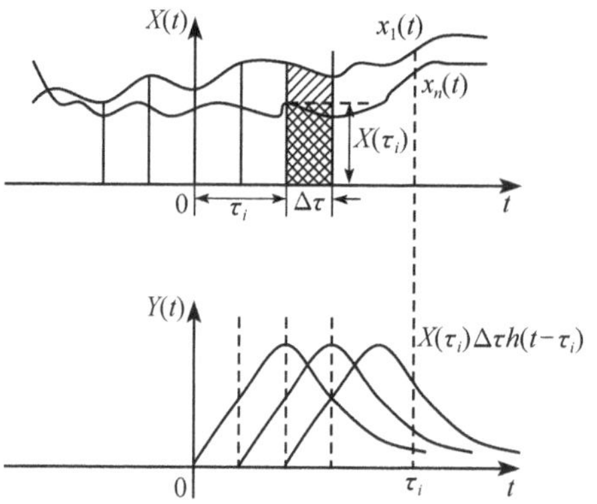

Fig. 2.15: Impulse response method of the random process.

The above equations show that when the input $X(t)$ is the normal random process, the output process $Y(t)$ is the weighted sum of the countless normal random variables. Using the conclusion that the probability distribution of the sum of multiple normal random variables still obeys the normal distribution, we know that for any time t, $Y(t)$ is always a normal random variable, so the output process $Y(t)$ is a normal random process.

We can also prove that when the input of the linear system is a normal process, with the output it is a joint normal process.

2.5.2 Input is a non-normal process of a broadband (relative to system's passband), and output is an approximate normal process

Given the non-normal process input $X(t)$, with spectral width Δf_X and correlation time τ_{0X}, and the passband Δf of the linear system, using the same method as above, we can regard $X(t)$ as the sum of a lot of narrow pulses (the pulse width is $\Delta\tau$) with random amplitude values, then make them go through the linear system, to finally obtain the following output process:

$$Y(t) \approx \sum_{i=1}^{n} X(\tau_i)h(t-\tau_i)\Delta\tau \tag{2.229}$$

where $X(\tau_i)$ is a non-normal random variable.

The central limit theorem shows that the distribution of the sum of a large number of independent random variables approximately obeys the normal distribution. Here, the amplitude of each narrow pulse is a random variable, as long as they satisfy the condition that the input process $X(t)$ can be divided into many narrow impulses (i.e., the number n is large enough) and the amplitude of each narrow pulse is statistically independent, according to the theorem, we can say the output process $Y(t)$ will approximate the normal distribution.

Choose the narrow pulse for width $\Delta\tau$. If the condition is satisfied:

$$\tau_{0X} \ll \Delta\tau \tag{2.230}$$

Then, there is no correlation between the amplitude of two adjacent narrow pulses, which can be regarded as statistical independence approximately. However, when choosing $\Delta\tau$, the number n of independent random variables should also be large enough. If the chosen $\Delta\tau$ is very large, which leads to dissatisfying the central limit theorem due to the too small number n, the output will still be a non-normal process. Its physical interpretation is: if $\Delta\tau$ is larger than the setup time t_r of the linear system, when the response output of the narrow pulses reaches a stable amplitude, the output process remains a non-normal process. Only when $\Delta\tau$ is much smaller than the setup time t_r of the system, after every narrow pulse through the linear system, can the probability distribution of the output narrow pulse amplitude change due to a serious of waveform distortion. As a result, the input non-normal distribution will change into normal distribution. According to eq. (2.227), the output process $Y(t)$ is a normal process, which is the weighted sum of a large number of normal random variables at any given moment.

Combining the above two conditions, when the input is a normal process, we can obtain the condition that makes the output become a nearly normal process:

$$\tau_{0X} \ll \Delta\tau \ll t_r \tag{2.231}$$

Since the correlation time is inversely proportional to the spectrum width, we have:

$$\tau_{0X} \propto \frac{1}{\Delta f_X} \tag{2.232}$$

According to the analysis of the linear system with certain signal through it, it can be seen that the setup time t_r is inversely proportional to the system passband Δf, i.e.,

$$t_r \propto \frac{1}{\Delta f} \tag{2.233}$$

So, the condition can be rewritten as:

$$\frac{1}{\Delta f_X} \ll \frac{1}{\Delta f} \tag{2.234}$$

or

$$\frac{\Delta f_X}{\Delta f} \gg 1 \tag{2.235}$$

The above qualitative analysis indicates that as long as the spectrum width Δf_X of the input random process is much larger than the system passband Δf, i.e., the so-called "broadband input" situation, regardless of what form the probability distribution of the input process is, the output process of the system is approximately the normal distribution. In this situation, it is called the "normalization" of the non-normal process.

The degree of normalization has the relationship ratio $\Delta f_X/\Delta f$ (or the number n of independent random variables in the central limit theorem). Moreover, if the distribution of the input signal obeys Rayleigh distribution, because it is a little closer to the normal distribution, we only need to set $\Delta f_X/\Delta f > 2 \sim 3$. If the distribution of the input process deviates far from the normal distribution, the ratio $\Delta f_X/\Delta f$ is required to be bigger. According to the actual requirements of engineering, this ratio is generally 3~5 times or 7~10 times.

2.5.3 Input is white noise and output of the limited bandwidth system is an approximate normal process

White noise can be a normal process or a non-normal process, but white noise is a wideband random process with spectral width $\Delta f_X \to \infty$ and correlation time $\tau_0 \to 0$. Moreover, the passband Δf of a linear system is always finite with nonzero and finite setup time t_r. Therefore, according to eq. (2.231), when the system input is white noise, the output $Y(t)$ is approximate to the normal process.

This conclusion has important practical value, providing a way to obtain the approximate normal process. As long as the wideband random process of any distribution rate passes through a narrowband linear system, the output is approximate to the normal process.

The intermediate frequency amplifier of radar is a narrowband linear system, and the output of the internal noise is the normal process. To interfere with this radar, it is best to make the interference of the jammer be a normal process after transmitted through the radar. However, it is difficult for the jammer to produce strict normal noise interference. Generally, it is normal distribution with limiting amplitude. According to the above analysis, as long as the spectral width of the normal noise interference with limiting amplitude is far greater than that of the radar, it can also obtain the effect of normal noise interference.

Example 2.5.1. A two-level cascaded linear network is shown in Fig. 2.16. Given the input stationary normal white noise $X(t)$, with power spectral density $G(\omega) = N_0/2$, the impulse response of network I is:

$$h(t) = \begin{cases} ae^{-at}, & t \geq 0 \\ 0, & t < 0 \end{cases} \tag{2.236}$$

Try to solve the probability density $p_z(z)$ of the output process $Z(t)$ of the cascade network.

Solution: A two-level cascaded linear network is still a linear system, so when the input is a normal process, the output is still a normal process. After the mean and the variance of the process $Z(t)$ have been obtained, we can write the $p_z(z)$ expression.

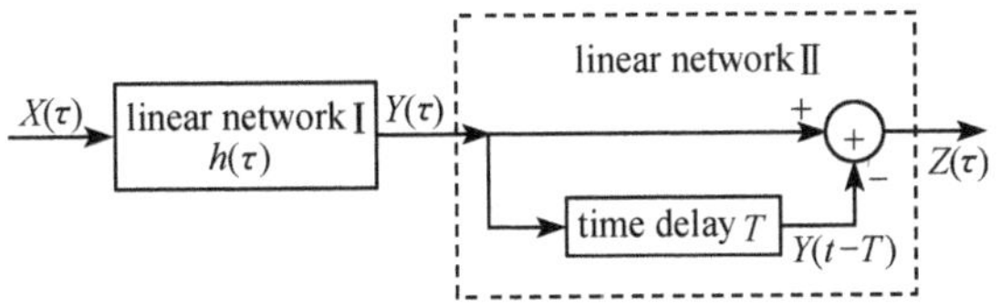

Fig. 2.16: Cascade of two linear networks.

Because the input $X(t)$, with zero mean, is stationary, the mean of process $Y(t)$ is:

$$m_Y(t) = E\left[\int_{-\infty}^{\infty} X(t)h(t-\tau)\,\mathrm{d}\tau\right] = \int_{-\infty}^{\infty} E[X(t)]h(t-\tau)\,\mathrm{d}\tau = 0 \tag{2.237}$$

In the same way:

$$m_X(t-T) = 0 \quad \text{and} \quad Z(t) = Y(t) - Y(t-T) \tag{2.238}$$

Therefore, we have:

$$m_Z(t) = m_Y(t) - m_Y(t-T) = 0 \tag{2.239}$$

The transfer function of linear network I is:

$$H_X(\omega) = \int_{-\infty}^{\infty} h(t)\mathrm{e}^{-\mathrm{j}\omega\tau}\,\mathrm{d}\tau = \int_{0}^{\infty} a\mathrm{e}^{-(a+\mathrm{j}\omega)\tau}\,\mathrm{d}\tau = \frac{a}{a+\mathrm{j}\omega} \tag{2.240}$$

The power spectral density of process $Y(t)$ is:

$$G_Y(\omega) = G_X(\omega)\,|H_X(\omega)|^2 = \frac{N_0}{2}\cdot\left|\frac{a}{a+\mathrm{j}\omega}\right|^2 = \frac{N_0/2}{1+(\omega/a)^2} \tag{2.241}$$

The corresponding covariance function is:

$$C_Y(\tau) = R_Y(\tau) = \frac{1}{\pi}\int_{0}^{\infty} R_Y(\omega)\cos\omega\tau\,\mathrm{d}\omega = \frac{N_0}{2\pi}\int_{0}^{\infty}\frac{\cos\omega\tau}{1+(\omega/a)^2}\,\mathrm{d}\omega = \frac{N_0 a}{4}\mathrm{e}^{-a|\tau|} \tag{2.242}$$

The variance of process $Z(t)$ is:

$$\begin{aligned}\sigma_Z^2 &= E\{[\mathring{Z}(t)]^2\} = E\{[Z(t)]^2\} = E\{Y^2(t) + Y^2(t-T) - 2Y(t)Y(t-T)\}\\ &= 2R_Y(0) - 2R_Y(T) = \frac{N_0 a}{2}(1-\mathrm{e}^{-a|\tau|})\end{aligned} \tag{2.243}$$

The probability density expression of $Z(t)$ is:

$$p_z(z) = \frac{1}{\sqrt{2\pi}\sigma_Z}\exp\left[-\frac{z^2}{2\sigma_Z^2}\right] \tag{2.244}$$

Exercises

2.1 Assume $\{X_n\}$ $(n = 1, 2, 3, \ldots)$ is a sequence of independent random variables. Given random variable X_n with two possible values 0 or n, we have:

$$P\{X_n = n\} = \frac{1}{n^2}, \quad P\{X_n = 0\} = 1 - \frac{1}{n^2} \tag{2.245}$$

Prove: $\{X_n\}$ converges to zero in probability but does not converge to zero in the mean square.

2.2 The linear system is shown in Fig. 2.17. Input two stationary processes $X_1(t)$ and $X_2(t)$, where $X_1(t)$ is outputted directly, and $X_2(t)$ is outputted after the differential transformation. Try to find the relational expression of the function $R_{Y_1Y_2}(\tau)$ of the output and the function $R_{X_1X_2}(\tau)$ of the input.

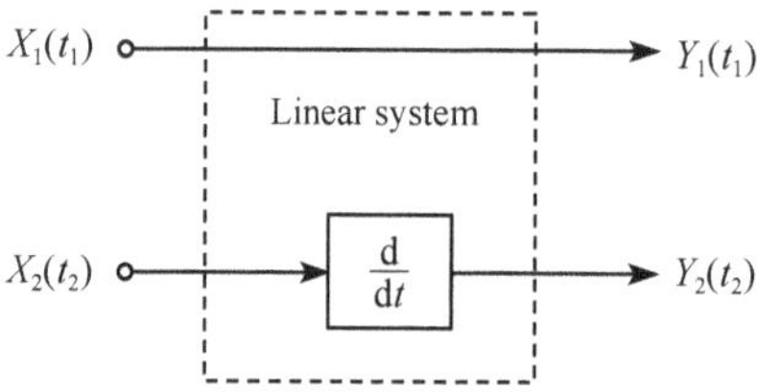

Fig. 2.17: Linear system.

2.3 The LR circuit is shown in Fig. 2.18. Assuming that the initial e.m.f. (electromotive force) of the inductance is zero, given the output stationary random process $X(t)$ with zero mean and the correlation function $R_{X_1X_2}(t_1, t_2) = \sigma^2 e^{-\beta|t_2-t_1|}$, try to solve: (1) the random differential equation and (2) the mean and the autocorrelation function of the output random process $Y(t)$.

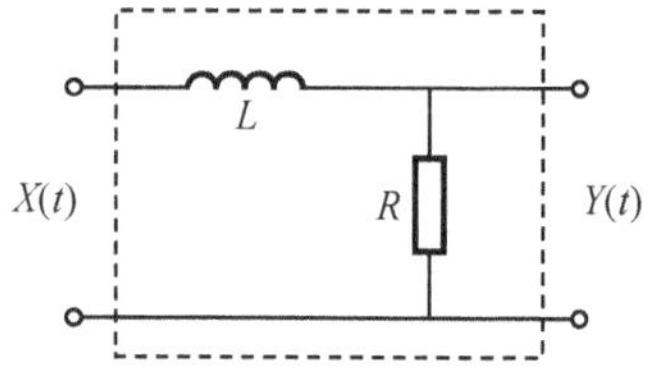

Fig. 2.18: LR circuit.

2.4 Assume the random process $Y(t) = \int_{-\infty}^{t} X(\tau)\,d\tau$. Given the power spectral density $G_X(\omega)$ of the input random process X(t), try to solve the power spectral density $G_Y(\omega)$ of $Y(t)$.

2.5 There is a stationary random process $X(t)$, where $m_X(t) = 1$, $R_X(\tau) = 1 + e^{-2|\tau|}$. Try to find the mean and the variance of the random variable $Y = \int_0^1 X(t)\,dt$.

2.6 The RC integral circuit is shown in Fig. 2.19. Assuming that the initial charge on the capacitor is zero, given the input stationary random process $X(t)$ with the zero mean and the correlation function $R_X(\tau) = \sigma^2 e^{-\beta|\tau|}$, the impulse response of the circuit is:

$$h(t) = \begin{cases} \alpha e^{-\alpha t}, & t \geq 0 \\ 0, & t < 0 \end{cases} \tag{2.246}$$

where $\alpha = 1/RC$. Try to find the output function $R_Y(t_1, t_2)$

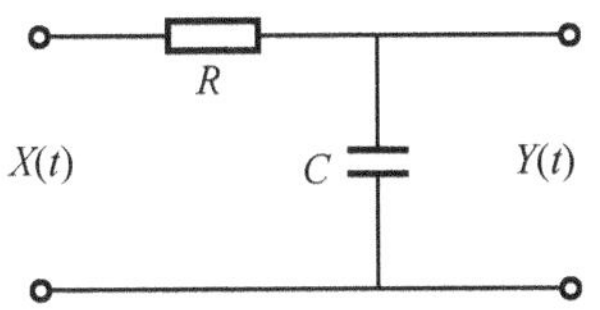

Fig. 2.19: RC integral circuit.

2.7 Try to solve the power spectral density $G_Y(\omega)$, Correlation function $R_Y(\tau)$, the relevant time τ_0 and noise passband $\Delta\omega_n$ after the white noise $X(t)$ through the RC differential circuit.

2.8 White noise $X(t)$ pass through the impulse response $h(t)$ of the linear system. Prove that the cross-correlation function between input and output is $R_{XY}(\tau) = N_0/2h(\tau)$.

2.9 In the linear system shown in Fig. 2.20, if input $X(t)$ is a stationary process, try to prove that the power spectral density of $Y(t)$ is $G_Y(\omega) = 2G_X(\omega)\,[1 + \cos\omega\tau]$.

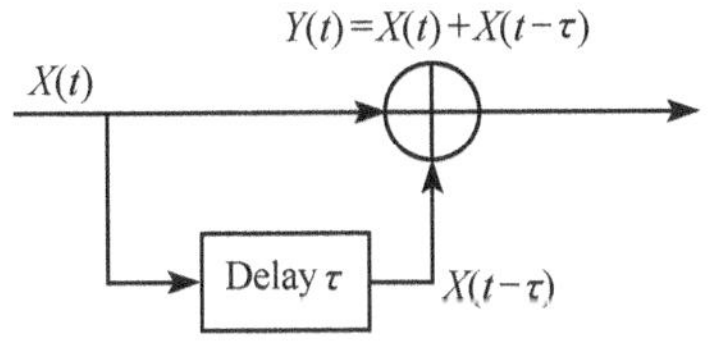

Fig. 2.20: A delay additive circuit.

2.10 Assuming the correlation function of white noise is $N_0/2\delta(\tau)$, it passes through the ideal bandpass amplifier with the amplitude-frequency feature as shown in Fig. 2.21. Try to find the total noise power of the amplifier output.

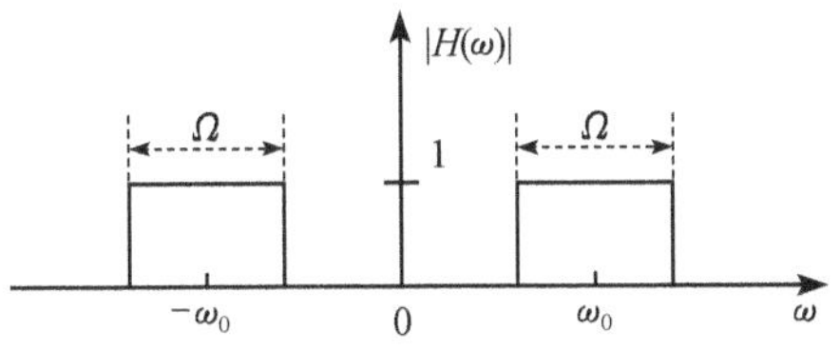

Fig. 2.21: Ideal bandpass.

2.11 The LR circuit is shown in Fig. 2.22. Given the power spectral density $N_0/2$ of the input $X(t)$, try to find the autocorrelation function $R_Y(\tau)$ on the output side.

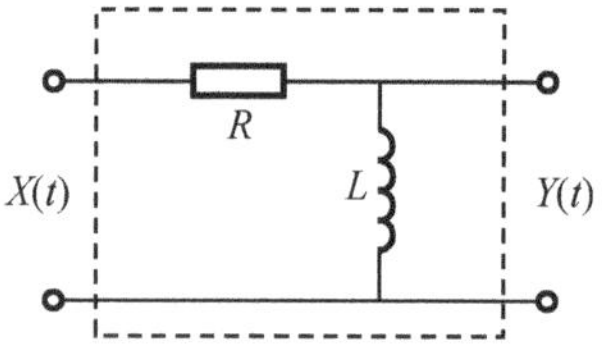

Fig. 2.22: LR circuit.

2.12 Assume that the power spectral density of the stationary Gaussian white noise is $N_0/2$. It is transmitted by a filter with transfer function $H(\omega) = \frac{1}{1+j(\omega/\omega_0)}$. Try to solve the one-dimensional probability density of the output noise.

2.13 Assuming the zero mean stationary random process $X(t)$ is inputted to a linear filter, the impulse response of the filter is as follows:

$$h(t) = \begin{cases} \alpha e^{-\alpha t}, & t \geq 0 \\ 0, & t < 0 \end{cases} \tag{2.247}$$

(1) Prove that the output power spectral density of the filter is:

$$\frac{\alpha^2}{\alpha^2 + \omega^2} G_X(\omega)$$

(2) If the filter impulse response is only one part of the index, that is,

$$h(t) = \begin{cases} \alpha e^{-\alpha t}, & 0 \leq t \leq T \\ 0, & \text{others} \end{cases} \tag{2.248}$$

prove that the output power spectral density is shown as:

$$\frac{\alpha^2}{\alpha^2 + \omega^2}(1 - 2e^{-\alpha T}\cos\omega T + e^{-2\alpha T})G_X(\omega)$$

2.14 Assuming the stationary white noise $X(t)$ passes through the linear system as shown in Fig. 2.23, try to solve the function $R_{Y_1Y_2}(\tau)$ and draw its graph.

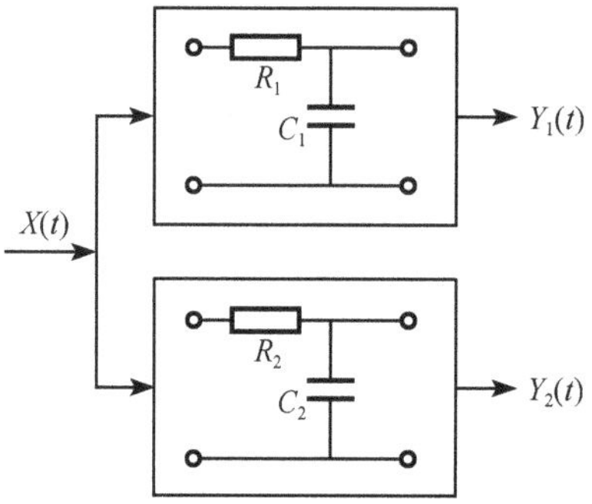

Fig. 2.23: Given linear system.

2.15 Assuming that the stationary normal white noise $X(t)$ passes through an LR circuit and the delay additive circuit as shown in Fig. 2.24, try to solve the one-dimensional probability density $p(y)$ of the output random process $Y(t)$.

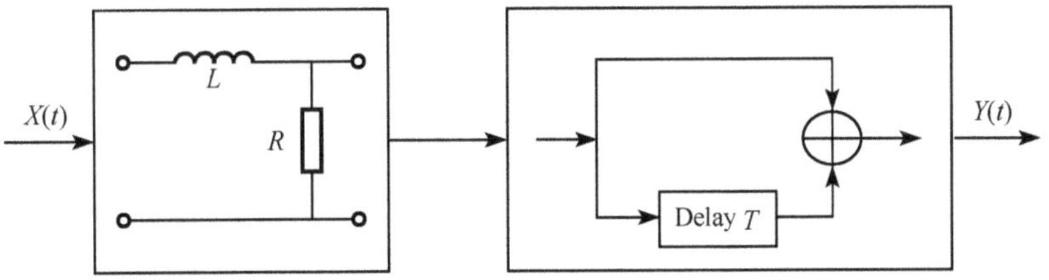

Fig. 2.24: LR and AD circuit.

2.16 Assuming that the stationary white noise $X(t)$ passes through the system as shown in Fig. 2.25, try to find the correlation function $R_Y(\tau)$, the variance σ_Y^2 and the mean m_Y of the output process $Y(t)$.

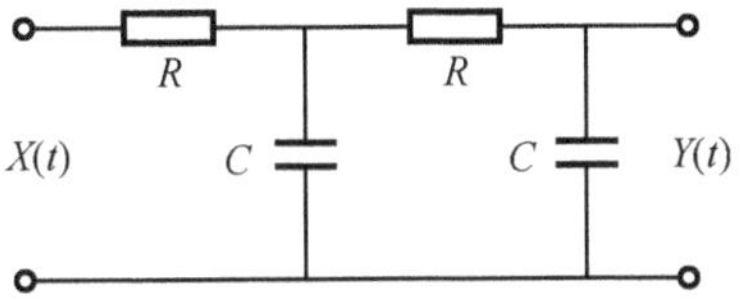

Fig. 2.25: Cascade RC circuit.

2.17 Assuming the linear system is as shown in Fig. 2.26, where $h_1(t)$ and $h_2(t)$ are impulse response functions of each subsystem, respectively, try to solve the cross-correlation function $R_{Y_1Y_2}(\tau)$ and the power spectral density $G_{Y_1Y_2}(\omega)$

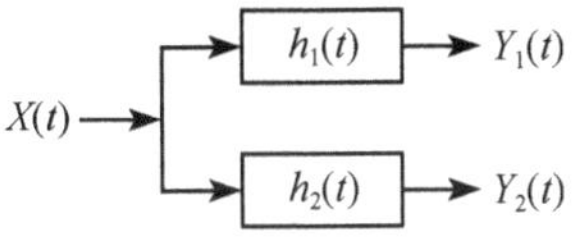

Fig. 2.26: Given linear system.

2.18 Assuming that the mean square of the stationary normal process $X(t)$ is differentiable with the mean m_X, the variance σ_X^2 and the autocorrelation function $R_X(\tau)$, try to solve the joint probability density $p_2\,(x, \dot{x}; t_1)$ of the random variable $X(t)$ and its derivative $X'(t)$ at a fixed time t_1

2.19 Machine problem: using the MATLAB program to design a composite signal of sinusoidal signal with Gaussian white noise. (1) analyze the power spectral density and amplitude distribution characteristics of composite signals; (2) analyze the power spectral density and the corresponding amplitude distribution characteristics of the composite signal through the RC integral circuit; and (3) analyze the power spectral density and the corresponding amplitude distribution characteristics of the composite signal through the ideal lowpass system.

3 Stationary and narrowband random processes

3.1 Narrowband random processes represent quasi-sinusoidal oscillation

3.1.1 Formation and characteristics of narrowband stochastic processes

In general radio receivers, there are usually high-frequency amplifiers or intermediate frequency amplifiers, whose passband is often far less than the central frequency f_0. That is:

$$\frac{\Delta f}{f_0} \ll 1 \quad \text{or} \quad \frac{\Delta\omega}{\omega_0} \ll 1 \tag{3.1}$$

This linear system is called a narrowband (linear) system.

If the input process of narrowband system $X(t)$ is white noise or broadband noise (as shown in Fig. 3.1), the output process $Y(t)$ is based on the system's pass-through selection feature. The power spectrum is shown in Fig. 3.2. Power spectrum $G_Y(\omega)$ distributes in the central frequency $\pm\omega_0$.The noise spectrum is wide in the narrowband range near $\Delta\omega_n \ll \omega_0$. This narrowband noise is called a narrowband stochastic process or a narrowband process.

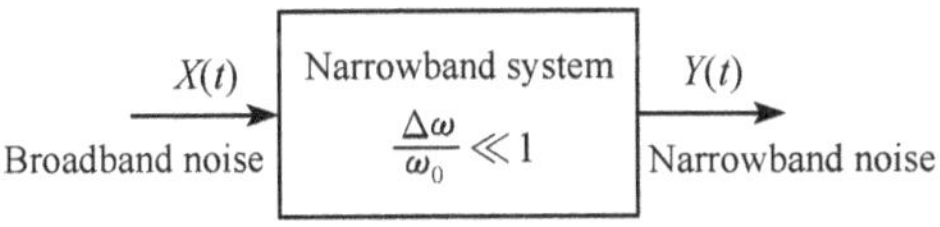

Fig. 3.1: Formation of narrowband random processes.

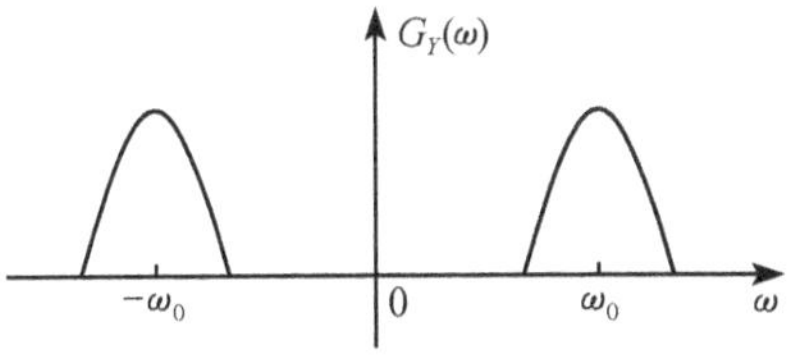

Fig. 3.2: Power spectra of narrowband random processes.

The output process $Y(t)$, is much like a modulated sinusoidal oscillation; the small oscillations of its carrier frequency ω near the central frequency ω_0 are not strictly amplitude modulated sinusoidal oscillations, hence the quasi-sinusoidal oscillation.

$Y(t)$ is an assemblage with large number of samples, and it can be expressed as:

$$Y(t) = A(t)\cos[\omega_0 t + \varphi(t)] \tag{3.2}$$

Here, $A(t)$, a random envelope, is a time function that changes slowly with respect to ω_0. The rate of change of the envelope is much slower than the instantaneous rate

https://doi.org/10.1515/9783110593808-004

of change for process $Y(t)$; $\varphi(t)$ is a random phase, relative to the frequency ω_0. It is also a slow change time function. Therefore, the narrowband stochastic process can be approximated as the quasi-sinusoidal oscillation of the amplitude.

The narrowband random process as a quasi-sinusoidal oscillation does not only have a clear physical meaning, but also facilitates analysis and application. For example, in order to research the output process of the narrowband through the envelope detector, it is only necessary to study the envelope transformation, without considering its carrier frequency and phase, because the output voltage of the envelope detector is proportional to the envelope of the input voltage and the carrier component is filtered.

3.1.2 Expression of narrowband stochastic processes

Intercept a section of 0- T from the waveform of the narrowband process $Y(t)$. For the period T, the interception waveform is repeated on the left and right, and the Fourier series of extension is used for the analysis. In the interval $(0, T)$, the process $Y(t)$ can be considered as the sum of an infinite number of simple harmonic oscillators with random enveloping and phase, namely:

$$Y(t) = \sum_{i=1}^{\infty} [a_i \cos \omega_i t + b_i \sin \omega_i t] \tag{3.3}$$

where:

$$a_i = \frac{2}{T} \int_0^T Y(t) \cos \omega_i t \, \mathrm{d}tr$$

$$b_i = \frac{2}{T} \int_0^T Y(t) \sin \omega_i t \, \mathrm{d}tr$$

$$\omega_i = i\omega_1 = i\frac{2\pi}{T}, \quad i = 1, 2, \ldots$$

Set the input $X(t)$ as a process of the zero mean stationary process. The mean of process $Y(t)$ is zero, so the above series expression does not contain a_0. Parameters a_i, b_i are random variables that vary with the sample, as shown in Fig. 3.3.

Setting $\omega_i = (\omega_i - \omega_0) + \omega_0$, the above formula can be rewritten as

$$\begin{aligned} Y(t) = & \sum_{i=1}^{\infty} [a_i \cos(\omega_i - \omega_0)t + b_i \sin(\omega_i - \omega_0)t] \cos \omega_0 t \\ & + \sum_{i=1}^{\infty} [b_i \cos(\omega_i - \omega_0)t - a_i \sin(\omega_i - \omega_0)t] \sin \omega_0 t \end{aligned} \tag{3.4}$$

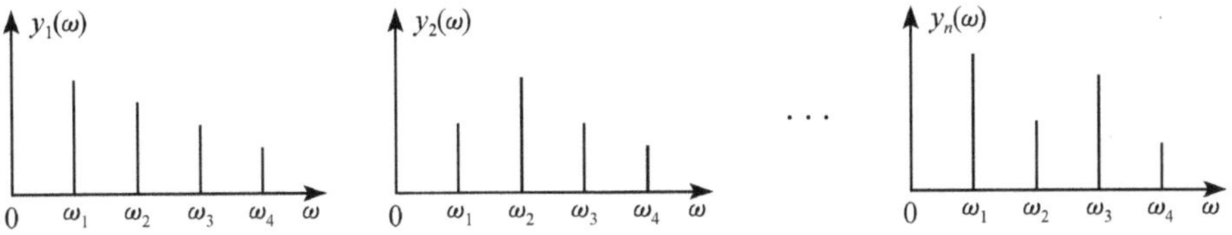

Fig. 3.3: The co-wave analysis of the random process.

Marking the amplitude of the cosine term in the above formula as $A_c(t)$ and writing the amplitude of the sine term is as $-A_s(t)$, the narrowband process $Y(t)$ can be expressed as

$$Y(t) = A_c(t)\cos\omega_0 t - A_s(t)\sin\omega_0 t \tag{3.5}$$

If

$$\begin{cases} A_c(t) = A(t)\cos\varphi(t) \\ A_s(t) = A(t)\sin\varphi(t) \end{cases} \tag{3.6}$$

The above expression can also be rewritten as

$$\begin{aligned} Y(t) &= A(t)\left[\cos\omega_0 t \cdot \cos\varphi(t) - \sin\omega_0 t \cdot \sin\varphi(t)\right] \\ &= A(t)\cos\left[\omega_0 t + \varphi(t)\right] \end{aligned} \tag{3.7}$$

where:

$$\begin{cases} A(t) = \sqrt{A_c^2(t) + A_s^2(t)} \\ \varphi(t) = \arctan\frac{A_s(t)}{A_c(t)} \end{cases} \tag{3.8}$$

Because both $A(t)$ and $\varphi(t)$ are time functions of slow change, the two orthogonal components $A_c(t)$ and $A_s(t)$ of the envelope $A(t)$ are also time functions of slow change. The statistical features of $A(t)$ as well as of $A_c(t)$, $A_s(t)$ will be discussed in detail later.

3.2 Analytic signals and Hilbert transforms

3.2.1 Complex signals of sinusoidal signals

The sinusoidal signal is

$$s(t) = A\cos(\omega_0 t + \varphi) \tag{3.9}$$

where the amplitude A, the angular frequency ω_0 and phase φ are constants.

The usual complex signals are expressed as complex exponentials or plural forms. The complex signals corresponding to eq. (3.9) are as follows:

$$\tilde{s}(t) = A\mathrm{e}^{\mathrm{j}(\omega_0 t + \varphi)} = \tilde{A}\mathrm{e}^{\mathrm{j}\omega_0 t} \tag{3.10}$$

where $\tilde{A} = Ae^{j\varphi}$ in the plural, called complex envelope and $e^{j\omega_0 t} = \cos\omega_0 t + j\sin\omega_0 t$ called complex carrier. With eq. (3.9) corresponding to the complex signals (plural) as follows:

$$\tilde{s}(t) = s(t) + j\hat{s}(t) \tag{3.11}$$

where $s(t) = A\cos(\omega_0 t + \varphi) = \mathrm{Re}[\tilde{s}(t)]$

This is the original real signal;

$$\hat{s}(t) = A\sin(\omega_0 t + \varphi) = \mathrm{Im}\,[\tilde{s}(t)] \tag{3.12}$$

It is a real function that is introduced for convenience.

The complex signal $\tilde{s}(t)$ whose envelope and phase angle are, respectively,

$$|\tilde{s}(t)| = \sqrt{s^2(t) + \hat{s}^2(t)} \tag{3.13}$$

$$\varphi(t) = \omega_0 t + \varphi = \arctan\frac{\hat{s}(t)}{s(t)} \tag{3.14}$$

The following analysis shows the relationship between the two spectral signals when the real signal is represented as a plural signal. With the Fourier transform to eq. (3.12), when $\varphi = 0$, the spectrum of $s(t)$ is:

$$S(\omega) = \pi A\,[\delta(\omega + \omega_0) + \delta(\omega - \omega_0)] \tag{3.15}$$

With the Fourier transform to eq. (3.13) and we can obtain the spectrum of the $j\hat{s}(t)$ as follows:

$$j\hat{S}(\omega) = -\pi A\,[\delta(\omega + \omega_0) - \delta(\omega - \omega_0)] \tag{3.16}$$

Therefore, from eq. (3.11), we can obtain the spectrum of the $\tilde{s}(t)$ as:

$$\tilde{S}(\omega) = S(\omega) + j\hat{S}(\omega) = 2\pi A\delta(\omega - \omega_0) \tag{3.17}$$

The $S(\omega)$ and $j\hat{S}(\omega)$ in the negative-frequency domain are the positive and negative cancellation, and they overlay the same number in the positive-frequency domain; the result $\tilde{S}(\omega)$ contains only positive spectral components and it is the original signal spectrum $S(\omega)$ two times the normal-frequency domain components, so there is a relationship:

$$\tilde{S}(\omega) = \begin{cases} 2S(\omega)\,, & \omega \geq 0 \\ 0\,, & \omega < 0 \end{cases} \tag{3.18}$$

Fig. 3.4 Shows the spectrum of sinusoidal signals.

3.2.2 Complex signals of high-frequency narrowband signals

Set the high-frequency narrowband real signal as:

$$s(t) = A(t)\cos\left[\omega_0 t + \varphi(t)\right] \tag{3.19}$$

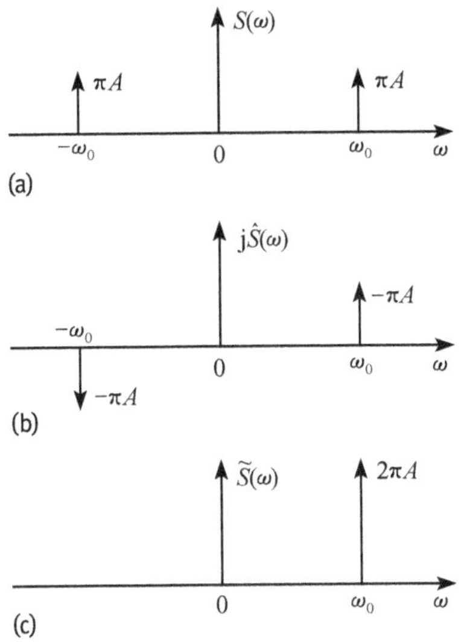

Fig. 3.4: Spectrum of sinusoidal signals.

where relative to the central frequency ω_0, the amplitude modulation signal $A(t)$ and the phase modulation signal $\varphi(t)$ are the time functions of low frequency that change slowly. According to the previous definition (3.19), the complexe signal can be obtained:

$$\tilde{s}(t) = A(t)e^{j[\omega_0 t+\varphi(t)]} = \tilde{A}(t)e^{j\omega_0 t} \tag{3.20}$$

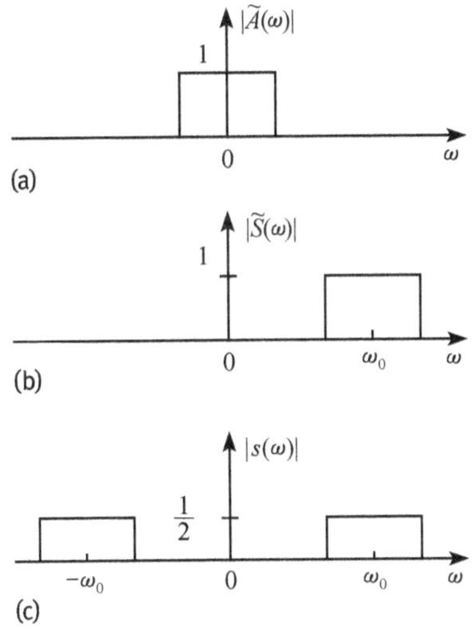

Fig. 3.5: Amplitude spectrum of high frequency narrowband signals.

In the formula, we have the complex envelope $\tilde{A}(t) = A(t)e^{j\varphi(t)}$, which contains all the information about the low-frequency modulation signal.

The envelope detector or phase detector are used to demodulate the high-frequency narrowband signal so that we can obtain a signal containing complex envelope part that keeps the information only and deletes the complex carrier part that has no information for signal processing. The spectra of the high-frequency narrowband signal can also be analyzed with the previous method.

(1) Find the spectrum $\tilde{S}(w)$ of the complex signal $\tilde{s}(t)$. Using the Fourier transform $\tilde{A}(t) \leftrightarrow \tilde{A}(\omega)$ and the frequency shift characteristics:

$$e^{j\omega_0 t} \leftrightarrow 2\pi\delta(\omega - \omega_0) \tag{3.21}$$

and taking advantage of the multiplication properties of Fourier transformation:

$$\tilde{A}(t)e^{j\omega_0 t} \leftrightarrow \frac{1}{2\pi}[\tilde{A}(\omega) \otimes 2\pi\delta(\omega - \omega_0)] \tag{3.22}$$

That is,

$$\tilde{s}(t) \leftrightarrow \tilde{S}(\omega) \tag{3.23}$$

Using the convolution property of the δ-function,

$$\tilde{S}(\omega) = \tilde{A}(\omega - \omega_0) \tag{3.24}$$

(2) The spectrum $S(\omega)$ of the real signal $s(t)$ is

$$s(t) = \text{Re}\,[\tilde{s}(t)] = \frac{1}{2}\,[\tilde{s}(t) + \tilde{s}^*(t)] \tag{3.25}$$

where $\tilde{s}^*(t)$ is the complex conjugate of $\tilde{s}(t)$.

Using the conjugate properties of Fourier transform: if $\tilde{s}(t) \leftrightarrow \tilde{S}(\omega)$, $\tilde{s}^*(t) \leftrightarrow \tilde{S}^*(-\omega)$. The Fourier transform of the formula is available:

$$S(\omega) = \frac{1}{2}[\tilde{S}(\omega) + \tilde{S}^*(-\omega)] = \frac{1}{2}[\tilde{A}(\omega - \omega_0) + \tilde{A}^*(-\omega - \omega_0)] \tag{3.26}$$

The spectrum is distributed symmetrically in positive and negative-frequency domains. Figure 3.5 shows the amplitude spectrum of the high-frequency narrowband signal.

Comparing eqs. (3.24) and (3.26), we know that there is a relationship between $\tilde{S}(\omega)$ and $S(\omega)$, as shown in eq. (3.18).

When the high-frequency narrow band real signal $s(t)$ has been transformed to the complex signal $\tilde{s}(t)$, we just need to focus on the low-frequency part of the modulation information – the complex envelope $\tilde{A}(t)$ or its spectrum $\tilde{A}(\omega)$. The above analysis shows that the spectrum $S(\omega)$ is known, and we can double the components of the normal-frequency domain of $S(\omega)$ to obtain $\tilde{S}(\omega)$, and letting $\tilde{S}(\omega)$ move left by ω_0, along axial ω, $\tilde{A}(\omega)$ appears. Therefore, it is necessary to perform operations on high-frequency signals, which can now be converted into operations of low-frequency signals to simplify the analysis and processing.

3.2.3 Analytic signals and the Hilbert transform

Writing the above two kinds of real signals $s(t)$ as a complex signal $\tilde{s}(t)$, we can obtain the relationship between $\tilde{S}(w)$ and $S(w)$ as follows:

$$\tilde{S}(\omega) = \begin{cases} 2S(\omega)\,, & \omega \geq 0 \\ 0\,, & \omega < 0 \end{cases} \tag{3.27}$$

A complex signal with this type of one-sided spectral characteristic $\tilde{s}(t) = s(t) + \mathrm{j}\hat{s}(t)$ is called the analytic signal of the real signal $s(t)$.

According to the unilateral spectrum of the signal shown in the formula, spectrum $\tilde{S}(\omega) = S(\omega) + \mathrm{j}\hat{S}(\omega)$ corresponds with the analytic signal $\tilde{s}(t) = s(t) + \mathrm{j}\hat{s}(t)$, in which $S(\omega)$ and $\mathrm{j}\hat{S}(\omega)$ have the following characteristics:

(1) The spectrum $S(\omega)$ of the real signal $s(t)$ is an even function of ω:

$$s(t) \leftrightarrow \begin{cases} S(\omega), & \omega \geq 0 \\ S(\omega), & \omega < 0 \end{cases} \tag{3.28}$$

(2) The spectrum $\mathrm{j}\hat{S}(\omega)$ of $\mathrm{j}\hat{s}(t)$ is an odd function of ω:

$$\mathrm{j}\hat{s}(t) \leftrightarrow \mathrm{j}\hat{S}(\omega) = \begin{cases} S(\omega)\,, & \omega \geq 0 \\ -S(\omega)\,, & \omega < 0 \end{cases} \tag{3.29}$$

The right-hand side of the above eq. (3.29) can be rewritten as

$$\mathrm{j}\hat{S}(\omega) = S(\omega)\,\mathrm{sgn}(\omega) \tag{3.30}$$

where

$$\mathrm{sgn}(\omega) = \begin{cases} 1\,, & \omega \geq 0 \\ -1\,, & \omega < 0 \end{cases} \tag{3.31}$$

The above function is the symbol function in the frequency domain, and eq. (3.27) can be rewritten as:

$$\tilde{S}(\omega) = S(\omega) + \mathrm{j}\hat{S}(\omega) = S(\omega) + S(\omega)\,\mathrm{sgn}(\omega) = S(\omega)[1 + \mathrm{sgn}(\omega)] = 2S(\omega)U(\omega) \tag{3.32}$$

where

$$U(\omega) = \frac{1}{2}[1 + \mathrm{sgn}(\omega)] = \begin{cases} 1\,, & \omega \geq 0 \\ 0\,, & \omega < 0 \end{cases} \tag{3.33}$$

The above is a unit step function in the frequency domain.

Now let us work out the relationship between the real part $s(t)$ and the imaginary part $\tilde{s}(t)$ in the analytic signal $\tilde{s}(t) = s(t) + \mathrm{j}\hat{s}(t)$.

By Fourier transformation on both sides, $\mathrm{j}\hat{s}(t) \leftrightarrow S(\omega)\,\mathrm{sgn}(\omega)$ and $s(t) \leftrightarrow S(\omega)$, $\mathrm{j}(1/\pi t) \leftrightarrow \mathrm{sgn}(\omega)$, we can obtain:

$$\mathrm{j}\hat{s}(t) = s(t) \otimes \mathrm{j}\frac{1}{\pi t} = \mathrm{j}\frac{1}{\pi}\int_{-\infty}^{\infty}\frac{s(\tau)}{t-\tau}\,\mathrm{d}\tau \tag{3.34}$$

or

$$\hat{s}(t) = \frac{1}{\pi}\int_{-\infty}^{\infty}\frac{s(\tau)}{t-\tau}\,\mathrm{d}\tau \tag{3.35}$$

This formula is the transformation $s(t)$ to $\hat{s}(t)$, called the Hilbert transform. On the contrary, if we know $\hat{s}(t)$ and want $s(t)$, the Hilbert inverse transform can be used:

$$s(t) = -\frac{1}{\pi}\int_{-\infty}^{\infty}\frac{\hat{s}(\tau)}{t-\tau}\,\mathrm{d}\tau \tag{3.36}$$

The Hilbert transform is a linear transformation between the real signals $s(t)$ and $\hat{s}(t)$ and plays an important role in signal analysis. Any real signal $s(t)$, by the Hilbert transform, can be transformed to $\hat{s}(t)$, and then we have analytic signal $\tilde{s}(t) = s(t) + \mathrm{j}\hat{s}(t)$. The analytic signal can also be defined according to the Hilbert transform.

The complex signal $\tilde{s}(t) = s(t) + \mathrm{j}\hat{s}(t)$, whose real part $s(t)$ and imaginary part $\hat{s}(t)$ are connected by the Hilbert transform, can be called the analytic signal (pre-envelope signal).

The Hilbert transform has some unique properties, and the main ones are described below, so as to help us to understand the role of the Hilbert transform.

The definition of the Hilbert transform indicates that working out $\hat{s}(t)$ by $s(t)$ actually gives the output $s(t)$ through impulse response $h(t) = 1/\pi t$; as is shown in Fig. 3.6.

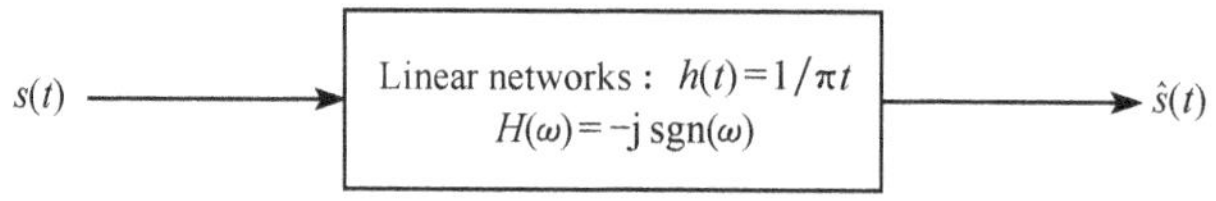

Fig. 3.6: The Hilbert transform.

From $H(\omega) \leftrightarrow h(t)$, the transfer function of this linear network is:

$$H(\omega) = \int_{-\infty}^{\infty} h(t)\mathrm{e}^{-\mathrm{j}\omega t}\,\mathrm{d}t = \int_{-\infty}^{\infty}\frac{1}{\pi t}\,[\cos\omega t - \mathrm{j}\sin\omega t]\,\mathrm{d}t = \begin{cases}-j, & \omega \geq 0\\ j, & \omega < 0\end{cases} \tag{3.37}$$

or

$$H(\omega) = -\mathrm{j}\,\mathrm{sgn}(\omega) \tag{3.38}$$

The amplitude-frequency characteristic $|H(\omega)|$ and the phase-frequency characteristics $\varphi(\omega)$ are shown in Fig. 3.7; it is clear that the Hilbert transform is equal to a late 90° phase shifter in the entire frequency domain.

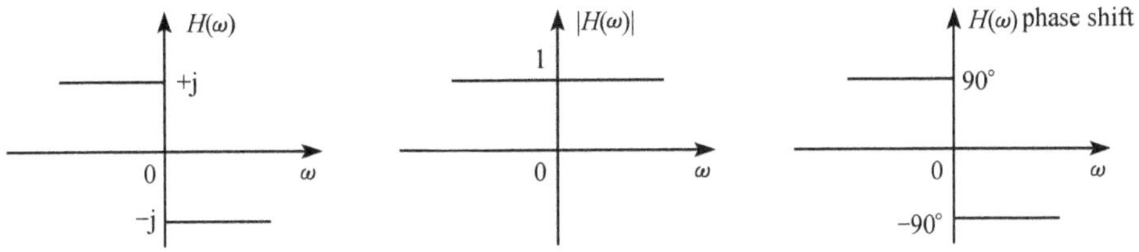

Fig. 3.7: Frequency characteristics of the Hilbert transform.

Example 3.2.1. Real signal $s(t) = \cos\omega_0 t$ known. Find the analytic signal $\tilde{s}(t)$.

Solution: From eq. (3.35):

$$\hat{s}(t) = \frac{1}{\pi}\int_{-\infty}^{\infty} \frac{\cos\omega_0\tau}{t-\tau}\,\mathrm{d}\tau = \frac{1}{\pi}\int_{-\infty}^{\infty} \frac{(t+\tau)\cos\omega_0\tau}{t^2-\tau^2}\,\mathrm{d}\tau \tag{3.39}$$

Make variable substitution $\tau = tx$ then $\mathrm{d}\tau = t\,\mathrm{d}x$, with:

$$\hat{s}(t) = \frac{2}{\pi}\int_{0}^{\infty} \frac{\cos\omega_0 tx}{1-x^2}\,\mathrm{d}x = \sin\omega_0 t \tag{3.40}$$

So, the analytic signal is:

$$\tilde{s}(t) = s(t) + \mathrm{j}\hat{s}(t) = \cos\omega_0 t + \mathrm{j}\sin\omega_0 t = \mathrm{e}^{-\mathrm{j}\omega_0 t} \tag{3.41}$$

Example 3.2.2. Real signal $s(t) = \cos\omega_0 t$ known. Find the signal $s_o(t)$ that changes by two Hilbert transforms.

Solution: According to the definition of the Hilbert transform, set the spectrum of $s_o(t)$ as $S(\omega)$ and the spectrum of $s_o(t)$ as $S_o(\omega)$. Then, we have:

$$S_o(\omega) = S(\omega)\,[-\mathrm{j}\,\mathrm{sgn}(\omega)]^2 = -S(\omega) \tag{3.42}$$

The inverse Fourier transform for eq. (3.42) is:

$$s_o(t) = -s(t) \tag{3.43}$$

After two Hilbert transformations, $s_o(t)$ has the reverse phase of the original signal $s(t)$. Making $s(t) = \cos\omega_0 t$, then $s_o(t) = -\cos\omega_0 t$.

In fact, for the second Hilbert transform, the relationship between the input and the output is:

$$s_o(t) = -s(t) = \hat{s}(t) \otimes \frac{1}{\pi t} \tag{3.44}$$

or:

$$s(t) = -\frac{1}{\pi}\int_{-\infty}^{\infty} \frac{\hat{s}(\tau)}{t-\tau}\,\mathrm{d}\tau \tag{3.45}$$

This is the inverse Hilbert transform that we defined previously.

3.3 Analytic complex stochastic process

The random process that is produced by the general method is the real function of time t, called the real random process. For example, the narrow band stochastic process described in Section 3.1, $Y(t) = A(t)\cos[\omega_0 t + \varphi(t)]$, is one. However, as described in Section 3.2, the complex stochastic process is commonly used to simplify the analysis and processing. Since stochastic processes are random variables that change over time, the statistical properties of complex stochastic processes need to be analyzed and the complex random variables discussed first.

3.3.1 Complex random variables

If X and Y are both real random variables, then:

$$Z = X + \mathrm{j}Y \tag{3.46}$$

Where Z is called a complex random variable.

The mathematical expectation, variance and covariance matrix of the real random variables extend to the complex random variable, with the requirements. (1) When the variable $Y = 0$, in the case that Z is a real random variable. The moment of the complex random variable Z should be equal to the real random variable X, in the moment. (2) The characteristics of random variable moments should be maintained; for example, the variance should be non-negative real value.

(1) The mathematical expectation of complex random variables Z is:

$$m_Z = E[Z] = E[X] + \mathrm{j}E[Y] = m_X + \mathrm{j}m_Y \tag{3.47}$$

If $Y = 0$, then $m_Z = m_X$, which meets the aforementioned requirements.

(2) Define the variance of complex random variable Z:

$$\sigma_Z^2 = D[Z] = E[|\mathring{Z}|^2] \tag{3.48}$$

where $\mathring{Z} = Z - m_Z$, the central complex random variable.

Because:

$$\mathring{Z} = X + \mathrm{j}Y - (m_X + \mathrm{j}m_Y) = \mathring{X} + \mathrm{j}\mathring{Y} \tag{3.49}$$

so:

$$D[Z] = E[\mathring{Z}^* \mathring{Z}] = E[\mathring{X}^2 + \mathring{Y}^2] = E[\mathring{X}^2] + E[\mathring{Y}^2] = D[X] + D[Y] \tag{3.50}$$

If $Y = 0$, then we have $D[Z] = D[X]$, which meets the aforementioned requirements. However, if the definition is $D[Z] = E[\mathring{Z}^2]$, the variance will be complex and not meet its characteristic requirements.

(3) If there are two complex random variables:

$$Z_1 = X_1 + \mathrm{j}Y_1\,,$$
$$Z_2 = X_2 + \mathrm{j}Y_2 \tag{3.51}$$

The covariance of complex random variables Z_1 and Z_2 is:

$$C_{Z_1Z_2} = E[\mathring{Z}_1^* \mathring{Z}_2] \tag{3.52}$$

where $\mathring{Z}_1^*$ is the complex conjugate of Z_1. Thus:

$$C_{Z_1Z_2} = E[(\mathring{X}_1 - \mathrm{j}\mathring{Y}_1)(\mathring{X}_2 + \mathrm{j}\mathring{Y}_2)] = C_{X_1X_2} + C_{Y_1Y_2} + \mathrm{j}[C_{X_1Y_2} + C_{Y_1X_2}] \tag{3.53}$$

If $Y_1 = Y_2 = 0$, then $C_{Z_1Z_2} = C_{X_1X_2}$, which meets the aforementioned requirements. However, if the definition is $C_{Z_1Z_2} = E[\mathring{Z}_1\mathring{Z}_2]$, then, when $Z_1 = Z_2 = Z$, the variance will be complex, which does not meet its characteristic requirements.

Here we show the uncorrelated, orthogonal and statistical independence of two complex random variables.

(1) If there are complex random variables Z_1 and Z_2, their covariance is:

$$C_{Z_1Z_2} = E[\mathring{Z}_1^* \mathring{Z}_2] = 0 \tag{3.54}$$

Complex variables Z_1 and Z_2 are uncorrelated.

(2) If there are complex random variables Z_1 and Z_2, we have:

$$E[Z_1^* Z_2] = 0 \tag{3.55}$$

Complex variables Z_1 and Z_2 are orthogonal.

(3) If the complex random variables are: $Z_1 = X_1 + \mathrm{j}Y_1$, $Z_2 = X_2 + \mathrm{j}Y_2$,

$$p(x_1, y_1; x_2, y_2) = p(x_1, y_1)p(x_2, y_2) \tag{3.56}$$

The complex variables Z_1 and Z_2 are statistically independent.

3.3.2 Complex random processes

If $X(t)$ and $Y(t)$ are both real and stochastic processes, then the complex random process is as follows:

$$Z(t) = X(t) + \mathrm{j}Y(t) \tag{3.57}$$

When the mathematical expectation, variance and covariance of the real stochastic process are generalized to the complex stochastic process, the requirements are as above: (1) when the process is taken as $Y(t) = 0$, the moment function of the complex process $Z(t)$ should be equal to the moment function of the real random process $X(t)$; (2) the characteristics of the random process moment function should be kept, for example, the variance should be a non-negative time function.

(1) Define the mathematical expectation of the complex stochastic process $Z(t)$:

$$m_Z(t) = E[Z(t)] = E[X(t)] + \mathrm{j}E[Y(t)] = m_X(t) + \mathrm{j}m_Y(t) \tag{3.58}$$

(2) Define the variance of the complex stochastic process $Z(t)$:

$$\sigma_Z^2(t) = D[Z(t)] = E[|\mathring{Z}(t)|^2] \tag{3.59}$$

where $\mathring{Z}(t) = Z(t) - m_Z(t) = \mathring{X}(t) + \mathrm{j}\mathring{Y}(t)$.

According the previous knowledge:

$$D[Z(t)] = D[X(t)] + D[Y(t)] \tag{3.60}$$

(3) Define the autocorrelation function of the complex stochastic process $Z(t)$ as:

$$R(t_1, t_2) = E\left[Z^*(t_1)Z(t_2)\right] \tag{3.61}$$

where $t_2 = t_1 + \tau$, $Z^*(t_1)$ as the complex conjugate of the $Z(t_1)$.

Note: Define $R(t_1, t_2) = E[Z(t_1)Z^*(t_2)]$, in which $t_2 = t_1 - \tau$ in some books.

(4) If there are two complex random processes $Z_1(t) = X_1(t) + \mathrm{j}Y_1(t)$ and $Z_2(t) = X_2(t)+\mathrm{j}Y_2(t)$, we define the cross-correlation function of the complex process $Z_1(t)$ and $Z_2(t)$ as:

$$R_{Z_1Z_2}(t_1, t_2) = E[Z_1^*(t_1)Z_2(t_2)] \tag{3.62}$$

where $t_2 = t_1 + \tau$, $Z_1^*(t_1)$ is the complex conjugate of the $Z_1(t_1)$.

Note: Define $R_{Z_1Z_2}(t_1, t_2) = E[Z_1(t_1)Z_2^*(t_2)]$, in which $t_2 = t_1 - \tau$in some books.

Example 3.3.1. A complex random process $V(t)$, with N complex signals, is as follows:

$$V(t) = \sum_{n=1}^{N} A_n \mathrm{e}^{\mathrm{j}(\omega_0 t + \varphi_n)} \tag{3.63}$$

where ω_0 is the constant; A_n and φ_n are the first n random amplitude and random phase of a signal, respectively. The random variable A_n is statistical independent of φ_n and φ_n and obeys uniform distribution in $(0, 2\pi)$. Find the autocorrelation function $V(t)$.

Solution: Defined by the complex stochastic process autocorrelation function, the autocorrelation function of $V(t)$ is:

$$\begin{aligned} R_V(t, t+\tau) = E[V^*(t)V(t+\tau)] &= E\left[\sum_{n=1}^{N} A_n \mathrm{e}^{-\mathrm{j}\omega_0 t - \mathrm{j}\varphi_n} \sum_{m=1}^{N} A_m \mathrm{e}^{\mathrm{j}\omega_0 t + \mathrm{j}\omega_0 \tau + \varphi_m}\right] \\ &= \sum_{n=1}^{N}\sum_{m=1}^{N} \mathrm{e}^{\mathrm{j}\omega_0\tau} E[A_n A_m \mathrm{e}^{\mathrm{j}(\varphi_m - \varphi_n)}] = R_V(\tau) \end{aligned} \tag{3.64}$$

Because A_n and φ_n are statistically independent:

$$R_V(\tau) = \mathrm{e}^{\mathrm{j}\omega_0\tau} \sum_{n=1}^{N} \sum_{m=1}^{N} E[A_n A_m] E[\mathrm{e}^{\mathrm{j}(\varphi_m - \varphi_n)}] \tag{3.65}$$

While

$$\begin{aligned} E[\mathrm{e}^{\mathrm{j}(\varphi_m-\varphi_n)}] &= E[\cos(\varphi_m - \varphi_n)] + \mathrm{j}E[\sin(\varphi_m - \varphi_n)] \\ &= \int_0^{2\pi}\int_0^{2\pi} \frac{1}{(2\pi)^2}[\cos(\varphi_m - \varphi_n) + \mathrm{j}\sin(\varphi_m - \varphi_n)]\,\mathrm{d}\varphi_m\,\mathrm{d}\varphi_n \\ &= \begin{cases} 0\,, & m \neq n \\ 1\,, & m = n \end{cases} \end{aligned} \tag{3.66}$$

Then

$$R_V(\tau) = \mathrm{e}^{\mathrm{j}\omega_0\tau} \sum_{n=1}^{N} \overline{A_n^2} \tag{3.67}$$

3.3.3 Correlation function and power spectral density of complex stochastic processes

The real random process $X(t)$ and the Hilbert transform $\hat{X}(t)$ are jointly stationary; we have the relationship:

$$R_{\hat{X}X}(\tau) = R_{X\hat{X}}(-\tau) \tag{3.68}$$

Now, we need to find the correlation function $R_{\tilde{X}}(\tau)$ and power spectral density $G_{\tilde{X}}(\omega)$ of the complex stochastic process $\tilde{X}(t) = X(t) + \mathrm{j}\hat{X}(t)$.

According to the definition of the autocorrelation function of the complex random process:

$$\begin{aligned} R_{\tilde{X}}(\tau) &= E[\tilde{X}^*(t)\tilde{X}(t+\tau)] \\ &= E\{[X(t) - \mathrm{j}\hat{X}(t)][X(t+\tau) + \mathrm{j}\hat{X}(t+\tau)]\} \\ &= R_X(\tau) + R_{\hat{X}}(\tau) + \mathrm{j}[R_{X\hat{X}}(\tau) - R_{\hat{X}X}(\tau)] \end{aligned} \tag{3.69}$$

Because the Hilbert transform is a linear transformation, the processes $X(t)$ and $\hat{X}(t)$ are jointly stationary. From previous knowledge, when finding the autocorrelation function $R_{\hat{X}}(\tau)$ and the cross-correlation functions $R_{X\hat{X}}(\tau)$ and $R_{\hat{X}X}(\tau)$, the following transformation relationship can be used.

This was found before for the case when the shock response of linear system is $h(\tau)$, the output is $Y(t)$ and we have:

$$R_{XY}(\tau) = R_X(\tau) \otimes h(\tau) \tag{3.70}$$

$$G_Y(\omega) = G_X(\omega)\,|H(\omega)|^2 \tag{3.71}$$

When making the Hilbert transform, we have $h(\tau) = 1/\pi\tau$, $|H(\omega)|^2 = 1$. Therefore,

$$R_{X\hat{X}}(\tau) = R_X(\tau) \otimes h(\tau) = \frac{1}{\pi} \int_{-\infty}^{\infty} \frac{R_X(a)}{\tau - a} \mathrm{d}a \tag{3.72}$$

$$G_{\hat{X}}(\omega) = G_X(\omega) \tag{3.73}$$

The Fourier transform of these two formulas is obtained as:

$$G_{X\hat{X}}(\omega) = -\mathrm{j}G_X(\omega) \cdot \mathrm{sgn}(\omega) \tag{3.74}$$

$$R_{\hat{X}}(\tau) = R_X(\tau) \tag{3.75}$$

Making $\tau = -\tau'$, eq. (3.72), namely

$$R_{X\hat{X}}\left(-\tau'\right) = -\frac{1}{\pi} \int_{-\infty}^{\infty} \frac{R_X(a)}{\tau' + a} \mathrm{d}a \tag{3.76}$$

Making a variable substitution $a = -a'$, then

$$R_{X\hat{X}}\left(-\tau'\right) = -\frac{1}{\pi} \int_{-\infty}^{\infty} \frac{R_X\left(-a'\right)}{\tau' - a'} \mathrm{d}a' = -\frac{1}{\pi} \int_{-\infty}^{\infty} \frac{R_X\left(a'\right)}{\tau' - a'} \mathrm{d}a' = -R_{X\hat{X}}\left(\tau'\right) \tag{3.77}$$

The above can be rewritten as

$$R_{X\hat{X}}(-\tau) = -R_{X\hat{X}}(\tau) \tag{3.78}$$

This formula shows the cross-correlation function $R_{X\hat{X}}(\tau)$ is an odd function of τ.

From this formula and eq. (3.78), the relationship between the two cross-correlation functions is:

$$R_{\hat{X}X}(\tau) = -R_{X\hat{X}}(\tau) \tag{3.79}$$

When $\tau = 0$, we have:

$$R_{\hat{X}X}(0) = -R_{X\hat{X}}(0) = 0 \tag{3.80}$$

This formula shows that the random variables $X(t)$ and $\hat{X}(t)$ are uncorrelated at the same time. If $X(t)$ is a normal process, then $\hat{X}(t)$ is also a normal process. Variables $X(t)$ and $\hat{X}(t)$ are statistically independent now.

Substituting eqs. (3.75) and (3.79) into eq. (3.69), we have:

$$R_{\tilde{X}}(\tau) = 2\left[R_X(\tau) + \mathrm{j}R_{X\hat{X}}(\tau)\right] \tag{3.81}$$

Taking the Fourier transform of eq. (3.81) and using eq. (3.74), we obtain:

$$\begin{aligned} G_{\tilde{X}}(\omega) &= 2\left[G_X(\omega) + \mathrm{j}G_{X\hat{X}}(\omega)\right] \\ &= 2G_X(\omega)\left[1 + \mathrm{sgn}(\omega)\right] \\ &= 4G_X(\omega)U(\omega) \end{aligned} \tag{3.82}$$

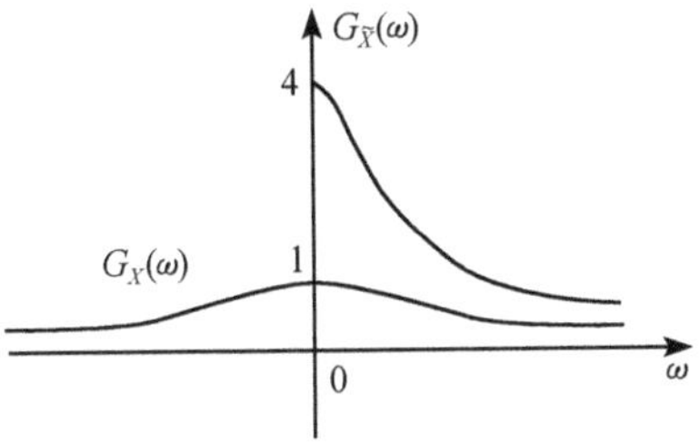

Fig. 3.8: Relationship between $G(x)$ and $G(w)$.

That is,

$$G_{\tilde{X}}(\omega) = \begin{cases} 4G_X(\omega)\,, & \omega \geq 0 \\ 0\,, & \omega < 0 \end{cases} \tag{3.83}$$

The relationship of $G_{\tilde{X}}(\omega)$ and $G_X(\omega)$ is shown in Fig. 3.8.

Example 3.3.2. $R_{X\hat{X}}(\tau) = \hat{R}_X(\tau)$

Proof. The Hilbert transform of $X(t)$ is

$$\hat{X}(t) = \frac{1}{\pi}\int_{-\infty}^{\infty} \frac{X(a)}{\tau - a}\,\mathrm{d}a \tag{3.84}$$

By definition, we have:

$$R_{X\hat{X}}(-\tau) = E[X(t)\hat{X}(t+\tau)] = E\left[X(t)\frac{1}{\pi}\int_{-\infty}^{\infty}\frac{X(a)}{t+\tau-a}\,\mathrm{d}a\right] = \frac{1}{\pi}\int_{-\infty}^{\infty}\frac{E[X(t)X(a)]}{t+\tau-a}\,\mathrm{d}a \tag{3.85}$$

Doing variable substitution: $\beta = a - t$:

$$R_{X\hat{X}}(\tau) = \frac{1}{\pi}\int_{-\infty}^{\infty}\frac{E\,[X(t)X(t+\beta)]}{\tau-\beta}\,\mathrm{d}\beta = \frac{1}{\pi}\int_{-\infty}^{\infty}\frac{R_X(\beta)}{\tau-\beta}\,\mathrm{d}\beta = \hat{R}_X(\tau) \tag{3.86}$$

□

3.3.4 Complex envelope and statistical properties of narrowband random processes

At the beginning, we obtained the representation of the narrowband reality process, namely:

$$Y(t) = A(t)\cos\left[\omega_0 t + \varphi(t)\right] \tag{3.87}$$

The complex process is

$$\tilde{Y}(t) = Y(t) + \mathrm{j}\hat{Y}(t) \tag{3.88}$$

Writing $\hat{Y}(t)$ as the Hilbert transform of the $Y(t)$, because $A(t)$ is a slow change function, we obtain:

$$\hat{Y}(t) = A(t)\sin\left[\omega_0 t + \varphi(t)\right] \tag{3.89}$$

Thus,

$$\tilde{Y}(t) = A(t)\mathrm{e}^{\mathrm{j}[\omega_0 t+\varphi(t)]} = \tilde{A}(t)\mathrm{e}^{\mathrm{j}\omega_0 t} \tag{3.90}$$

with $\tilde{A}(t) = A(t)\mathrm{e}^{\mathrm{j}\varphi(t)}$ as the complex envelope for the narrowband process.

Now, let us find the statistical properties of $\tilde{Y}(t)$ and the complex envelope $\tilde{A}(t)$.

3.3.4.1 Statistical characteristics of complex processes

According to the definition of the autocorrelation function of a complex process, the correlation function of $\tilde{Y}(t)$ is:

$$\begin{aligned} R_{\tilde{Y}}(\tau) &= E[\tilde{Y}^*(t)\tilde{Y}(t+\tau)] = E[\tilde{A}^*(t)\tilde{A}(t+\tau)]\mathrm{e}^{-\mathrm{j}\omega_0 t}\mathrm{e}^{\mathrm{j}\omega_0(t+\tau)} \\ &= R_{\tilde{A}}(\tau)\mathrm{e}^{\mathrm{j}\omega_0\tau} \end{aligned} \tag{3.91}$$

The statistical character of $\tilde{Y}(t)$ depends on the correlation function $R_{\tilde{A}}(\tau)$ of the complex envelope $\tilde{A}(t)$. $R_{\tilde{Y}}(\tau)$ is expressed by the cross-correlation function of the real part, and the imaginary part of the complex process $\tilde{Y}(t)$, that is,

$$\begin{aligned} R_{\tilde{Y}}(\tau) &= E[\tilde{Y}^*(t)\tilde{Y}(t+\tau)] = E\{[Y(t) - \mathrm{j}\hat{Y}(t)][Y(t+\tau) + \mathrm{j}\hat{Y}(t+\tau)]\} \\ &= R_Y(\tau) + R_{\hat{Y}}(\tau) + \mathrm{j}[R_{Y\hat{Y}}(\tau) - R_{\hat{Y}Y}(\tau)] \end{aligned} \tag{3.92}$$

Using the conclusion of the analytic complex stochastic process, we have:

$$R_{\hat{Y}}(\tau) = R_Y(\tau) \tag{3.93}$$

$$R_{\hat{Y}Y}(\tau) = -R_{Y\hat{Y}}(\tau) \tag{3.94}$$

Substituting the eqs. (3.93) and (3.94) into eq. (3.92), we have

$$R_{\tilde{Y}}(\tau) = 2[R_Y(\tau) + \mathrm{j}R_{Y\hat{Y}}(\tau)] \tag{3.95}$$

Taking the positive Fourier transform of the above eq. (3.95), the power spectral density is:

$$G_{\tilde{Y}}(\omega) = 2[G_Y(\omega) + \mathrm{j}G_{Y\hat{Y}}(\omega)] \tag{3.96}$$

Using the conclusion of the analytic complex stochastic process, we have:

$$G_{Y\hat{Y}}(\omega) = -\mathrm{j}G_Y(\omega)\cdot\mathrm{sgn}(\omega) \tag{3.97}$$

Substituting eq. (3.97) into the eq. (3.96), we have

$$\begin{aligned} G_{\tilde{Y}}(\omega) &= 2\left[G_Y(\omega) + G_Y(\omega)\cdot\mathrm{sgn}(\omega)\right] \\ &= 4G_Y(\omega)U(\omega) \end{aligned} \tag{3.98}$$

The relationship of $G_{\tilde{Y}}(\omega)$ with $G_Y(\omega)$ is shown in Fig. 3.9 (a) and (b).

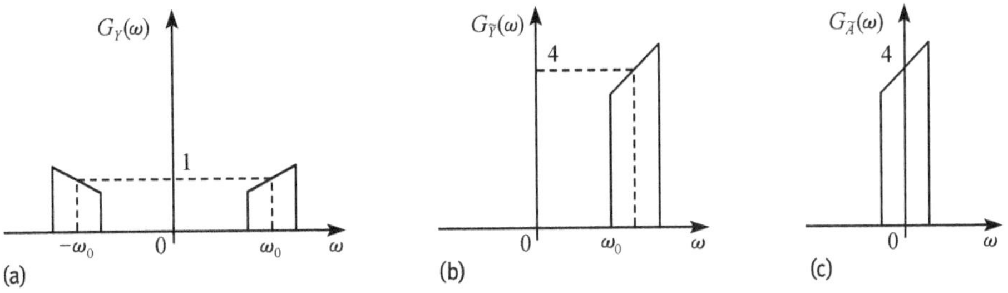

Fig. 3.9: Power spectrum relation of the narrowband real process and analytic signal.

Taking the Fourier transform of eq. (3.91), we obtain:

$$G_{\tilde{Y}}(\omega) = G_{\tilde{A}}(\omega - \omega_0) \tag{3.99}$$

This formula indicates that the $G_{\tilde{Y}}(\omega)$ can be solved by shifting the power spectrum $G_{\tilde{A}}(\omega)$ of the complex envelope $\tilde{A}(t)$ to the right by ω_0. This can be seen in Fig. 3.9 (b) and (c).

We now deduce the general expression of the correlation function of the stable narrowband process. Suppose that process $Y(t)$ has zero mean and power spectral density $G_Y(\omega)$. Because $G_Y(\omega)$ is an even function of w, the correlation function of $Y(t)$ is:

$$R_Y(\tau) = \frac{1}{\pi} \int_0^{\infty} G_Y(\omega) \cos \omega\tau \cdot d\omega \tag{3.100}$$

Doing variable substitution $\omega = \omega_0 + \Omega$ (i.e., the unilateral power spectrum $G_{\tilde{Y}}(\omega)$ is obtained by $G_Y(\omega)$, then shifted to the left by ω_0), we have:

$$R_Y(\tau) = \frac{1}{\pi} \int_{-\omega_0}^{\infty} G_Y(\omega_0 + \Omega) \cos(\omega_0 + \Omega)\tau \cdot d\Omega \tag{3.101}$$

Because $G_{\tilde{A}}(\Omega) = 4G_Y(\omega_0 + \Omega)$, the correlation function of the stationary narrowband process is generally expressed as:

$$\begin{aligned} R_Y(\tau) &= \frac{1}{4\pi} \int_{-\omega_0}^{\infty} G_{\tilde{A}}(\Omega)\,[\cos \Omega\tau \cdot \cos \omega_0\tau - \sin \Omega\tau \cdot \sin \omega_0\tau] \cdot d\Omega \\ &= A(\tau) \cos \omega_0\tau - B(\tau) \sin \omega_0\tau \end{aligned} \tag{3.102}$$

Write:

$$A(\tau) = \frac{1}{4\pi} \int_{-\omega_0}^{\infty} G_{\tilde{A}}(\Omega) \cos \Omega\tau \cdot d\Omega \tag{3.103}$$

$$B(\tau) = \frac{1}{4\pi} \int_{-\omega_0}^{\infty} G_{\tilde{A}}(\Omega) \sin \Omega\tau \cdot d\Omega \tag{3.104}$$

or:

$$R_Y(\tau) = A_0(\tau) \cos [\omega_0 \tau + \nu(\tau)] \tag{3.105}$$

Write:

$$A_0(\tau) = \sqrt{A^2(\tau) + B^2(\tau)}\,, \quad \nu(\tau) = \arctan \frac{B(\tau)}{A(\tau)} \tag{3.106}$$

Because $Y(t)$ is a stationary process, $R_Y(\tau)$ is the even function of τ, by $R_Y(\tau) = A(\tau) \cos \omega_0 \tau - B(\tau) \sin \omega_0 \tau$, $A(\tau)$ must be the even function of τ, and $B(\tau)$ must be the odd function of τ.

If the narrowband process $Y(t)$ has a symmetrical power spectrum (relative to the central frequency $\pm\omega_0$), then $G_{\tilde{A}}(\Omega)$ is the even function of Ω and $B(\tau) = 0$ [i.e., we have $\nu(\tau) = 0$] and therefore, the correlation function is:

$$R_Y(\tau) = A(\tau) \cos \omega_0 \tau \tag{3.107}$$

Example 3.3.3. Set zero mean stable and narrowband noise $Y(t)$ with symmetric power spectral density. The correlation function is shown in the above equation. Try to find the correlation function $R_{\hat{Y}}(\tau)$, $R_{\tilde{Y}}(\tau)$ and variance $\sigma^2_{\hat{Y}}$ and $\sigma^2_{\tilde{Y}}$.

Solution:

$$R_{\hat{Y}}(\tau) = R_Y(\tau) = A(\tau) \cos \omega_0 \tau \tag{3.108}$$

Choosing $\tau = 0$ in the above, the variance is:

$$\sigma^2_{\hat{Y}} = \sigma^2_Y = A(0) \tag{3.109}$$

According to Example 3.3.2, we have

$$R_{Y\hat{Y}}(\tau) = \hat{R}_Y(\tau) = A(\tau) \sin \omega_0 \tau \tag{3.110}$$

Substituting formulas $R_Y(\tau)$ and $R_{Y\hat{Y}}(\tau)$ in eq. (3.95), the correlation function of complex stochastic process $\tilde{Y}(t)$ is:

$$R_{\tilde{Y}}(\tau) = 2A(\tau) e^{j\omega_0 \tau} \tag{3.111}$$

Therefore, the variance is:

$$\sigma^2_{\tilde{Y}} = 2A(0) \tag{3.112}$$

3.3.4.2 Statistical properties of complex envelopes

For the narrowband process $Y(t)$, its general expression can also use the orthogonal expression, namely:

$$Y(t) = A_c(t) \cos \omega_0 t - A_s(t) \sin \omega_0 t \tag{3.113}$$

The Hilbert transform for the above formula is:

$$\hat{Y}(t) = A_c(t) \sin \omega_0 t + A_s(t) \cos \omega_0 t \tag{3.114}$$

The simultaneous solution can be obtained as:

$$A_c(t) = Y(t)\cos\omega_0 t + \hat{Y}(t)\sin\omega_0 t \tag{3.115}$$

$$A_s(t) = -Y(t)\sin\omega_0 t + \hat{Y}(t)\cos\omega_0 t \tag{3.116}$$

If $Y(t)$ is a zero mean stationary normal process, it is known that the Hilbert transform is a linear transformation, and $\hat{Y}(t)$ is also the normal process of the zero mean stationary process. At this moment, $A_c(t)$ and $A_s(t)$ are a linear combination of these two normal processes and zero mean stationary normal processes.

By the above expression of $A_c(t)$, the correlation function of $A_c(t)$ is:

$$\begin{aligned} R_c(\tau) &= E[A_c(t)A_c(t+\tau)] \\ &= E\{[Y(t)\cos\omega_0 t + \hat{Y}(t)\sin\omega_0 t][Y(t+\tau)\cos\omega_0(t+\tau) + \hat{Y}(t+\tau)\sin\omega_0(t+\tau)]\} \\ &= R_Y(\tau)\cos\omega_0 t\cdot\cos\omega_0(t+\tau) + R_{\hat{Y}}(\tau)\sin\omega_0 t\cdot\sin\omega_0(t+\tau) \\ &\quad + R_{Y\hat{Y}}(\tau)\cos\omega_0 t\cdot\sin\omega_0(t+\tau) + R_{\hat{Y}Y}(\tau)\sin\omega_0 t\cdot\cos\omega_0(t+\tau) \end{aligned} \tag{3.117}$$

Substituting $R_{\hat{Y}}(\tau) = R_Y(\tau)$, $R_{\hat{Y}Y}(\tau) = -R_{Y\hat{Y}}(\tau)$ in eq. (3.117), we have:

$$R_c(\tau) = R_Y(\tau)\cos\omega_0 t + R_{Y\hat{Y}}(\tau)\sin\omega_0 t \tag{3.118}$$

With this method, the correlation function of $A_s(t)$ is:

$$R_s(\tau) = R_Y(\tau)\cos\omega_0 t + R_{Y\hat{Y}}(\tau)\sin\omega_0 t \tag{3.119}$$

Hence, the relationship is:

$$R_c(\tau) = R_s(\tau) \tag{3.120}$$

The expression shows that the correlation functions of $A_c(t)$ and $A_s(t)$ are equal, so their power spectral density is the same.

Choosing $\tau = 0$ in the correlation function of $A_c(t)$ and $A_s(t)$, we have

$$R_c(0) = R_s(0) = R_Y(0) \tag{3.121}$$

This formula shows that both average power of two low-frequency processes $A_c(t)$ and $A_s(t)$ are equal to the average power of the narrowband process $Y(t)$. Because of the zero mean, the variances of the three correlation functions are the same.

By the correlation functions of $A_c(t)$ and $A_s(t)$, it is known that any two horizontal components $A_c(t)$ and $A_c(t+\tau)$ or vertical components $A_s(t)$ and $A_s(t+\tau)$ with interval τ are all correlative random variables.

Copying the derivation above, the cross-correlation function of $A_c(t)$ and $A_s(t)$ is:

$$\begin{aligned} R_{cs}(\tau) &= E\left[A_c(t)A_s(t+\tau)\right] \\ &= -R_Y(\tau)\cos\omega_0 t\cdot\sin\omega_0(t+\tau) + R_{\hat{Y}}(\tau)\sin\omega_0 t\cdot\cos\omega_0(t+\tau) \\ &\quad + R_{Y\hat{Y}}(\tau)\cos\omega_0 t\cdot\cos\omega_0(t+\tau) - R_{\hat{Y}Y}(\tau)\sin\omega_0 t\cdot\sin\omega_0(t+\tau) \\ &= -R_Y(\tau)\sin\omega_0 t + R_{Y\hat{Y}}(\tau)\cos\omega_0 t \end{aligned} \tag{3.122}$$

The same method can be used for the cross-correlation function:

$$R_{sc}(\tau) = E\left[A_s(t)A_c(t+\tau)\right] = R_Y(\tau)\sin\omega_0 t - R_{Y\hat{Y}}(\tau)\cos\omega_0 t \tag{3.123}$$

The relationship is as follow:

$$R_{cs}(\tau) = -R_{sc}(\tau) \tag{3.124}$$

This expression indicates that the cross-correlation function of $A_c(t)$ and $A_s(t)$ is an odd function of τ. When $\tau = 0$, we have

$$R_{cs}(0) = -R_{sc}(0) = 0 \tag{3.125}$$

We can see that the two components $A_c(t)$ and $A_s(t)$ are orthogonal at the same time, and that they are uncorrelated random variables.

Example 3.3.4. Set zero mean stationary and narrowband noise $Y(t)$, with the symmetrical power spectrum, its correlation function is $R_Y(\tau) = A(\tau)\cos\omega_0\tau$. Try to find the correlation functions $R_c(\tau)$ and $R_s(\tau)$,variances σ_c^2 and σ_s^2, and autocorrelation functions $R_{cs}(\tau)$ and $R_{sc}(\tau)$.

Solution: From the above example, we have:

$$R_{\hat{Y}}(\tau) = R_Y(\tau) = A(\tau)\cos\omega_0\tau \tag{3.126}$$

$$R_{Y\hat{Y}}(\tau) = \hat{R}_Y(\tau) = A(\tau)\sin\omega_0\tau \tag{3.127}$$

Substituting $R_c(\tau)$, $R_s(\tau)$ in it:

$$R_c(\tau) = R_s(\tau) = A(\tau) \tag{3.128}$$

Substituting $R_{cs}(\tau)$, $R_{sc}(\tau)$ in it:

$$R_{cs}(\tau) = -R_{sc}(\tau) = 0 \tag{3.129}$$

This formula shows that any two low-frequency components $A_c(t)$ and $A_s(t+\tau)$ or $A_s(t)$ and $A_c(t+\tau)$ with the spacing τ are all orthogonal and they are all uncorrelated random variables.

Because the mean is zero, choose $\tau = 0$ in $R_c(\tau) = R_s(\tau) = A(\tau)$. The variance is:

$$\sigma_c^2 = \sigma_s^2 = A(0) \tag{3.130}$$

3.4 Probability distribution of narrowband normal process envelopes and phase

The intermediate frequency amplifier of the receiver is usually a narrowband linear system, and its input noise is broadband noise. According to previous knowledge, the output noise of the system is the normal process. However, the narrowband normal process of the medium output needs to be demodulated by the envelope detector or phase detector, so it is necessary to study the probability distribution (envelope and phase) of the normal process. In this section, first, the situation when noise exists only but no signal is discussed, and then we study the situation when both the signal and superimposed noise exist. Last, the detection system of narrowband signals is introduced briefly.

3.4.1 Probability distribution of the narrowband normal noise envelope and phase

3.4.1.1 One-dimensional distribution

Narrowband noise $n(t)$ is the stationary normal process, the mean is zero and the variance is σ^2; the expression is as follows:

$$n(t) = A(t)\cos[\omega_0 t + \varphi(t)] = A_c(t)\cos\omega_0 t - A_s(t)\sin\omega_0 t \tag{3.131}$$

From the previous section we know that $A_c(t)$ and $A_s(t)$ are both stationary normal processes, the mean is zero and the variance is σ^2, $A_c(t)$ and $A_s(t)$,orthogonal components, are not relevant at the same time. For normal processes, this means that they are statistically independent. Ignoring the time variable t, random variables $A_c(t)$ and $A_s(t)$ can be shorted to A_c and A_s. Then the probability density is:

$$p(A_c) = \frac{1}{\sqrt{2\pi}\sigma}\exp\left[-\frac{A_c^2}{2\sigma^2}\right], \quad -\infty < A_c < \infty \tag{3.132}$$

$$p(A_s) = \frac{1}{\sqrt{2\pi}\sigma}\exp\left[-\frac{A_s^2}{2\sigma^2}\right], \quad -\infty < A_s < \infty \tag{3.133}$$

The joint probability density is:

$$p_2(A_c, A_s) = p(A_c)p(A_s) = \frac{1}{2\pi\sigma^2}\exp\left[-\frac{A_c^2 + A_s^2}{2\sigma^2}\right], \quad -\infty < (A_c, A_s) < \infty \tag{3.134}$$

Now, we transform the binary random variable (A_c, A_s) into a new binary random variable (A, φ). Among these A and φ represent the envelope and phase of the narrow band process $n(t)$, respectively, at the corresponding moment. Obviously, time variable t is hidden.

According to the previous results, the relationships between envelope with phase and the orthogonal components are as follows, respectively:

$$\begin{cases} A = \sqrt{A_c^2 + A_s^2}\,, & A \ge 0 \\ \varphi = \arctan\frac{A_s}{A_c}\,, & 0 \le \varphi \le 2\pi \end{cases} \tag{3.135}$$

The antifunction is

$$\begin{cases} A_c = A\cos\varphi\,, & -\infty < A_c < \infty \\ A_s = A\sin\varphi\,, & -\infty < A_s < \infty \end{cases} \tag{3.136}$$

According to the function transformation of the random variable, the joint probability density is:

$$p_2(A, \varphi) = |J|\, p_2(A_C, A_S) \tag{3.137}$$

Write the Jacobean determinant as:

$$J = \frac{\partial(A_C, A_S)}{\partial(A, \varphi)} = \begin{vmatrix} \frac{\partial A_c}{\partial A} & \frac{\partial A_c}{\partial \varphi} \\ \frac{\partial A_s}{\partial A} & \frac{\partial A_s}{\partial \varphi} \end{vmatrix} = \begin{vmatrix} \cos\varphi & -A\sin\varphi \\ \sin\varphi & A\cos\varphi \end{vmatrix} = A \tag{3.138}$$

Substituting eq. (3.138) into eq. (3.137):

$$p_2(A, \varphi) = \begin{cases} \frac{A}{2\pi\sigma^2} e^{-\frac{A^2}{2\sigma^2}}, & A \geq 0, 0 \leq \varphi \leq 2\pi \\ 0, & \text{other} \end{cases} \tag{3.139}$$

The one-dimensional distribution of the envelope A and phase φ is the result of the integral of random variables A and φ in the domain:

$$p(A) = \int_0^{2\pi} p_2(A, \varphi)\, d\varphi = \frac{A}{\sigma^2} e^{-\frac{A^2}{2\sigma^2}}, \quad A \geq 0 \tag{3.140}$$

$$p(\varphi) = \int_0^{\infty} p_2(A, \varphi)\, dA = \frac{1}{2\pi}, \quad 0 \leq \varphi \leq 2\pi \tag{3.141}$$

The two modes show that the envelope of the narrowband normal noise obeys the Rayleigh distribution, while the phase obeys uniform distribution. By the above three types, we have the following equation:

$$p_2(A, \varphi) = p(A)p(\varphi) \tag{3.142}$$

This formula shows that the random variables $A_c(t)$ and $\varphi(t)$ are statistically independent at the same time. However, the following analysis of bivariate distribution at two different moments shows that the random processes of the envelope and phase, $A_c(t)$ and $\varphi(t)$, are not statistically independent.

3.4.1.2 Two-dimensional distribution

The derivation step is the same as the above: find the four-dimensional probability density $p_4(A_C, A_S, A_{CT}, A_{ST})$ first. Ten transform it to the new four-dimensional probability density $p_4(A, A_T, \varphi, \varphi_T)$. Finally, work out the marginal probability densities $p_2(A, A_T)$ and $p_2(\varphi, \varphi_T)$.

(1) Finding $p_4(A_C, A_S, A_{CT}, A_{ST})$

The value of the normal process $n(t)$ is a random variable at the same time t, which can be divided into two statistically independent orthogonal components A_c and A_S. Similarly, for the same moment $t + \tau$, normal variable $n(t + \tau)$ can also be divided into two orthogonal components A_{CT} and A_{ST}. These four random variables, A_C, A_S, A_{CT}, A_{ST}, are the normal distribution, with zero mean values and the variances σ^2. They form the four-dimensional joint normal random variables A_C, A_S, A_{CT}, A_{ST}. Therefore, the probability density of the four-dimensional joint normal random variable is:

$$p_4(x_1, x_2, x_3, x_4) = \frac{1}{4\pi^2|C|^{\frac{1}{2}}} \exp\left[-\frac{1}{2|C|}\sum_{i=1}^{4}\sum_{j=1}^{4}|C_{ij}|(x_i - m_i)(x_j - m_j)\right] \tag{3.143}$$

If $m_i = m_j = 0$, $C(\tau) = R(\tau)$, according to the results of Example 3.3.4, there is a stationary and narrowband process with symmetrical power spectrum $R_c(\tau) = R_s(\tau) = A(\tau)$, $R_{cs}(\tau) = -R_{sc}(\tau) = 0$. The determinant of the covariance matrix C is:

$$|C| = |R| = \begin{vmatrix} R_c(0) & R_{cs}(0) & R_c(\tau) & R_{cs}(\tau) \\ R_{sc}(0) & R_s(0) & R_{sc}(\tau) & R_s(\tau) \\ R_c(-\tau) & R_c(-\tau) & R_c(0) & R_{cs}(0) \\ R_{sc}(-\tau) & R_s(-\tau) & R_{sc}(0) & R_s(0) \end{vmatrix} = \begin{vmatrix} \sigma^2 & 0 & A(\tau) & 0 \\ 0 & \sigma^2 & 0 & A(\tau) \\ A(\tau) & 0 & \sigma^2 & 0 \\ 0 & A(\tau) & 0 & \sigma^2 \end{vmatrix}$$
$$= [\sigma^4 - A^2(\tau)]^2 \tag{3.144}$$

The cofactors are as follows:

$$|C_{11}| = |C_{22}| = |C_{33}| = |C_{44}| = \sigma^2|R|^{\frac{1}{2}} \tag{3.145}$$

$$|C_{13}| = |C_{31}| = |C_{24}| = |C_{42}| = -A(\tau)|R|^{\frac{1}{2}} \tag{3.146}$$

All the rest are zero. Therefore, the four-dimensional joint probability density can be obtained as:

$$p_4\left(A_c, A_s, A_{CT}, A_{ST}\right)$$
$$= \frac{1}{4\pi^2\,|R|^{\frac{1}{2}}} \exp\left\{-\frac{1}{2\,|R|^{\frac{1}{2}}}\left[\sigma^2\left(A_c^2 + A_s^2 + A_{CT}^2 + A_{CT}^2\right) - 2A(\tau)\left(A_c A_{CT} + A_s A_{ST}\right)\right]\right\} \tag{3.147}$$

(2) Find $p_4(A, A_\tau, \varphi, \varphi_\tau)$
According to the function transformation of the random variable, the joint probability density is:

$$p_4(A, A_\tau, \varphi, \varphi_\tau) = |J| p_4(A_c, A_s, A_{CT}, A_{ST}) \tag{3.148}$$

We have the relationships:

$$\begin{cases} A_c = A\cos\varphi \\ A_s = A\sin\varphi \end{cases}, \quad \begin{cases} A_{CT} = A_\tau\cos\varphi_\tau \\ A_{ST} = A_\tau\sin\varphi_\tau \end{cases} \tag{3.149}$$

The Jacobean determinant is obtained following the preceding method:

$$J = \frac{\partial\left(A_c, A_s, A_{CT}, A_{ST}\right)}{\partial\left(A, A_\tau, \varphi, \varphi_\tau\right)} = AA_\tau \tag{3.150}$$

Substituting the above two formulas in the formula that calculates the joint probability density, we get:

$$p_4\left(A, A_\tau, \varphi, \varphi_\tau\right) = \begin{cases} \frac{AA_\tau}{4\pi^2|R|^{\frac{1}{2}}} \exp\left\{-\frac{1}{2|R|^{\frac{1}{2}}}\left[\sigma^2\left(A^2+A_\tau^2\right) - 2A(\tau)AA_\tau\cos(\varphi_\tau-\varphi)\right]\right\}, & A,\ A_\tau > 0 \\ & 0 \geq \varphi, \varphi_\tau \geq 2\pi \\ 0, & \text{otherwise} \end{cases} \tag{3.151}$$

(3) Calculate $p_2(A, A_\tau)$ and $p_2(\varphi, \varphi_\tau)$

Taking the integrals of eq. (3.151) of the variables φ and φ_τ,

$$\begin{aligned} p_2\left(A, A_\tau\right) &= \int_0^{2\pi}\int_0^{2\pi} p_4\left(A, A_\tau, \varphi, \varphi_\tau\right) \mathrm{d}\varphi\, \mathrm{d}\varphi_\tau \\ &= \left\{\frac{1}{4\pi^2}\int_0^{2\pi}\int_0^{2\pi} \exp\left[\frac{AA_\tau}{|R|^{\frac{1}{2}}} A(\tau)\cos\left(\varphi_\tau-\varphi\right)\right] \mathrm{d}\varphi\, \mathrm{d}\varphi_\tau\right\} \\ &\cdot \frac{AA_\tau}{|R|^{\frac{1}{2}}} \exp\left\{-\frac{\sigma^2\left(A^2+A_\tau^2\right)}{2\,|R|^{\frac{1}{2}}}\right\} \end{aligned} \tag{3.152}$$

The integral in the curly braces can be transformed as follows by the variable substitution $\varphi = \varphi_\tau - \varphi$:

$$\frac{1}{2\pi}\int_0^{2\pi} \mathrm{d}\varphi \cdot \frac{1}{2\pi}\int_0^{2\pi} \exp\left[\frac{AA_\tau}{|R|^{\frac{1}{2}}} A(\tau)\cos\varphi\right] \mathrm{d}\varphi = I_0\left[\frac{AA_\tau A(\tau)}{|R|^{\frac{1}{2}}}\right] \tag{3.153}$$

where $I_0(x) = 1/2\pi \int_0^{2\pi} \exp(x\cos\varphi)\,\mathrm{d}\varphi$ is called the first class of the zero-order modified Bessel function. Therefore, the joint probability density of A and A_τ is:

$$p_2\left(A, A_\tau\right) = \begin{cases} \frac{AA_\tau}{|R|^{\frac{1}{2}}} \cdot I_0\left[\frac{AA_\tau A(\tau)}{|R|^{\frac{1}{2}}}\right] \cdot \exp\left\{-\frac{\sigma^2\left(A^2+A_\tau^2\right)}{2|R|^{\frac{1}{2}}}\right\}, & A,\ A_\tau \geq 0 \\ 0, & \text{otherwise} \end{cases} \tag{3.154}$$

Because $A(\tau) = \sigma^2 r(\tau)$, among them $r(\tau)$ is the correlation coefficient. By formula $|C| = |R| = [\sigma^4 - A^2(\tau)]^2$, we can obtain:

$$|R|^{\frac{1}{2}} = \sigma^4\left[1 - r^2(\tau)\right] \tag{3.155}$$

Substitute eq. (3.155) into eq. (3.154), and then:

$$p_2\left(A, A_\tau\right) = \frac{AA_\tau}{\sigma^4\left[1-r^2(\tau)\right]} \cdot I_0\left\{\frac{AA_\tau A(\tau)}{\sigma^4\left[1-r^2(\tau)\right]}\right\} \cdot \exp\left\{-\frac{\left(A^2+A_\tau^2\right)}{2\sigma^2\left[1-r^2(\tau)\right]}\right\}, \quad A,\ A_\tau \geq 0 \tag{3.156}$$

Integrating the $p_4(A, A_\tau, \varphi, \varphi_\tau)$ with respect to A and A_τ, we have:

$$\begin{aligned} p_2(\varphi, \varphi_\tau) &= \int_0^\infty \int_0^\infty p_4(A, A_\tau, \varphi, \varphi_\tau)\, \mathrm{d}A\, \mathrm{d}A_\tau \\ &= \int_0^\infty \int_0^\infty \frac{AA_\tau}{4\pi^2 |R|^{\frac{1}{2}}} \exp\left\{-\frac{1}{2|R|^{\frac{1}{2}}}\left[\sigma^2\left(A^2 + A_\tau^2\right)\right.\right. \\ &\qquad \left.\left. -2A(\tau)AA_\tau \cos(\varphi_\tau - \varphi)\right]\right\} \mathrm{d}A\, \mathrm{d}A_\tau \end{aligned} \tag{3.157}$$

As proved in Appendix 2, when $0 \le \varphi \le 2\pi$, the integral formula is as follows:

$$\int_0^\infty \int_0^\infty Z_1 Z_2 \exp\left[-(Z_1^2 + Z_2^2 + 2Z_1 Z_2 \cos\varphi)\right] \mathrm{d}Z_1\, \mathrm{d}Z_2 = \frac{1}{4}\csc^2\varphi(1 - \varphi \operatorname{arccot}\varphi) \tag{3.158}$$

Using this formula to compute the integral for $p_2(\varphi, \varphi_\tau)$. the variables are substituted first:

$$Z_1 = \frac{\sigma A}{\sqrt{2}\,|R|^{\frac{1}{2}}}, \quad Z_2 = \frac{\sigma A_\tau}{\sqrt{2}\,|R|^{\frac{1}{2}}} \tag{3.159}$$

Write:

$$\cos\varphi = -\frac{A(\tau)}{\sigma^2}\cos(\varphi_\tau - \varphi) = -r(\tau)\cos(\varphi_\tau - \varphi) \tag{3.160}$$

Then

$$\begin{aligned} p_2(\varphi, \varphi_\tau) &= \frac{|R|^{\frac{1}{2}}}{\pi^2\sigma^4} \int_0^\infty \int_0^\infty Z_1 Z_2 \exp\left[-\left(Z_1^2 + Z_2^2 + 2Z_1 Z_2 \cos\varphi\right)\right] \mathrm{d}Z_1\, \mathrm{d}Z_2 \\ &= \frac{|R|^{\frac{1}{2}}}{4\pi^2\sigma^4}\csc^2\varphi\,(1 - \varphi \operatorname{arccot}\varphi) \end{aligned} \tag{3.161}$$

Because:

$$\csc^2\varphi = \frac{1}{1 - \cos^2\varphi}, \quad \cot\varphi = \frac{\cos\varphi}{(1 - \cos^2\varphi)^{\frac{1}{2}}} \tag{3.162}$$

Finally, the joint probability density of φ and φ_τ is:

$$p_2(\varphi, \varphi_\tau) = \begin{cases} \frac{|R|^{\frac{1}{2}}}{4\pi^2\sigma^4}\left[\frac{(1-\cos^2\varphi)^{\frac{1}{2}} - \varphi\cos\varphi}{(1-\cos^2\varphi)^{\frac{3}{2}}}\right], & 0 \le \varphi, \varphi_\tau \le 2\pi \\ 0, & \text{otherwise} \end{cases} \tag{3.163}$$

Write: $\varphi = \cos^{-1}[-r(\tau)\cos(\varphi_\tau - \varphi)]$

By the expression of $p_4(A, A_\tau, \varphi, \varphi_\tau)$, $p_2(A, A_\tau)$ and $p_2(\varphi, \varphi_\tau)$ it is not hard to see that:

$$p_4(A, A_\tau, \varphi, \varphi_\tau) \ne p_2(A, A_\tau)\, p_2(\varphi, \varphi_\tau) \tag{3.164}$$

This formula shows that the envelope and phase of the narrowband normal process are not statistically independent random processes.

3.4.2 Probability distribution of the envelope and phase of a narrowband normal noise plus sine (type) signal

In the same way, narrowband noise $n(t)$ is a stationary normal process, whose mean is zero and variance is σ^2. The expression is:

$$n(t) = A_n(t)\cos\left[\omega_0 t + \varphi_n(t)\right] = n_c(t)\cos\omega_0 t - n_s(t)\sin\omega_0 t \tag{3.165}$$

Supposing signal is a random initial trust number:

$$s(t) = A\cos(\omega_0 t + \theta) = A\cos\theta\cos\omega_0 t - A\sin\theta\sin\omega_0 t \tag{3.166}$$

Amplitude A and angular frequency ω_0 are known, and the phase θ obeys uniform distribution in $[0, 2\pi]$.

The synthesis process is:

$$Y(t) = s(t) + n(t) = [A\cos\theta + n_c(t)]\cos\omega_0 t - [A\sin\theta + n_s(t)]\sin\omega_0 t \tag{3.167}$$

Set:

$$\begin{cases} A_c(t) = A\cos\theta + n_c(t) \\ A_s(t) = A\sin\theta + n_s(t) \end{cases} \tag{3.168}$$

Then:

$$Y(t) = A_c(t)\cos\omega_0 t - A_s(t)\sin\omega_0 t = R(t)\cos\left[\omega_0 t + \varphi(t)\right] \tag{3.169}$$

where $R(t) = \sqrt{A_c^2(t) + A_s^2(t)}$ is the envelope of the synthesis process; $\varphi(t) = \arctan\frac{A_s(t)}{A_c(t)}$ is the phase of the synthesis process.

If there is no signal, then the components $n_c(t)$ and $n_s(t)$ are both normal random variables and statistically independent with zero mean and variance σ^2. Now, there is a signal, as long as θ is given, $A\cos\theta$ and $A\sin\theta$ are determined. From the expressions of $A_c(t)$ and $A_s(t)$, we know that the components $A_c(t)$ and $A_s(t)$ are still normal random variables with statistical independence, but the mean and variance are:

$$E[A_c(t)] = A\cos\theta \tag{3.170}$$

$$E[A_s(t)] = A\sin\theta \tag{3.171}$$

$$D[A_c(t)] = D[A_s(t)] = \sigma^2 \tag{3.172}$$

Now, canceling the time variable t, we have random variables $A_c(t)$, $A_s(t)$, $R(t)$ and $\varphi(t)$, respectively, shorthand for A_c, A_s, R, φ. Under the condition that signal phase θ is certain, the probability densities of A_c and A_s are:

$$p(A_c|\,\theta) = \frac{1}{\sqrt{2\pi}\sigma}\exp\left[-\frac{(A_c - A\cos\theta)^2}{2\sigma^2}\right] \tag{3.173}$$

$$p(A_s|\,\theta) = \frac{1}{\sqrt{2\pi}\sigma}\exp\left[-\frac{(A_s - A\sin\theta)^2}{2\sigma^2}\right] \tag{3.174}$$

The joint probability density is:

$$p_2(A_c, A_s|\,\theta) = \frac{1}{2\pi\sigma^2}\exp\left[-\frac{(A_c - A\cos\theta)^2 + (A_s - A\sin\theta)^2}{2\sigma^2}\right] \tag{3.175}$$

The process $Y(t)$ can be synthesized by the same transformation as the previously unsigned condition, and the joint probability density of the envelope and phase is:

$$p_2(R,\varphi|\theta) = \frac{R}{2\pi\sigma^2}\exp\left[-\frac{R^2 + A^2 - 2RA\cos(\varphi - \theta)}{2\sigma^2}\right], \quad R \geq 0,\ 0 \leq \varphi \leq 2\pi \tag{3.176}$$

The probability densities of the envelope and phase of the synthesis process are $p(R)$ and $p(\varphi)$.

3.4.2.1 Probability density $p(R)$ of the envelope

Integrating $p_2(R,\varphi|\theta)$ to φ, the conditional probability density of envelope R is:

$$\begin{aligned} p(R|\theta) &= \int_0^{2\pi} p_2(R,\varphi|\theta)\,\mathrm{d}\varphi \\ &= \frac{R}{\sigma^2}\exp\left(-\frac{R^2 + A^2}{2\sigma^2}\right)\cdot\frac{1}{2\pi}\int_0^{2\pi}\exp\left[-\frac{RA\cos(\varphi - \theta)}{\sigma^2}\right]\mathrm{d}\varphi \\ &= \frac{R}{\sigma^2}\exp\left(-\frac{R^2 + A^2}{2\sigma^2}\right)\cdot I_0\left(\frac{RA}{\sigma^2}\right), \quad R \geq 0 \end{aligned} \tag{3.177}$$

The right-hand side of eq. (3.177) has nothing to do with θ, so it can be rewritten as:

$$p(R) = \frac{R}{\sigma^2}\exp\left(-\frac{R^2 + A^2}{2\sigma^2}\right)\cdot I_0\left(\frac{RA}{\sigma^2}\right), \quad R \geq 0 \tag{3.178}$$

If there is no signal, then $A = 0$ and $I_0(0) = 1$ and the above equation becomes the Rayleigh distribution of $p(R) = \frac{R}{\sigma^2}\exp(-\frac{R^2}{2\sigma^2})$, $R \geq 0$. This is known as generalized Rayleigh distribution or Rice distribution.

Setting $v = \frac{R}{\sigma}$, $q = \frac{A}{\sigma}$, then the normalized form of the above equation is:

$$p(v) = ve^{-\frac{v^2+q^2}{2}} I_0(vq)\,, \quad v \geq 0 \tag{3.179}$$

As is shown in Fig. 3.10, if there is no signal, then $q = 0$ and $p(v)$ is the Rayleigh distribution at this time. With an increase of the signal-to-noise ratio q, the peak point of the curve $p(v)$ moves to the right, and the curve tends to be a normal distribution. This feature can be used to make amplitude detection by comparing the output value v with the fixed threshold v_0. Then the signal can be found from the noise background.

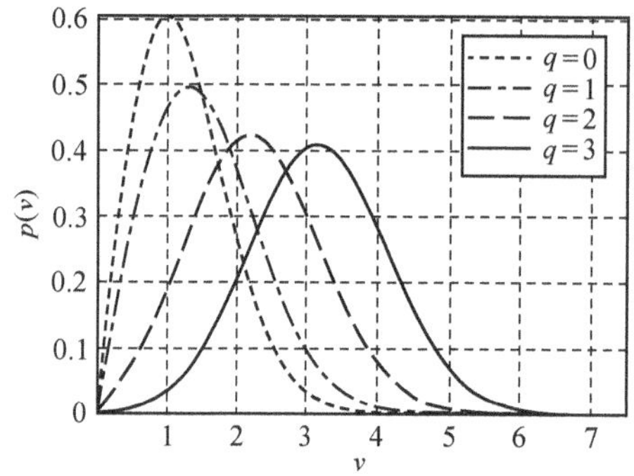

Fig. 3.10: Probability density of generalized Rayleigh distribution.

The asymptotic properties of generalized Rayleigh distribution are discussed below.

The modified Bessel function $I_0(x)$ can be expanded into the sum of the infinite series:

$$I_0(x) = \sum_{k=0}^{\infty} \frac{x^{2k}}{2^{2k}(k!)^2} \tag{3.180}$$

When $x \ll 1$, we have:

$$I_0(x) = 1 + \frac{x^2}{2^2} + \frac{x^4}{2^4 2^2} + \cdots \approx e^{\frac{x^2}{4}} \tag{3.181}$$

When $x \gg 1$, we have:

$$I_0(x) = \frac{e^x}{\sqrt{2\pi x}}\left\{1 + \frac{1}{1!(8x)} + \frac{9}{2!(8x)^2} + \cdots\right\} \approx \frac{e^x}{\sqrt{2\pi x}} \tag{3.182}$$

Therefore, the generalized Rayleigh distribution has the following asymptotic properties.

(1) With low signal-to-noise ratio ($vq \ll 1$), there is $I_0(vq) \approx 1$, which can be obtained by eq. (3.179):

$$p(v) \approx v e^{-\frac{v^2}{2}}, \quad v \geq 0 \tag{3.183}$$

This shows that the generalized Rayleigh distribution tends to the Rayleigh distribution.

(2) With large signal-to-noise ratio ($vq \gg 1$) $R \approx A, v \approx q$, using eqs. (3.182) and (3.179):

$$p(v) \approx v e^{-\frac{v^2+q^2}{2}} \frac{e^{vq}}{\sqrt{2\pi vq}} \approx \frac{1}{\sqrt{2\pi}} e^{-\frac{(v-q)^2}{2}} \tag{3.184}$$

This indicates that the generalized Rayleigh distribution tends to be normal distribution.

3.4.2.2 Probability density of the phase $p(\varphi)$

The integral of $p_2(R,\varphi|\theta)$ to the R, the conditional probability density of the phase φ is:

$$p(\varphi|\theta) = \int_0^\infty p_2(R,\varphi|\theta)\,\mathrm{d}R = \exp\left(-\frac{A^2}{2\sigma^2}\right)\cdot\int_0^\infty \frac{R}{2\pi\sigma^2}\exp\left[-\frac{R^2-2RA\cos(\varphi-\theta)}{2\sigma^2}\right]\mathrm{d}R \tag{3.185}$$

After a simple operation, we have the following:

$$p(\varphi|\theta) = \exp\left(-\frac{A^2\sin^2(\varphi-\theta)}{2\sigma^2}\right)\cdot\int_0^\infty \frac{R}{2\pi\sigma^2}\exp\left\{-\frac{[R-A\cos(\varphi-\theta)]^2}{2\sigma^2}\right\}\mathrm{d}R \tag{3.186}$$

Making variable substitution $z = \frac{R-A\cos(\varphi-\theta)}{\sigma}$, the integral in the above formula becomes:

$$\int_{-\frac{A\cos(\varphi-\theta)}{\sigma}}^{\infty} \frac{z\sigma + A\cos(\varphi-\theta)}{2\pi\sigma}\mathrm{e}^{-\frac{z^2}{2}}\,\mathrm{d}z$$
$$= \frac{1}{2\pi}\int_{-\frac{A\cos(\varphi-\theta)}{\sigma}}^{\infty} z\mathrm{e}^{-\frac{z^2}{2}}\,\mathrm{d}z + \frac{A\cos(\varphi-\theta)}{2\pi\sigma}\int_{-\frac{A\cos(\varphi-\theta)}{\sigma}}^{\infty}\mathrm{e}^{-\frac{z^2}{2}}\,\mathrm{d}z$$
$$= \frac{1}{2\pi}\exp\left[-\frac{A^2\cos^2(\varphi-\theta)}{2\sigma^2}\right] + \frac{A\cos\varphi(\varphi-\theta)}{2\pi\sigma}\left[\frac{1}{2}+\frac{1}{\sqrt{2\pi}}\int_0^{\frac{A\cos(\varphi-\theta)}{\sigma}}\mathrm{e}^{-\frac{z^2}{2}}\,\mathrm{d}z\right]$$

Substituting $p(\varphi|\theta)$ into the above formula, we have:

$$p(\varphi|\theta) = \frac{1}{2\pi}\exp\left(-\frac{A^2}{2\sigma^2}\right) + \frac{A\cos(\varphi-\theta)}{\sqrt{2\pi}\sigma}\exp\left\{-\frac{A^2\sin^2(\varphi-\theta)}{2\sigma^2}\right\}\left\{\frac{1}{2}+\Psi\left[\frac{A\cos(\varphi-\theta)}{\sigma}\right]\right\} \tag{3.187}$$

where $\Psi(x) = \frac{1}{\sqrt{2\pi}}\int_0^x \mathrm{e}^{-z^2/2}\,\mathrm{d}z$ is the Laplace function. The error function is:

$$\mathrm{erf}(x) = \frac{2}{\sqrt{\pi}}\int_0^x \mathrm{e}^{-z^2}\,\mathrm{d}z \tag{3.188}$$

There is relationship between $\Psi(x)$ and $\mathrm{erf}(x)$:

$$\Psi(x) = \frac{1}{2}\mathrm{erf}\left(\frac{x}{\sqrt{2}}\right) \tag{3.189}$$

The power signal-to-noise ratio is $Q^2 = \frac{A^2}{2\sigma^2} = q^2/2$, and the expression of the above $p(\varphi|\theta)$ can be rewritten as:

$$p(\varphi|\theta) = \frac{e^{-Q^2}}{2\pi} + \frac{Q\cos(\varphi-\theta)}{2\sqrt{\pi}} e^{-Q^2\sin^2(\varphi-\theta)}\{1 + \operatorname{erf}[Q\cos(\varphi-\theta)]\} \tag{3.190}$$

If there is no signal, then $Q = 0$ and the result from the above is:

$$p(\varphi|\theta) = \frac{1}{2\pi} \tag{3.191}$$

The error function that has the following asymptotic properties, when $x \gg 1$ is:

$$\operatorname{erf}(x) = 1 - \frac{e^{-x^2}}{\sqrt{\pi}x}\left[1 - \frac{1}{2x^2} + \frac{1\cdot 3}{2^2x^4} - \frac{1\cdot 3\cdot 5}{2^3x^6} + \cdots\right] \approx 1 - \frac{e^{-x^2}}{\sqrt{\pi}x} \tag{3.192}$$

So, the large signal-to-noise ratio ($Q \gg 1$), which can be obtained from the presentation of the type $p(\varphi|\theta)$ is:

$$p(\varphi|\theta) \approx \frac{Q\cos(\varphi-\theta)}{\sqrt{\pi}} \exp[-Q^2\sin^2(\varphi-\theta)] \tag{3.193}$$

This expression indicates that $p(\varphi|\theta)$ is the even function of $(\varphi - \theta)$; when $\varphi = \theta$, we have the peak $Q/\sqrt{\pi}$, when φ deviates from θ, $p(\varphi|\theta)$ reduces rapidly, as is shown in Fig. 3.11. This shows that the synthesis process $Y(t)$ is mainly focused on the near signal phase θ when the signal-to-noise ratio is large. When this feature is used for phase detection by comparing the output value φ with the fixed threshold φ_0, the signal can be found from the background noise.

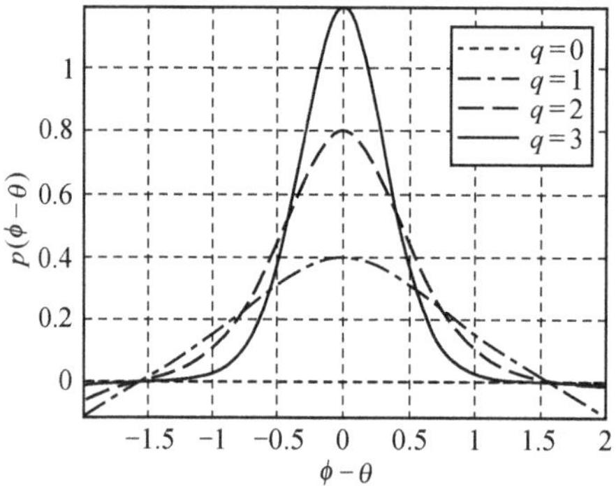

Fig. 3.11: Probability density of phase in the synthesis process.

3.5 Probability distribution of narrowband random process enveloping squares

Envelope detection is the most commonly used detection method, not only because it is easier to implement than phase detection, but also because incoherent signals, which initial phase random changes and where phase detection is impossible, can only use envelope detection. Envelope detection requires an envelope detector, and according to the detection characteristic, it can be divided into linear (law) geophones and square geophones. The output voltage (or current) $u(t)$ of a linear detector is proportional to the envelope $R(t)$ of the input signal. Practical application is complicated, but the theoretical analysis is even more complicated. The output voltage (or current) $u(t)$ of a square law detector is proportional to the square of the envelope of the input $R(t)$ of the input signal. Practical application of the square is less easy, but the theoretical analysis is easier.

Next, we discuss the situation when there is no signal and subsequently discuss the situation when the signal is superimposed. Finally, the presented χ^2 distribution and noncentered χ^2 distribution is discussed in the case of video signal making accumulation after square law detection.

3.5.1 Probability distribution of narrowband normal noise enveloping squares

The envelope of narrowband normal noise was obtained as follows:

$$A = \sqrt{A_c^2 + A_s^2} \tag{3.194}$$

The probability density obeys the Rayleigh distribution.

$$p(A) = \frac{A}{\sigma^2} \mathrm{e}^{-\frac{A^2}{2\sigma^2}}, \quad A \geq 0 \tag{3.195}$$

The envelope detector has the characteristic of half-wave square law detection, and the output voltage is $u(t) = A^2(t)$, for short:

$$u = A^2, \quad u, \quad A \geq 0 \tag{3.196}$$

as is shown in Fig. 3.12.

According to the transformation of the random variable function, the probability density of the envelope squared u is

$$p(u) = \left|\frac{\mathrm{d}A}{\mathrm{d}u}\right| p(A) = \frac{1}{2A} \cdot \frac{A}{\sigma^2} \mathrm{e}^{-\frac{A^2}{2\sigma^2}} = \frac{1}{2\sigma^2} \mathrm{e}^{-\frac{u}{2\sigma^2}}, \quad u \geq 0 \tag{3.197}$$

The upper expression indicates that the square of the narrowband normal noise envelope obeys the exponential distribution.

The normalized random variable $v = u/\sigma^2$, and the probability density is:

$$p(v) = \left|\frac{\mathrm{d}u}{\mathrm{d}v}\right| p(u) = \frac{1}{2} \mathrm{e}^{-\frac{v}{2}}, \quad v \geq 0 \tag{3.198}$$

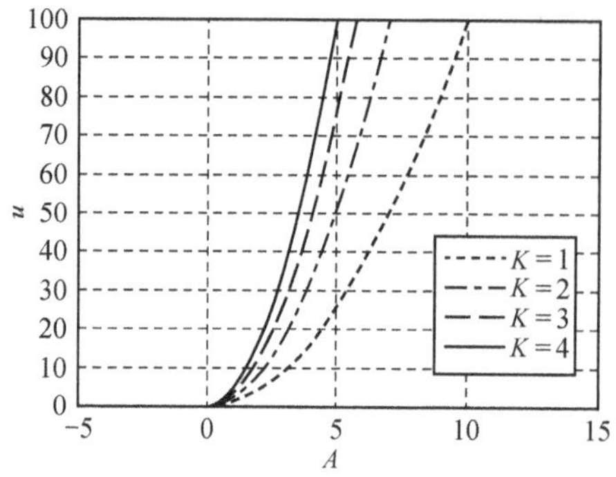

Fig. 3.12: Half-wave square law detection characteristics.

3.5.2 Probability distribution of synthesis process enveloping squares in narrowband normal noise plus sine (type) signals

According to the previous analysis, the synthesis process is:

$$Y(t) = s(t) + n(t) = R(t)\cos\left[\omega_0 t + \varphi(t)\right] \tag{3.199}$$

The probability density of the envelope $R(t)$ obeys the generalized Rayleigh distribution:

$$p(R) = \frac{R}{\sigma^2}\exp\left(-\frac{R^2 + A^2}{2\sigma^2}\right)\cdot I_0\left(\frac{RA}{\sigma^2}\right), \quad R \geq 0 \tag{3.200}$$

Still, the detection characteristic is the half-wave square law, namely:

$$u = R^2, \quad u, \quad R \geq 0 \tag{3.201}$$

In accordance with the previous transformation, the probability density of the envelope square u is

$$p(u) = \left|\frac{\mathrm{d}R}{\mathrm{d}u}\right| p(R) = \frac{1}{2\sigma^2}\exp\left(-\frac{u + A^2}{2\sigma^2}\right)\cdot I_0\left(\frac{\sqrt{u}A}{\sigma^2}\right), \quad u \geq 0 \tag{3.202}$$

Setting $v = u/\sigma^2$, the probability density is:

$$p(v) = \left|\frac{\mathrm{d}u}{\mathrm{d}v}\right| p(u) = \frac{1}{2}\exp\left(-\frac{v + \frac{A^2}{\sigma^2}}{2}\right)\cdot I_0\left(\frac{\sqrt{v}A}{\sigma}\right), \quad v \geq 0 \tag{3.203}$$

3.5.3 χ^2 Distribution and noncentered χ^2 distribution

In order to improve the detection performance of periodic signals in the noise background with the envelope detection method, so-called video signal accumulation is usually adopted; the composition of the detection system is shown in Fig. 3.13.

The diagram shows that the envelope process $R^2(t) = A_c^2(t) + A_s^2(t)$ is obtained by the envelope detection of the narrowband normal process $Y(t)$ through the square law

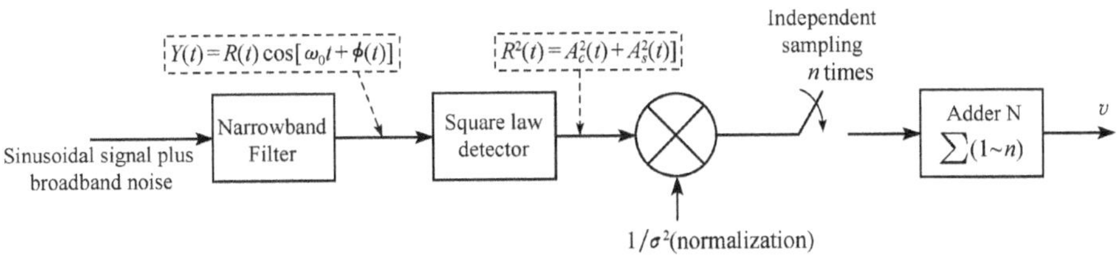

Fig. 3.13: A block diagram of the detection system for video signal accumulation.

detector. According to what was said previously, whether or not there is a signal, $A_c(t)$ and $A_s(t)$ are in normal distribution. These two low-frequency (video) process multiplied by $1/\sigma^2$, then normalization, and do n times independent sampling to obtain $2n$ square values of the statistical independent standard normal random variables (A_c and A_s), then output their sum with the adder to make a statistical decision.

The pulse radar is commonly used in this way, and the echo signal is a series of video narrow pulses after detection that take periodic (radar plus cycle) sampling. The sample obtained may keep noise only or keep noise mixed with signals. As long as the samples are compared with the threshold level, the presence of a signal can be determined. The following analysis shows that the output value v of this system obeys the χ^2 distribution or noncentered χ^2 distribution.

3.5.3.1 χ^2 Distribution

There are n statistical independent standard normal random variables, $X_1, X_2, \ldots,$ X_n, obeying $N(0, 1)$ distribution (zero mean, unit variance). The sum of the squares of these random variables is expressed as:

$$\chi^2 = \sum_{i=1}^{n} X_i^2 \tag{3.204}$$

They are called χ^2 variables with n degrees of freedom, and the probability distribution is called χ^2 distribution.

When deriving the expression of the χ^2 distribution, for the sake simplicity, we abbreviate the variable χ^2 as v, namely $v = \chi^2$.

X_i, as a standard normal variable, has the probability density:

$$p(x_i) = \frac{1}{\sqrt{2\pi}} e^{-\frac{x_i^2}{2}} \tag{3.205}$$

After the transformation of the full-wave squared law $y = x_i^2$ as is shown in Fig. 3.14, the probability density of the new variable y is (note the number 2, because $y = x_i^2$ is symmetric about the longitudinal axis):

$$p(y) = 2\left|\frac{dx_i}{dy}\right| p(x_i) = \frac{1}{\sqrt{2\pi y}} e^{-\frac{y}{2}}, \quad y \geq 0 \tag{3.206}$$

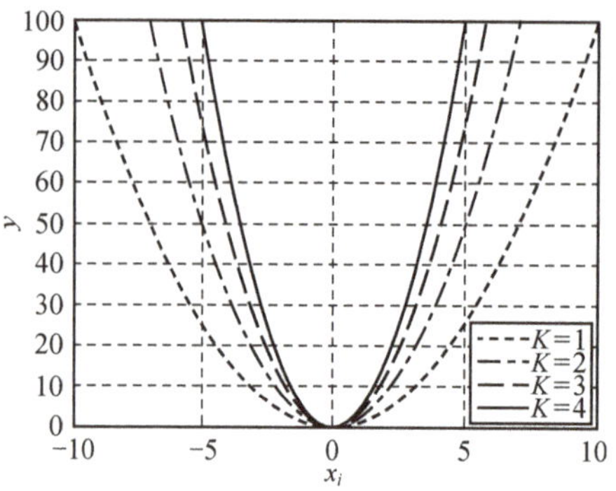

Fig. 3.14: Full-wave square law detection characteristics ($y = k^* x_i^2$).

The corresponding characteristic function is the Fourier transform of the probability density, namely:

$$\Phi_Y(\lambda) = \int_{-\infty}^{\infty} p(y) e^{j\lambda y}\, dy = \frac{1}{\sqrt{2\pi}} \int_0^{\infty} e^{-\frac{y}{2}} e^{-(\frac{1}{2} - j\lambda) y}\, dy = \frac{1}{(1 - j2\lambda)^{\frac{1}{2}}} \tag{3.207}$$

because the various X_i are statistically independent, and the same as X_i^2.

Properties of the characteristic function: the characteristic function of the sum of independent random variables is equal to the product of the characteristic function of the random variables. Namely, the characteristic function of $v = \chi^2$ is:

$$\Phi_v(\lambda) = \frac{1}{(1 - j2\lambda)^{\frac{n}{2}}} \tag{3.208}$$

Applying the inverse Fourier transform of the above equation, the expression of the χ^2 distribution is

$$p(v) = \frac{1}{2\pi} \int_{-\infty}^{\infty} \Phi_v(\lambda) e^{-j\lambda y}\, d\lambda = \frac{1}{2^{\frac{n}{2}} \Gamma\left(\frac{n}{2}\right)} v^{\frac{n}{2}-1} e^{-\frac{v}{2}}, \quad v \geq 0 \tag{3.209}$$

where $\Gamma(a) = \int_0^{\infty} t^{a-1} e^{-t}\, dt$.

The probability density curve of the χ^2 distribution is shown in Fig. 3.15. When the degrees of freedom $n = 2$,then:

$$p(v) = \frac{1}{2^{\frac{n}{2}} \Gamma\left(\frac{n}{2}\right)} v^{\frac{n}{2}-1} e^{-\frac{v}{2}}$$

becomes the exponential distribution, as is shown in Fig. 3.15 with the form $p(v) = \frac{1}{2} e^{-v/2}$.

If the mean of the independent normal random variables is zero, and the variance is σ^2, let $v_0 = \sigma^2 v$. Apply the transform to the expression of the probability density $p(v)$, which obeys χ^2 distribution. The probability density of the new variable v_0 is:

$$p(v_0) = \frac{1}{2^{\frac{n}{2}} \sigma^n \Gamma\left(\frac{n}{2}\right)} v_0^{\frac{n}{2}-1} e^{-\frac{v_0}{2\sigma^2}}, \quad v_0 \geq 0 \tag{3.210}$$

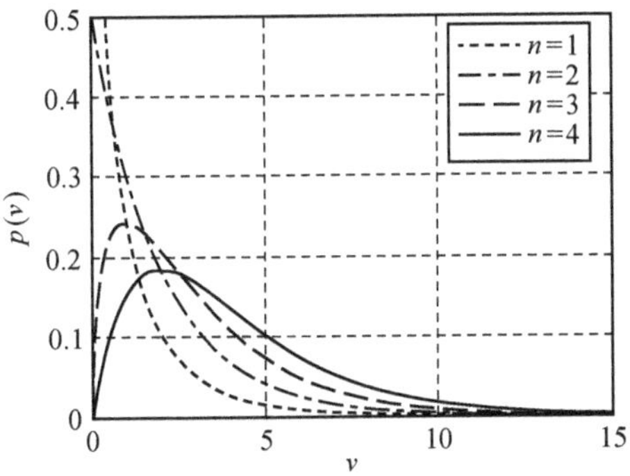

Fig. 3.15: Probability density curve of card square distribution.

χ^2 distribution has the following properties:

(1) The sum of the two statistically independent χ^2 variables is still χ^2 variable. If they have degrees of freedom n_1 and n_2, the degrees of freedom of the sum are $n = n_1 + n_2$.

(2) By the relationship between the characteristic function and the moment, we can obtain that the mean of the χ^2 distribution is $E[\chi^2] = n$, and the variance is $D[\chi^2] = 2n$.

3.5.3.2 Noncentered χ^2 distribution

There are n normal random variables, with $X_1, X_2, \ldots, X_n$, statistically independent. Their mean is zero and variance is σ^2; the sum of the normalized variables is:

$$V = \frac{1}{\sigma^2}\sum_{i=1}^{n}(X_i + A_i)^2 \tag{3.211}$$

It obeys noncentered distribution with n degrees of freedom; among them A_i is a non-random variable.

The physical model of the completed operation is still "the block diagram of the detection system for the implementation of video signal accumulation," but the narrow-band process in the input is the sum of the narrowband normal noise (the mean value is zero and the variance is σ^2) and the sinusoidal signals.

Derive the expression of the probability density which obeys noncentered χ_2 distribution. The random variable $X'_i = X_i + A_i$, the probability density of the variables X'_i is:

$$p\left(x'_i\right) = \frac{1}{\sqrt{2\pi}\sigma}\exp\left[-\frac{\left(x'_i - A_i\right)^2}{2\sigma^2}\right] \tag{3.212}$$

After the transformation of the full wave squared law $y_i = (x_i + A_i)^2$, the new variable y_i is available, and its probability density is:

$$p(y_i) = \frac{1}{2\left(2\pi\sigma^2 y_i\right)^{\frac{1}{2}}}\left\{\exp\left[-\frac{\left(\sqrt{y_i} - A_i\right)^2}{2\sigma^2}\right] + \exp\left[-\frac{\left(-\sqrt{y_i} - A_i\right)^2}{2\sigma^2}\right]\right\} \tag{3.213}$$

Expending the square of the exponent in eq. (3.213) and using the Euler formula: $e^z + e^{-z} = 2\text{ch}z$, we have

$$p(y_i) = \left(\frac{1}{2\pi\sigma^2 y_i}\right)^{\frac{1}{2}} \exp\left(-\frac{y_i + A_i^2}{2\sigma^2}\right) \text{ch}\left(A_i\sqrt{y_i}\right) \tag{3.214}$$

The corresponding characteristic function is:

$$\Phi_{Y_i}(\lambda) = \frac{1}{(1 - j2\sigma^2\lambda)^{\frac{1}{2}}} \exp\left(-\frac{A_i^2}{2\sigma^2}\right) \exp\left(\frac{\frac{A_i^2}{2\sigma^2}}{1 - j2\sigma^2\lambda}\right) \tag{3.215}$$

Set $Q = \sum_{i=1}^n Y_i^2$, the characteristic function of the variable Q as follows:

$$\Phi_Q(\lambda) = \prod_{i=1}^n \Phi_{Y_i}(\lambda) = \frac{1}{(1 - j2\sigma^2\lambda)^{\frac{n}{2}}} \exp\left(-\frac{\sum_{i=1}^n A_i^2}{2\sigma^2}\right) \exp\left(\frac{\sum_{i=1}^n \frac{A_i^2}{2\sigma^2}}{1 - j2\sigma^2\lambda}\right) \tag{3.216}$$

Applying the inverse Fourier transform for the above, the probability density of the variable Q can be obtained:

$$p(q) = \frac{1}{2\sigma^2}\left(\frac{q}{\lambda'}\right)^{\frac{n-2}{4}} \exp\left(\frac{\lambda' + q}{-2\sigma^2}\right) I_{\frac{n}{2}-1}\left(\frac{\sqrt{q\lambda'}}{\sigma^2}\right), \quad q \geq 0 \tag{3.217}$$

where $\lambda' = \sum_{i=1}^n A_i^2$ is called a noncentered parameter; $I_{n/2-1}(x)$ is the first category $n/2 - 1$-order modified Bessel function.

Because $v = Q/\sigma^2$ and transformation of eq. (3.217), the probability density expression of the noncentered χ_2 variable v is:

$$p(v) = \frac{1}{2}\left(\frac{v}{\lambda}\right)^{\frac{n-2}{4}} \exp\left(-\frac{\lambda + v}{2}\right) I_{\frac{n}{2}-1}\left(\sqrt{v\lambda}\right), \quad v \geq 0 \tag{3.218}$$

where $\lambda = 1/\sigma^2 \sum_{i=1}^n A_i^2$ is the accumulated power signal-to-noise ratio.

The probability density curve of the noncentral χ_2 distribution is shown in Fig. 3.16. The figure shows that if the input signal-to-noise ratio λ/n increases, the curve peak point moves to the right, which increases the probability of the variable v exceeding the fixed threshold and improves the detection performance. On the other hand, input signal-to-noise ratio λ/n is constant, and cumulative times n increases, the peak point is also moved to the right, indicating that the accumulation can improve the detection performance.

The noncentered χ^2 distribution has the following properties:

(1) The sum of the two statistically independent noncentral χ^2 variables is still a non central χ^2 variable. If it has degrees of freedom n_1 and n_2, the noncentral parameters are, respectively, λ_1 and λ_2 and then the sum of the degrees of freedom is $n = n_1 + n_2$. The noncentral parameter is $\lambda = \lambda_1 + \lambda_2$.

(2) The mean of the noncentral χ^2 variable is $E[v] = \lambda + n$, the variance is $D[v] = 4\lambda + 2n$.

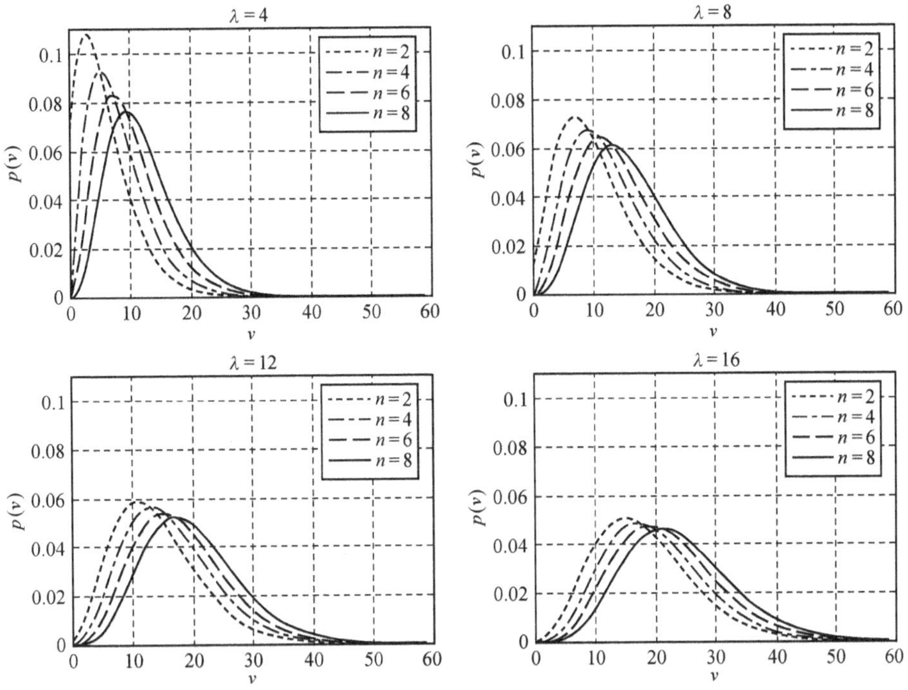

Fig. 3.16: Probability density curve of noncentral square distribution.

Example 3.5.1. Show the composition of the envelope detection system in Section 3.5.3 (Fig. 3.13). The input of the square law detector is the sum of the narrowband normal noise (zero mean and variance σ^2) and the sinusoidal signal, namely:

$$Y(t) = A\cos(\omega_0 t+\theta)+n(t) = [A\cos\theta+n_c(t)]\cos\omega_0 t-[A\sin\theta+n_s(t)]\sin\omega_0 t \quad (3.219)$$

After detection and normalization, apply m sampling independently. Which distribution will the added output value v obey? Also find the parameters.

Solution: Set the square law detector output voltage $u(t)$ as the square of envelope of the input process $R(t)$:

$$u(t) = R^2(t) = [A\cos\theta + n_c(t)]^2 + [A\sin\theta + n_s(t)]^2 \quad (3.220)$$

After m sampling, we have m statistically independent random variables, but each sampling value can be divided into two orthogonal components, so the output value of the adder is:

$$v = \frac{1}{\sigma^2}\sum_{i=1}^{m}(A\cos\theta + n_{ci})^2 + \frac{1}{\sigma^2}\sum_{i=1}^{m}(A\sin\theta + n_{si})^2 \quad (3.221)$$

From the definition of the noncentral χ^2 distribution, both sums in the above formula are noncentral χ^2 distribution, with m degrees freedom and the noncentral parameters are:

$$\lambda_1 = \frac{1}{\sigma^2}\sum_{i=1}^{m}(A\cos\theta)^2 = \frac{mA^2}{\sigma^2}\cos^2\theta \tag{3.222}$$

$$\lambda_2 = \frac{1}{\sigma^2}\sum_{i=1}^{m}(A\sin\theta)^2 = \frac{mA^2}{\sigma^2}\sin^2\theta \tag{3.223}$$

Because of the statistical independence of these two types, it is known that the sum is still a noncentral χ^2 distribution, but the degrees of freedom are $n = 2m$, the noncentral parameter is $\lambda = \lambda_1 + \lambda_2 = mA^2/\sigma^2$, and according to the formula of probability density expression of noncentral χ^2 distribution v, we obtain:

$$p(v) = \frac{1}{2}\left(\frac{v}{\lambda}\right)^{\frac{m-1}{2}}\exp\left(-\frac{\lambda+v}{2}\right)I_{m-1}\left(\sqrt{v\lambda}\right), \quad v \geq 0 \tag{3.224}$$

The ratio of the noncentral parameters to freedom is:

$$\frac{\lambda}{n} = \frac{A^2}{2\sigma^2} \tag{3.225}$$

This is the power signal-to-noise ratio on the input end of the detector.

The mean and variance of the variable v are:

$$E[v] = 2m\left(1 + \frac{A^2}{2\sigma^2}\right) \tag{3.226}$$

$$D[v] = 4m\left(1 + \frac{A^2}{\sigma^2}\right) \tag{3.227}$$

Exercises

3.1 Find the analytic signal $\tilde{X}(t)$ of the sine signal $X(t) = \sin\omega_0 t$.

3.2 Supposing $X(t) = \mathrm{Rect}(t/T)$, find the analytic signal $\tilde{X}(t)$ of $X(t)$.

3.3 Supposing $\tilde{X}(t) = X(t) + \mathrm{j}\hat{X}(t)$, among them $X(t)$ is the generalized stationary process. Define $R_X(\tau) = E\{X(t)X(t-\tau)\}$. Try to prove $R_{\hat{X}X}(\tau) = \hat{R}_X(\tau)$.

3.4 The stationary process with the symmetrical spectrum is assumed to be: $Y(t) = A_c\cos\omega_0 t - A_s\sin\omega_0 t$. Try to prove $E\{A_c(t)A_s(t+\tau)\} = 0$.

3.5 Suppose the stationary normal noise is: $n(t) = X(t)\cos\omega_0 t - Y(t)\sin\omega_0 t$. Try to prove that the autocorrelation function of $n(t)$ is: $R_n(\tau) = R_X(\tau)\cos\omega_0\tau - R_{XY}(\tau)\sin\omega_0\tau$.

3.6 Suppose the random initial signal is $X(t) = A\cos(\omega_0 t + \theta)$, and θ has uniform distribution in $(0, 2\pi)$ and A and ω_0 are both constants. The signal responds after the linear system with the impulse response $\omega_0/\pi t$. The output signal is $Y(t)$. Try to find the correlation functions $R_Y(\tau)$, $R_{XY}(\tau)$, $R_{YX}(\tau)$.

3.7 Supposing the random variable X obeys centralization χ^2 distribution; its probability density is:

$$p(x) = \begin{cases} \frac{1}{2^{\frac{n}{2}}\Gamma(\frac{n}{2})} e^{-\frac{x}{2}} x^{\frac{n}{2}-1}, & x > 0 \\ 0, & x \le 0 \end{cases} \quad (3.228)$$

where n is degree of freedom, $n = 1, 2, \ldots, k$.

(1) Try to prove that the characteristic function is of X is $\Phi_X(\lambda) = (1 - j2\lambda)^{-n/2}$.

(2) Try to prove that $E\{X\} = n$, $D\{X\} = 2n$.

3.8 The standard normal noise voltage of the narrowband is passed by the square law detector, and four independent samples are extracted and added together in the adder. Which distribution of the noise voltage u, outputted by the adder, is obeyed? Write the expression of the probability density $p(u)$ and find the mean $E\{u\}$ and variance $D\{u\}$.

3.9 Knowing that stochastic processes $Y(t) = s(t) + n(t)$, among them $s(t) = A_s \cos(\omega_0 t + \theta)$, while A_s, ω_0, θ are constants; $n(t) = A_n(t)\cos[\omega_0 t + \varphi(t)]$ is symmetrical spectral narrowband stationary noise with zero mean and variance σ_n^2. Try to find the joint probability density $p_2(A, \dot{A}; t_1)$ of the envelope $A(t_1)$ of the process $Y(t)$ and its derivative $\dot{A}(t_1)$ at fixed time t_1.

3.10 Suppose the white noise $X(t)$ of the autocorrelation function $N_0/2\delta(\tau)$. The frequency response characteristics of the two channels are, respectively, $H_1(\omega)$ and $H_2(\omega)$. Divide $X(t)$ into two paths through the symmetric spectrum narrowband system shown in Fig. 3.17.

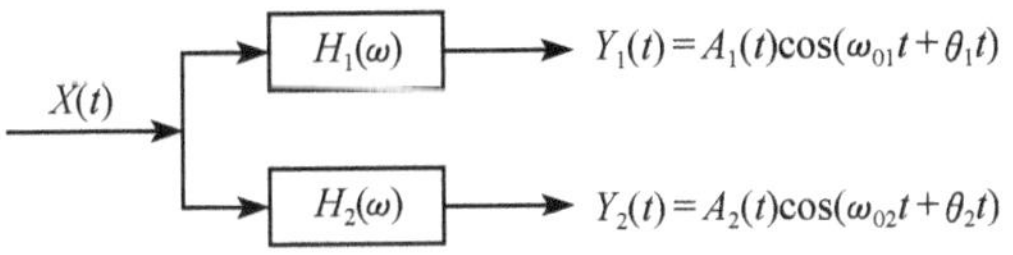

Fig. 3.17: Given narrowband system.

(1) In what conditions for $H_1(\omega)$ and $H_2(\omega)$ is cross-correlation function $R_{Y_1Y_1}(\tau)$ an even function?

(2) In what conditions for $H_1(\omega)$ and $H_2(\omega)$ are $Y_1(t)$ and $Y_1(t)$ statistically independent?

(3) When the central frequency meets $\omega_{01} = \omega_{02}$, find the cross-correlation functions $R_{\hat{Y}_1\hat{Y}_2}(\tau)$, $R_{Y_1\hat{Y}_2}(\tau)$, and $R_{\hat{Y}_1Y_2}(\tau)$.

3.11 On the machine: Design sine-type signals and Gaussian white noise signals by using MATLAB.

(1) Analyze the power spectral density and amplitude distribution of the sinusoidal signal, the Gaussian noise and the composite signal, respectively;

(2) Find the Hilbert transform of the three signals in (1) and compare the changes of the power spectrum and the amplitude distribution;

(3) Find the complex signals of the three signals in (1) and compare the changes of the power spectrum and the amplitude distribution;
(4) Analyze and observe the orthogonality between the three signals and their corresponding Hilbert transform signals – the three signals in (2).

3.12 Design and implement the accumulation of the video signal detection system shown in Fig. 3.13 by using MATLAB and analyze the frequency and time-domain characteristics of input and output signals from each module of this system, as well as the corresponding signal statistical characteristics.

4 The nonlinear transformation of stationary random processes

4.1 Nonlinear transformation overview

In general electronic devices, in addition to linear circuits, there are usually nonlinear circuits, such as detectors, limiters, modulators, frequency discriminators, logarithmic amplifiers and so on. The transmission characteristics of these nonlinear circuits (the relationship between the output value y and the input value x) have some kind of nonlinear function relationship:

$$y = f(x) \tag{4.1}$$

The actual transmission characteristics measured with the test method are generally not of enough regularization. In order to facilitate the theoretical analysis, it usually takes on an idealized approximation according to the required accuracy to make it become a simple function, such as broken lines, indexes, polynomials and so on. Figure 4.1 shows several common nonlinear functions (full-wave square law, half-wave linear law, one-way ideal limiting amplitude).

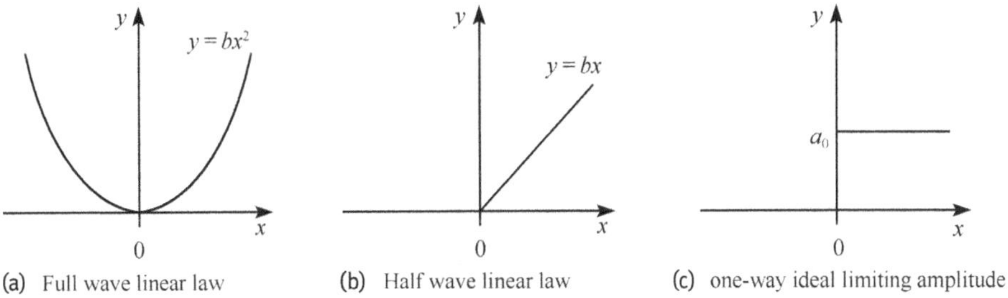

Fig. 4.1: Common nonlinear functions.

Compared with the linear circuit, the nonlinear circuit has the following characteristics:

(1) The superposition principle is no longer applicable. So, when signal and noise are through the nonlinear circuit together, we cannot separate their research like we do in linear circuits;
(2) Spectral transformation occurs. New spectrum components are generated, which do not appear in the input circuit, such as harmonics of the input signal. If there are multiple frequency components, the output will have their sum frequency components and difference frequency components.

It can be seen that the nonlinear transformation is more tedious and complicated.

https://doi.org/10.1515/9783110593808-005

Nonlinear circuits are classified into two types: inert nonlinear circuits and non-inert nonlinear circuits, depending on their presence of inertia (i.e., no memory or not). If the circuit contains an inert element (such as an inductor L or a capacitance C), it is an inert nonlinear circuit. If the circuit contains only resistive devices, it is a non-inert nonlinear circuit.

Actual nonlinear circuits of general electronic equipment, usually contain inert components and are inert nonlinear, and the front levels and the rear levels of the circuits are often linear circuits. If we strictly consider the impact of these inertial components in this stage of nonlinear circuits and the interaction among these stages, it is generally necessary to solve the nonlinear differential equations, and the analysis will be very tedious and complicated. In order to simplify the analysis, sometimes the influence of the inert components is merged into the linear circuit of the front or rear stages to make the current circuit a non-inert nonlinear circuit.

This chapter mainly analyzes non-inert nonlinear circuits. The random process is through the non-inert nonlinear circuit shown in Fig. 4.2. We assume that the nonlinear function of the circuit is $y = f(x)$, it does not change with time and is the time-invariance nonlinear system. If the input process is $X(t)$, the output process is:

$$Y(t) = f[X(t)] \tag{4.2}$$

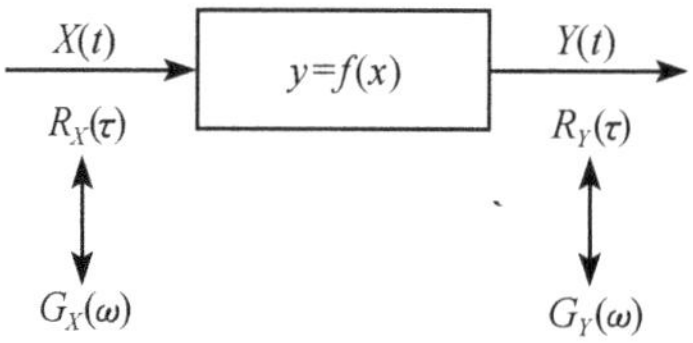

Fig. 4.2: The random process passes a non-inert nonlinear circuit.

Then the input process is $X(t+\varepsilon)$, and the output process is:

$$Y(t+\varepsilon) = f\,[X(t+\varepsilon)] \tag{4.3}$$

The so-called lack of inertias of nonlinear system (i.e., no memory), that is, when the moment $t = t_1$, the output value $Y(t_1)$ is only related to the input value $X(t_1)$ at the same moment t_1 but is irrelevant to the input values before the t_1 moment.

The research topics of nonlinear transformation of random processes are:

(1) transformation of the probability distribution,

(2) transformation of moment function and power spectral density.

The first problem is actually a function transformation of the random variable. As long as the probability density $p(x,t)$ and $p_2(x_1, x_2; t_1, t_2)$ of the input process $X(t)$ is known, the one-dimensional probability density of the output process $Y(t)$ is:

$$p(y,t) = |J|\,p(x,t) \tag{4.4}$$

where, $J = \mathrm{d}x/\mathrm{d}y$. If $y = f(x)$ is a nonmonotonic function, then:

$$p(y,t) = |J_1|\,p(x_1,t) + |J_2|\,p(x_2,t) + \cdots \tag{4.5}$$

where $J_1 = \mathrm{d}x_1/\mathrm{d}y$, $J_2 = \mathrm{d}x_2/\mathrm{d}y, \ldots$. The two-dimensional probability density of process $Y(t)$ is:

$$p_2(y_1,y_2;t_1,t_2) = |J|\,p_2(x_1,x_2;t_1,t_2) \tag{4.6}$$

Because $Y(t_1) = f[X(t_1)]$, $Y(t_2) = f[X(t_2)]$ are described as $y_1 = f(x_1)$, $y_2 = f(x_2)$ and the Jacobian determinant in the eq. (4.6) is:

$$J = \frac{\partial(x_1,x_2)}{\partial(y_1,y_2)} = \begin{vmatrix} \frac{\partial x_1}{\partial y_1} & \frac{\partial x_2}{\partial y_1} \\ \frac{\partial x_1}{\partial y_2} & \frac{\partial x_2}{\partial y_2} \end{vmatrix} \tag{4.7}$$

An example of a nonlinear transformation of probability density has been given, and an example will be provided later for analysis. This chapter focuses on the nonlinear transformation of moment functions and power spectral densities.

Knowing a nonlinear function $y = f(x)$ and the probability density $p(x,t)$ and $p_2(x_1,x_2;t_1,t_2)$ of the input process $X(t)$, we can obtain the mean of the output process $Y(t)$:

$$m_Y(t) = E[Y(t)] = \int_{-\infty}^{\infty} y p(y,t)\,\mathrm{d}y = \int_{-\infty}^{\infty} f(x)p(x,t)\,\mathrm{d}x \tag{4.8}$$

Similarly, the n-order moment function of process $Y(t)$ can be obtained as:

$$E\{[Y(t)]^n\} = \int_{-\infty}^{\infty} [f(x)]^n p(x,t)\,\mathrm{d}x \tag{4.9}$$

The correlation function of the process $Y(t)$ is:

$$R_Y(t_1,t_2) = E[Y(t_1)Y(t_2)] = \int_{-\infty}^{\infty}\int_{-\infty}^{\infty} f(x_1)f(x_2)p_2(x_1,x_2;t_1,t_2)\,\mathrm{d}x_1\,\mathrm{d}x_2 \tag{4.10}$$

If the input $X(t)$ is a stationary process, $p(x,t) = p(x)$ and $p_2(x_1,x_2;t_1,t_2) = p_2(x_1,x_2;\tau)$ $(\tau = t_2 - t_1)$. At this time, the mean and correlation functions of the process $Y(t)$ become:

$$m_Y = \int_{-\infty}^{\infty} f(x)p(x)\,\mathrm{d}x \tag{4.11}$$

$$R_Y(\tau) = \int_{-\infty}^{\infty}\int_{-\infty}^{\infty} f(x_1)f(x_2)p_2(x_1,x_2;\tau)\,\mathrm{d}x_1\,\mathrm{d}x_2 \tag{4.12}$$

It can be seen that at this time the output process, $Y(t)$ is still a stationary process. Having applied the Fourier transform to the formula above, we can obtain the power spectral density $G_Y(\omega)$ of the output process $Y(t)$. In the following, we mainly analyze the nonlinear transformation of stationary stochastic processes and focus on the two statistical parameters of the correlation function and the power spectral density.

Equation (4.12) is the basic relationship of the correlation function of the stationary process through the non-inert nonlinear system. According to the different methods used when solving the points, generally it is divided into two commonly used methods – the direct method and the transformation method.

The method of directly using this formula for double integral calculation is called as the direct method. If the nonlinear function relation, and the probability density expression are complex, then direct double integration will be difficult. So, this method is generally applicable only to cases where the function relation and the probability density expression are relatively simple (for example, square law or half-wave linear law, normal distribution, or exponential distribution).

Not directly using this formula for the double integral, but using the Fourier transform or the Laplace transform, transforming the nonlinear function into a so-called "transfer function," transforming the probability density into a characteristic function, and changing the integral form for the calculation, is called the transformation method. This method is available when the nonlinear function is complex, but its analytical calculations are still quite cumbersome.

When the stationary high-frequency narrowband process passes the envelope detector, a special method – the slowly changing envelope method – can be used, which only considers the nonlinear transformation of the envelope. Although the use of this method is limited, and the input process must be a narrowband process, it has the advantage of simple and convenient calculation.

4.2 Direct method of nonlinear transformation of random processes

In the following, we introduce the direct method through two typical cases (stationary normal noise, sinusoidal signals and stationary normal noise through the full-wave square-law device together). Then, we introduce a metamorphic form of the direct method: the Hermite polynomial method.

4.2.1 Stationary normal noise through full–wave square-law devices

Let the nonlinear function relation be:

$$y = bx^2 \tag{4.13}$$

where b is a constant. The known input $X(t)$ is the stationary normal noise whose mean is zero and whose variance is σ^2. The two-dimensional probability density is:

$$p_2(x_1, x_2, \tau) = \frac{1}{2\pi\sigma^2\sqrt{1-r^2}} \exp\left\{-\frac{x_1^2 + x_2^2 - 2rx_1x_2}{2\sigma^2(1-r^2)}\right\} \tag{4.14}$$

where $x_1 = X(t_1)$, $x_2 = X(t_2)$, $r = r_X(\tau)$.

The following are necessary for solving the correlation function $R_Y(\tau)$ and power spectral density $G_Y(f)$ of the output process $Y(t)$.

Substituting the two equations above into

$$R_Y(\tau) = \int_{-\infty}^{\infty}\int_{-\infty}^{\infty} f(x_1)f(x_2)p_2(x_1, x_2, \tau)\,dx_1\,dx_2$$

letting $r = \cos\theta$, and substituting the variables, $x_1 = \sigma\rho\cos(\theta/2+\varphi)$, $x_2 = \sigma\rho\cos(\theta/2-\varphi)$, $R_Y(\tau)$ can be obtained. However, using the fourth-order mixed origin moment of the zero-mean stationary normal process to solve the formula yields:

$$\overline{X_1X_2X_3X_4} = \overline{X_1X_2}\cdot\overline{X_3X_4} + \overline{X_1X_3}\cdot\overline{X_2X_4} + \overline{X_1X_4}\cdot\overline{X_2X_3} \tag{4.15}$$

It is easier to let $X_3 = X_1$, $X_4 = X_2$ and then:

$$\overline{X_1^2X_2^2} = \overline{X_1^2}\cdot\overline{X_2^2} + 2\left(\overline{X_1X_2}\right)^2 \tag{4.16}$$

Since the mean is zero, we have:

$$\overline{X_1^2} = \overline{X_2^2} = \sigma^2 = R_X(0) \tag{4.17}$$

and

$$\overline{X_1X_2} = R_X(0) \tag{4.18}$$

So, the relationship between the output correlation function and the input correlation function is:

$$R_Y(\tau) = b^2\overline{X_1^2X_2^2} = b^2\left[\sigma^4 + 2R_X^2(\tau)\right] \tag{4.19}$$

When τ=0, the total average power of the process $Y(t)$ is:

$$R_Y(0) = 3b^2\sigma^4 \tag{4.20}$$

When $\tau = \infty$, because $R_Y(\infty) = \sigma^2 r_X(\infty) = 0$, the DC power of the process $Y(t)$ is:

$$R_Y(\infty) = b^2\sigma^4 \tag{4.21}$$

Thus, the average undulation power of the process $Y(t)$ (the sum of low frequency and high frequency) is:

$$\sigma_Y^2 = R_Y(0) - R_Y(\infty) = 2b^2\sigma^4 \tag{4.22}$$

Use Fourier transform on the $R_Y(\tau)$ above. Because:

$$R_X(\tau) \cdot R_X(\tau) \leftrightarrow G_X(f) \otimes G_X(f) \tag{4.23}$$

the power spectral density of the process $Y(t)$ is:

$$\begin{aligned} G_Y(f) &= \int_{-\infty}^{\infty} R_Y(\tau) e^{-j2\pi f\tau} \,d\tau \\ &= \int_{-\infty}^{\infty} \left[b^2\sigma^4 + 2b^2 R_X^2(\tau)\right] e^{-j2\pi f\tau} \,d\tau \\ &= b^2\sigma^4\delta(f) + 2b^2 \int_{-\infty}^{\infty} G_X(f')G_X(f-f')\,df' \end{aligned} \tag{4.24}$$

On the right-hand side of eq. (4.24), the first term is the DC component, which is recorded as:

$$G_{Y=}(f) = b^2\sigma^4\delta(f) \tag{4.25}$$

The second term on the right-hand side (the convolution term) is the undulating component:

$$G_{Y\sim}(f) = 2b^2 \int_{-\infty}^{\infty} G_X(f')G_X(f-f')\,df' \tag{4.26}$$

If the power spectral density $G_X(f)$ of the input process $X(t)$ is given, then $G_{Y\sim}(f)$ can be obtained by the convolution integral of the formula above.

Example 4.2.1. Knowing that the nonlinear function relation is $y = bx^2$ and the input process $X(t)$ is stationary narrowband normal noise, the power spectral density is:

$$G_X(f) = \begin{cases} c\,, & f_0 - \frac{\Delta f}{2} < |f| < f_0 + \frac{\Delta f}{2} \\ 0\,, & \text{otherwise} \end{cases} \tag{4.27}$$

The graph is shown in Fig. 4.3 (a). Find the power spectral density of the output process $Y(t)$, and solve the power of its DC, low-frequency and high-frequency components.

Solution: From $G_X(f)$ in Fig. 4.3 (a) we can see that the input process $X(t)$ does not have DC components, and its variance is:

$$\sigma^2 = \int_{-\infty}^{\infty} G_X(f)\,df = 2c\Delta f \tag{4.28}$$

The DC component power spectral density of the output process $Y(t)$ is:

$$G_{Y=}(f) = b^2\sigma^4\delta(f) \tag{4.29}$$

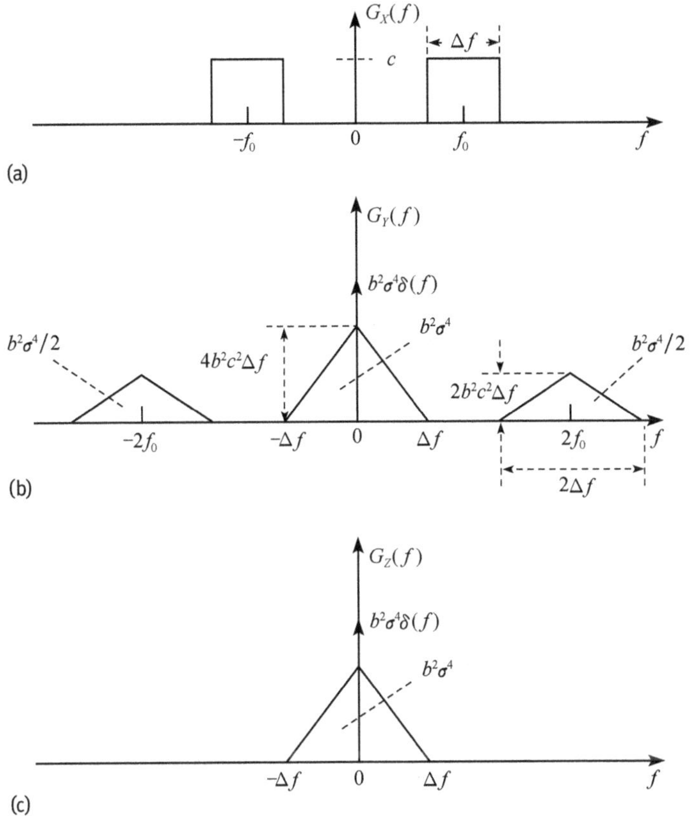

Fig. 4.3: Power spectral density of stationary narrow-band noise through full wave square law

From the convolution calculation of eq. (4.26), the available power spectral density is shown in Fig. 4.3 (b), i.e.,

$$G_{Y\sim}(f) = \begin{cases} 4b^2c^2(\Delta f - |f|)\,, & 0 < |f| < \Delta f \\ 2b^2c^2(\Delta f - \big||f| - 2f_0\big|)\,, & 2f_0 - \Delta f < |f| < 2f_0 + \Delta f \\ 0\,, & \text{elsewhere} \end{cases} \tag{4.30}$$

Figure 4.3 shows that, due to nonlinear transformation, the DC, low-frequency and high-frequency harmonic components, which are not obtained in the input, appear in the output power spectrum. From Fig. 4.3 (b), it can be calculated that the power of these three components is exactly equal to $b^2\sigma^4$.

It should be noted that Fig. 4.3 (b) shows the short-circuit current power spectrum in a nonlinear device when the output is shorted. If the output is connected to the ideal lowpass filter to form a square law envelope detector, the high-frequency components are completely filtered, and only low-frequency components are outputted (including DC). The graphics of power spectral density $G_Z(f)$ is shown in Fig. 4.3 (c).

4.2.2 Common noise and signals through full-wave square-law device

The input process is:

$$X(t) = s(t) + n(t) \tag{4.31}$$

where $n(t)$ is a stationary normal process whose mean is zero and variance is σ^2, and $s(t)$ is random initial phase signal:

$$s(t) = A\cos(\omega_0 t + \theta) \tag{4.32}$$

where θ is uniformly distributed in $[0, 2\pi]$. Signal and noise are statistically independent. The nonlinear function is still $y = bx^2$.

Because the input $X(t)$ is a stationary process, the output $Y(t)$ is still a stationary process. We obtain the correlation function of the process $Y(t)$:

$$R_Y(\tau) = \overline{YY_\tau} = b^2\overline{X^2X_\tau^2} = b^2\overline{(s+n)^2(s_\tau+n_\tau)^2} \tag{4.33}$$

expand the formula, and use the statistically independent relations and $\overline{n} = \overline{n_\tau} = 0$, $\overline{s} = \overline{s_\tau} = 0$ to obtain:

$$\begin{aligned} R_Y(\tau) &= b^2\left(\overline{s^2s_\tau^2} + \overline{n^2n_\tau^2} + \overline{s^2n_\tau^2} + \overline{s_\tau^2n^2} + 4\overline{ss_\tau}\cdot\overline{ss_\tau}\right) \\ &= R_{ss}(\tau) + R_{nn}(\tau) + R_{sn}(\tau) \end{aligned} \tag{4.34}$$

where, $R_{ss}(\tau) = b^2\overline{s^2s_\tau^2}$, the correlation function of signal s^2 and s_τ^2, referred to as the signal–signal component;

$R_{nn}(\tau) = b^2\overline{n^2n_\tau^2}$, the correlation function of noise n^2 and n_τ^2, referred to as the noise–noise component;

$R_{sn}(\tau) = b^2[\overline{s^2n_\tau^2} + \overline{s_\tau^2n^2} + 4\overline{ss_\tau}\cdot\overline{ss_\tau}]$, the cross-correlation function of signal and noise, referred to as the signal–noise component.

It is not difficult to obtain the following three components:

$$R_{ss}(\tau) = b^2\left[\frac{A^4}{4} + \frac{A^4}{8}\cos 2\omega_0\tau\right] \tag{4.35}$$

$$R_{nn}(\tau) = b^2\left[\sigma^4 + 2R_n^2(\tau)\right] \tag{4.36}$$

$$\begin{aligned} R_{sn}(\tau) &= b^2\left[\frac{A^2}{2}\sigma^2 + \frac{A^2}{2}\sigma^2 + 4\frac{A^2}{2}\cos\omega_0\tau\cdot R_n(\tau)\right] \\ &= b^2\left[A^2\sigma^2 + 2A^2R_n(\tau)\cos\omega_0\tau\right] \end{aligned} \tag{4.37}$$

Substitute the above three equations into eq. (4.34) to obtain:

$$R_Y(\tau) = b^2\left[\left(\frac{A^2}{2} + \sigma^2\right)^2 + 2R_n^2(\tau) + 2A^2R_n(\tau)\cos\omega_0\tau + \frac{A^4}{8}\cos 2\omega_0\tau\right] \tag{4.38}$$

Use the Fourier transform on $R_{ss}(\tau)$, $R_{nn}(\tau)$, $R_{sn}(\tau)$ and $R_Y(\tau)$ above to obtain the corresponding power spectral density as follows:

$$G_{ss}(f) = \frac{b^2A^4}{4}\delta(f) + \frac{b^2A^4}{16}\left[\delta(f-2f_0) + \delta(f+2f_0)\right] \tag{4.39}$$

$$G_{nn}(f) = b^2\sigma^4\delta(f) + 2b^2\int_{-\infty}^{\infty} G_n(f')G_n(f-f')\,\mathrm{d}f' \tag{4.40}$$

$$\begin{aligned} G_{sn}(f) &= b^2A^2\sigma^2\delta(f) + 2b^2A^2G_n(f) \otimes \frac{1}{2}\left[\delta(f-f_0) + \delta(f+f_0)\right] \\ &= b^2A^2\sigma^2\delta(f) + b^2A^2\left[G_n(f-f_0) + G_n(f+f_0)\right] \end{aligned} \tag{4.41}$$

$$\begin{aligned} G_Y(f) = {} & b^2\left(\frac{A^2}{2} + \sigma^2\right)^2\delta(f) + 2b^2\int_{-\infty}^{\infty} G_n(f')G_n(f-f')\,\mathrm{d}f' \\ & + b^2A^2\left[G_n(f-f_0) + G_n(f+f_0)\right] + \frac{b^2A^4}{16}\left[\delta(f-2f_0) + \delta(f+2f_0)\right] \end{aligned} \tag{4.42}$$

Letting $\tau = 0$, the total average power of the output be obtained by $R_Y(\tau)$ in eq. (4.38):

$$\begin{aligned} R_Y(0) &= b^2\left[\left(\frac{A^2}{2} + \sigma^2\right)^2 + 2\sigma^4 + 2A^2\sigma^2 + \frac{A^4}{8}\right] \\ &= 3b^2\left[\frac{A^4}{8} + A^2\sigma^2 + \sigma^4\right] \end{aligned} \tag{4.43}$$

Because the first term on the right-hand side of eq. (4.38) of $R_Y(\tau)$ represents the DC component, the output DC power is:

$$m_Y^2 = b^2\left(\frac{A^2}{2} + \sigma^2\right)^2 \tag{4.44}$$

So, the output average fluctuation power is:

$$\sigma_Y^2 = R_Y(0) - m_Y^2 = 2b^2\left[\frac{A^4}{16} + A^2\sigma^2 + \sigma^4\right] \tag{4.45}$$

Example 4.2.2. Noise $n(t)$ and signal $s(t)$ are common through the full-wave square-law device. The nonlinear function relation is $y = bx^2$. $n(t)$ is stationary normal noise whose mean is zero and variance is σ^2, and its power spectral density is:

$$G_n(f) = \begin{cases} c\,, & f_0 - \frac{\Delta f}{2} < |f| < f_0 + \frac{\Delta f}{2} \\ 0\,, & \text{otherwise} \end{cases} \tag{4.46}$$

$s(t) = A\cos(\omega_0 t + \theta)$ is a random initial phase signal, and the signal and noise are statistically independent. Find the power spectrum of output process $Y(t)$ and its respective spectrum component.

Solution: The correlation function of the input signal $s(t)$ is:

$$R_s(\tau) = \frac{A^2}{2} \cos \omega_0 \tau \tag{4.47}$$

Use Fourier transform to obtain the power spectral density of $s(t)$:

$$G_s(f) = \frac{A^2}{4} [\delta(f - f_0) + \delta(f + f_0)] \tag{4.48}$$

It is known that the power spectral density of the input process $X(t)= s(t)+ n(t)$ is:

$$G_X(f) = G_s(f) + G_n(f) = \frac{A^2}{4} [\delta(f - f_0) + \delta(f + f_0)] + \begin{cases} c\,, & f_0 - \frac{\Delta f}{2} < |f| < f_0 + \frac{\Delta f}{2} \\ 0\,, & \text{otherwise} \end{cases} \tag{4.49}$$

As is shown in Fig. 4.4 (a), it consists of two line spectrums, which are continuous spectrums. The power spectral density $G_Y(f)$ of output process $Y(t)$ consists of three component spectrums: the signal–signal component spectrum $G_{ss}(f)$, the noise–noise component spectrum $G_{nn}(f)$, and the signal–noise component spectrum $G_{sn}(f)$. The signal–signal component spectrum $G_{ss}(f)$ is:

$$G_{ss}(f) = \frac{b^2 A^4}{4} \delta(f) + \frac{b^2 A^4}{16} [\delta(f - 2f_0) + \delta(f + 2f_0)] \tag{4.50}$$

The graph is shown in Fig. 4.4 (b), which is a three-line spectrum.

The noise–noise component spectrum $G_{nn}(f)$ can be obtained by the formula of $G_{nn}(f)$ and the expression $G_n(f)$:

$$G_{nn}(f) = b^2 \sigma^4 \delta(f) + \begin{cases} 4b^2 c^2 (\Delta f - |f|)\,, & 0 < |f| < \Delta f \\ 2b^2 c^2 (\Delta f - \| f| - 2f_0|)\,, & 2f_0 - \Delta f < |f| < 2f_0 + \Delta f \\ 0\,, & \text{elsewhere} \end{cases} \tag{4.51}$$

where, $\sigma^2 = 2c\Delta f$. The graphics of $G_{nn}(f)$ are shown in Fig. 4.4 (c); it is a line spectrum consisting of three sections of continuous spectrums.

The signal–noise component spectrum $G_{sn}(f)$ can be obtained by the formula of $G_{sn}(f)$ and the expression $G_n(f)$:

$$G_{sn}(f) = b^2 A^2 \sigma^2 \delta(f) + \begin{cases} 2b^2 A^2 c^2\,, & 0 < |f| < \frac{\Delta f}{2} \\ b^2 A^2 c^2\,, & 2f_0 - \frac{\Delta f}{2} < |f| < 2f_0 + \frac{\Delta f}{2} \\ 0\,, & \text{elsewhere} \end{cases} \tag{4.52}$$

The graph is shown in Fig. 4.4 (d), which is a line spectrum consisting of three sections of continuous spectrums.

With the sum of the above three kinds, you can get the expression of the total power spectrum of process $Y(t)$. The total power spectrum $G_Y(f)$ is obtained by superimposing the three-component spectrums. The graph of $G_Y(f)$ is shown in Fig. 4.4 (e), which is a three-line spectrum consisting of three sections of continuous spectrums.

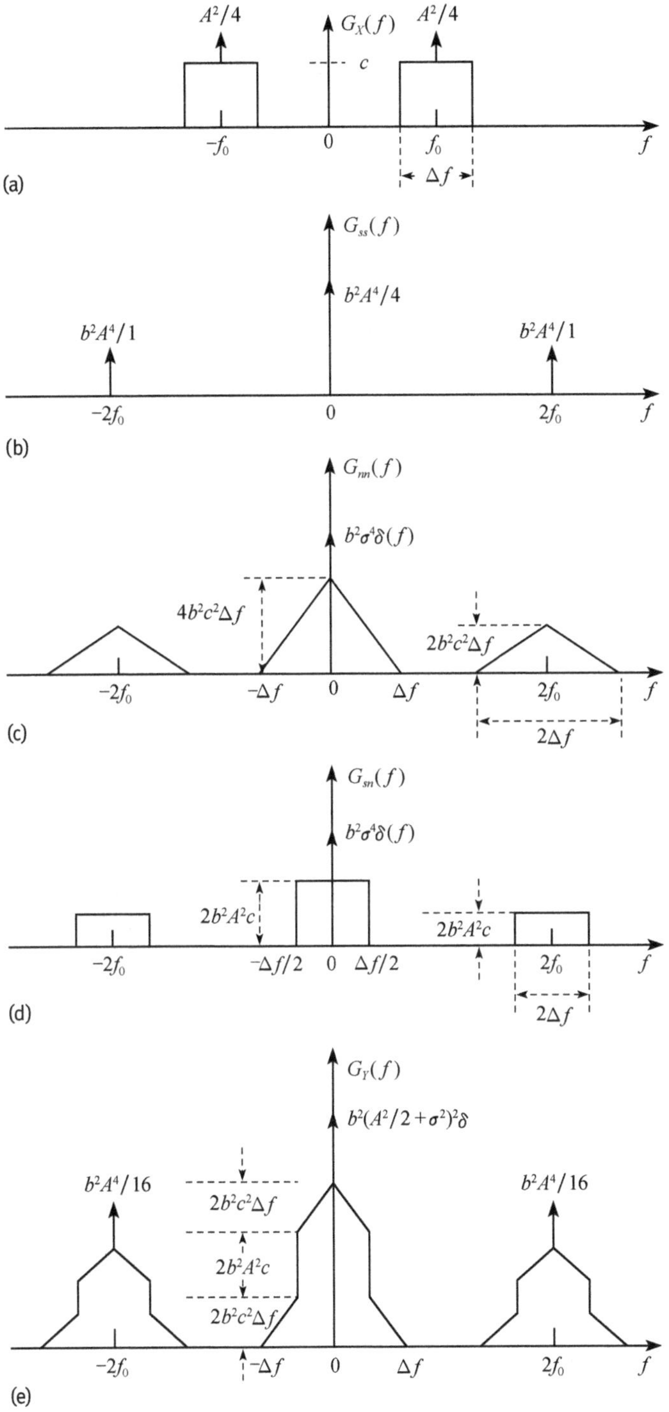

Fig. 4.4: Power spectral density of stationary narrowband noise and sinusoidal signals passing a square law device together

It should be noted that if the upper case circuit is a detector, after passing the ideal low-frequency-pass filter, the output consists of a line spectrum (a signal component) that is located in zero frequency and a continuous spectrum (a low-frequency fluctuation component). If the above case circuit is a second harmonic multiplier, then after passing through the ideal high-frequency bandpass filter, the output consists of two line spectrums (signal component) and two continuous spectrums (noise component) located in $\pm 2f_0$.

4.2.3 Determination of the output power spectrum with the difference beat method and the sum beat method

After the noise of the symmetric uniform spectrum has passed the nonlinear circuit, the output is a continuous triangle spectrum $G_{nn}(f)$, shown as Example 4.2.2. It is obtained from the convolution integral, but can we explain its physical meaning?

We know that when the voltage at two different frequencies f_1, f_2 commonly acts on the nonlinear device, due to nonlinear transformation, the output voltage will generally contain their DC, fundamental, and the harmonics. In addition, due to the principle of the difference beat method and the sum beat method, their beat frequency (difference frequency and frequency) will be produced, that is, $mf_1 \pm nf_2$, $m, n = 1, 2, \ldots$.

The continuous spectrum of input noise is divided into n components, whose width is δf and then each component spectrum is approximately a sine wave, but their frequencies are different. Due to the principle of the difference beat and the sum beat, it will produce a lot of difference frequency and sum frequency components. Obtaining the low-frequency continuous spectrum by the convolution integral is called the difference beat method and obtaining the high-frequency continuous spectrum by the convolution integral is called the sum beat method. The following is divided into two cases to be discussed.

4.2.3.1 High-frequency band noise input

We assume that the input high frequency band noise have the rectangular symmetrical power spectrum showed in Fig. 4.5 (a), whose spectral width is Δf. It is divided into n components, and any two components will produce a difference frequency component, of which the highest difference frequency is $\Delta f - \delta f$, and only one subfrequency component is produced by the difference beat of the two components numbered 1 and n. The second high-difference frequency is $\Delta f - 2\delta f$, and the difference frequency components are a total of 2, which are produced by the difference beat of the components numbered 1 and $n - 1$, 2 and n. The subsequent difference frequency is $\Delta f - 3\delta f$, and there is a total of three difference frequency components in which the lowest difference frequency is δf and its number is $n - 1$. It can be seen from the above, with the value of difference frequency increasing, the number of difference frequency components is a diminishing of the linear law. So, when $n \to \infty$ ($\delta f \to 0$), the low-frequency continuous spectrum of the output noise is a right-angled triangle, as shown in Fig. 4.5 (b).

Among the n components divided above, any two components will produce sum frequency components, in which the lowest is $2f_0 - \Delta f + 2\delta f$ and only one, produced by the sum beat of the two components numbered by 1 and 2. The sum frequency $2f_0 - \Delta f + 4\delta f$ has two components, which are, respectively, made by the sum beat of the components numbered by 1 and 4, and 2 and 3. The sum frequency $2f_0 - \Delta f + 6\delta f$ has three components, ..., and when the frequency is $2f_0$, the number of this sum frequency component is the largest and $n/2$. It can be seen that in this segment of the sum frequency range, with the sum frequency increasing, the number of sum frequency components is an increase of the linear law. On the contrary, with the sum frequency increasing, the number of sum frequency components higher than $2f_0$ is a decrease of the linear law. So, when $n \to \infty$ ($\delta f \to 0$), the high-frequency continuous spectrum of the output noise is isosceles triangular, as shown in Fig. 4.5 (b).

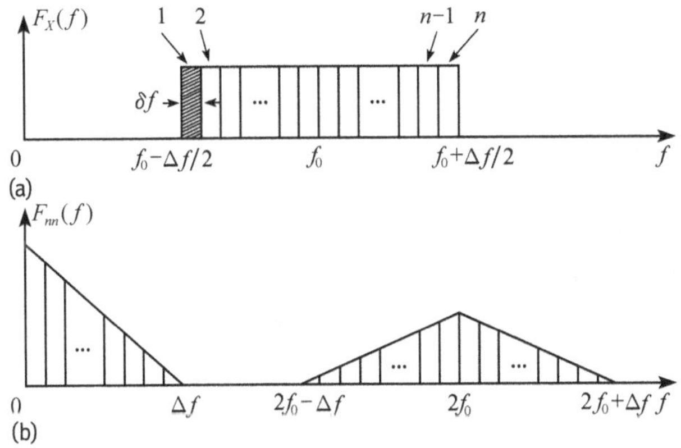

Fig. 4.5: The output power spectrum determined by beat frequency (only input noise).

4.2.3.2 Noise and signal input

We assume that the input noise is the same as described above, and the input signal is sine wave with a single frequency f_0 and in the center of the noise spectrum width Δf, as shown in Fig. 4.6 (a). The various components of the signal and the noise will turn the difference beat into signal–noise low-frequency components, and the number of the difference frequency components is equal in the spectrum width range of $0 \sim \Delta f/2$, as shown in Fig. 4.6 (a). The various components of the signal and the noise will turn the sum beat into signal–noise high-frequency components, and the number of sum frequency components is equal in the spectrum width range of $2f_0 - (\Delta f/2) \sim 2f_0 + (\Delta f/2)$, as shown in Fig. 4.6 (b).

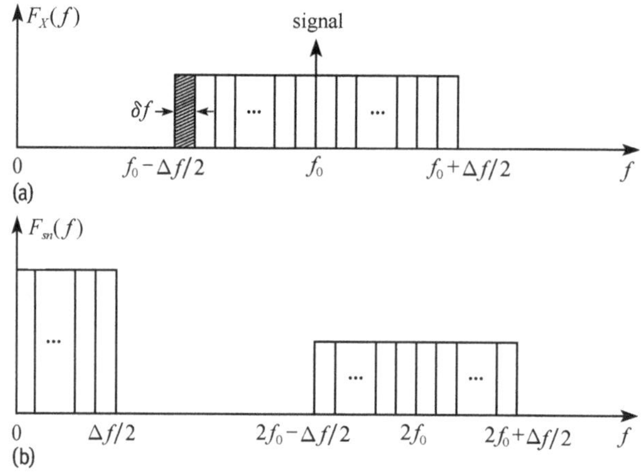

Fig. 4.6: The output power spectrum (when the signal noise is common input) is determined by the beat law.

When the high-frequency band noise and the signal are input together, the power spectrum of the noise–noise component $F_{nn}(f)$, as shown in Fig. 4.6 (b) in the previous section.

Example 4.2.3. The input power spectrum of the square-law detector (ignoring its load effect) is shown in Fig. 4.7 (a). The input signal is sinusoidal amplitude modulated, and the input noise is a narrowband rectangular power spectrum. Try to use the difference beat method to characterize the total power spectrum $F_Y(\omega)$ after ideally filtering through the lowpass detector and its graphics of the component spectrum $F_{\mathrm{ss}}(\omega)$, $F_{\mathrm{nn}}(\omega)$, $F_{\mathrm{sn}}(\omega)$.

Solution: The difference beat method can be used to qualitatively draw the various component spectrums (including DC) after lowpass filtering through the detectors $F_{ss}(\omega)$, $F_{nn}(\omega)$ and $F_{sn}(\omega)$, respectively, shown in Fig. 4.7 (b)–(d). The superimposed total power spectrum $F_Y(\omega)$ is shown in Fig. 4.7 (e), where, the respective spectrum components are shown by arrows, but the amplitude ratio is not indicated.

4.2.4 Hermite polynomial method

The Hermitian polynomials are defined as:

$$H_k(z) = (-1)^k \mathrm{e}^{\frac{z^2}{2}} \frac{\mathrm{d}^k}{\mathrm{d}z^k}\left(\mathrm{e}^{-\frac{z^2}{2}}\right) \tag{4.53}$$

where $k = 0, 1, 2, \ldots$, the order of the derivative.

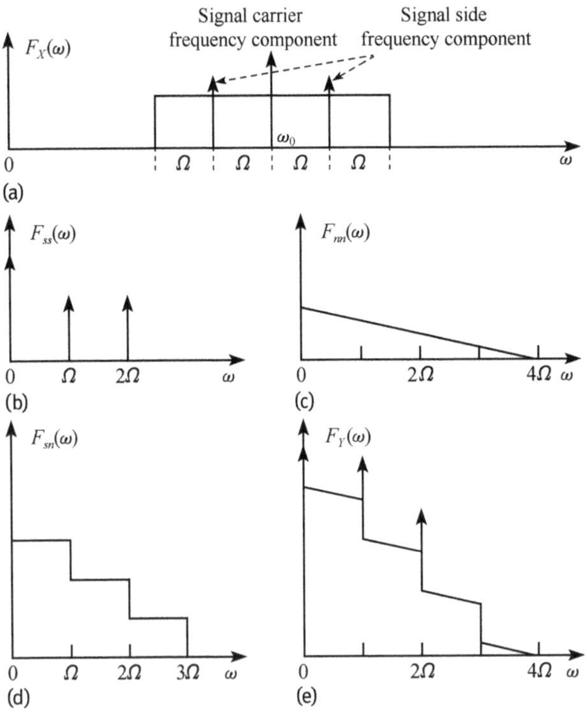

Fig. 4.7: Determine the output power spectrum (common input of signal and noise) by beat frequency

The first six items of Hermitian polynomials are as follows:

$$\begin{gathered}H_0(z) = 1\,,\quad H_1(z) = z\,,\quad H_2(z) = z^2 - 1\,,\quad H_3(z) = z^3 - 3z\,,\\ H_4(z) = z^4 - 6z^2 + 3\,,\quad H_5(z) = z^5 - 10z^3 + 15z\end{gathered} \tag{4.54}$$

The higher-order items can be obtained by the following recursive formula:

$$H_{k+1}(z) = z \cdot H_k(z) - k \cdot H_{k-1}(z) \tag{4.55}$$

From the above, a special value can be obtained when $z = 0$:

$$H_0(0) = 1\,,\quad H_{2k}(0) = (-1)^k(2k-1)!!\,,\quad H_{2k+1}(0) = 0 \tag{4.56}$$

It can be shown that the Hermitian polynomial has orthogonality:

$$\int_{-\infty}^{\infty} H_j(z)H_k(z)\mathrm{e}^{-\frac{z^2}{2}}\,\mathrm{d}z = \begin{cases} k!\sqrt{2\pi}\,, & j = k \\ 0\,, & j \neq k \end{cases} \tag{4.57}$$

When the nonlinear function relation and the two-dimensional probability density expression $p_2(x_1, x_2; \tau)$ are not complicated,

$$R_Y(\tau) = \int_{-\infty}^{\infty}\int_{-\infty}^{\infty} f(x_1)f(x_2)p_2(x_1, x_2; \tau)\,\mathrm{d}x_1\,\mathrm{d}x_2$$

can be made double integrals directly to solve the correlation function output. However, if it is a complex function, it is difficult to do integration operation directly. At this time, if the input random process is the normal distribution, you can expand $p_2(x_1, x_2; \tau)$ according to Maclaurin series into the Hermitian polynomial. Since the partial integral is easy to compute, and the Hermitian polynomial has orthogonality, the double integral operation can be simplified to a single integral. This method is clever, easy to use and has strong applicability (but is still limited to the input for the normal distribution), and is thus widely cited. The following is a detailed derivation.

Setting input $X(t)$ for a stationary normal process, whose mean is zero and variance is σ^2, the two-dimensional probability density is:

$$p_2(x_1, x_2; \tau) = \frac{1}{2\pi\sigma^2\sqrt{1-r^2}} \exp\left\{-\frac{x_1^2 + x_2^2 - 2rx_1x_2}{2\sigma^2(1-r^2)}\right\} \tag{4.58}$$

First substitute variables for the standardization of random variables. Letting $z_1 = x_1/\sigma$, $z_2 = x_2/\sigma$, then you can obtain:

$$R_Y(\tau) = \int_{-\infty}^{\infty}\int_{-\infty}^{\infty} f(\sigma z_1) f(\sigma z_2) p_2(z_1, z_2; \tau)\, dz_1\, dz_2 \tag{4.59}$$

where:

$$p_2(z_1, z_2; \tau) = \sigma^2 p_2(x_1, x_2; \tau) = \frac{1}{2\pi\sigma^2\sqrt{1-r^2}} \exp\left\{-\frac{z_1^2 + z_2^2 - 2rz_1z_2}{2(1-r^2)}\right\} \tag{4.60}$$

where $r = r_X(\tau)$.

$p_2(z_1, z_2; \tau)$ and the characteristic function $\Phi_Z(\lambda_1, \lambda_2; \tau)$ are a pair of two-dimensional Fourier transforms. It is known that the two-dimensional characteristic function of stationary normal processes (the mean is zero, the variance is 1 and the correlation coefficient is r) is:

$$\Phi_Z(\lambda_1, \lambda_2; \tau) = \exp\left\{-\frac{1}{2}\left(\lambda_1^2 + \lambda_2^2 - 2r\lambda_1\lambda_2\right)\right\} \tag{4.61}$$

So that:

$$\begin{aligned} p_2(z_1, z_2; \tau) = \frac{1}{(2\pi)^2} \int_{-\infty}^{\infty}\int_{-\infty}^{\infty} &\exp\left\{-\frac{1}{2}\left(\lambda_1^2 + \lambda_2^2 - 2r\lambda_1\lambda_2\right)\right\} \\ &\cdot \exp\{-j(\lambda_1 z_1 + \lambda_2 z_2)\}\, d\lambda_1\, d\lambda_2 \end{aligned} \tag{4.62}$$

Expand $\exp\{-r\lambda_1\lambda_2\}$ into Mark Lauryn series:

$$e^{-r\lambda_1\lambda_2} = \sum_{k=0}^{\infty} \frac{(-r)^k}{k!} (\lambda_1\lambda_2)^k \tag{4.63}$$

and substitute it into the formula above:

$$p_2(z_1, z_2; \tau) = \sum_{k=0}^{\infty} \frac{(-r)^k}{k!} \left[\frac{1}{2\pi}\int_{-\infty}^{\infty} \lambda_1^k e^{-\frac{\lambda_1^2}{2} - j\lambda_1 z_1}\, d\lambda_1\right] \cdot \left[\frac{1}{2\pi}\int_{-\infty}^{\infty} \lambda_2^k e^{-\frac{\lambda_2^2}{2} - j\lambda_2 z_2}\, d\lambda_2\right] \tag{4.64}$$

In the following, the integral formula in the square brackets should become the Hermitian polynomial.

One-dimensional probability density $p(z)$ and one-dimensional characteristic functions $\Phi_Z(\lambda)$ are a pair of Fourier transforms, and there is a relationship:

$$p(z) = \frac{1}{2\pi} \int_{-\infty}^{\infty} \Phi_Z(\lambda) \mathrm{e}^{-\mathrm{j}\lambda z} \, \mathrm{d}\lambda \tag{4.65}$$

for the standard normal process, whose mean is zero and whose variance is 1. From the equation above, we can obtain:

$$\frac{1}{\sqrt{2\pi}} \mathrm{e}^{-\frac{z^2}{2}} = \frac{1}{2\pi} \int_{-\infty}^{\infty} \mathrm{e}^{-\frac{\lambda^2}{2} - \mathrm{j}\lambda z} \, \mathrm{d}\lambda \tag{4.66}$$

On the both sides of the previous equation we can take the derivation of variables z for k times to obtain:

$$\frac{1}{\sqrt{2\pi}} \frac{\mathrm{d}^k}{\mathrm{d}z^k} \left(\mathrm{e}^{-\frac{z^2}{2}} \right) = \frac{(-\mathrm{j})^k}{2\pi} \int_{-\infty}^{\infty} \lambda^k \mathrm{e}^{-\frac{\lambda^2}{2} - \mathrm{j}\lambda z} \, \mathrm{d}\lambda \tag{4.67}$$

use the definition in eq. (4.53) of $H_k(z)$ to obtain:

$$\frac{1}{2\pi} \int_{-\infty}^{\infty} \lambda^k \mathrm{e}^{-\frac{\lambda^2}{2} - \mathrm{j}\lambda z} \, \mathrm{d}\lambda = \frac{1}{\sqrt{2\pi} \mathrm{j}^k} \mathrm{e}^{-\frac{z^2}{2}} H_k(z) \tag{4.68}$$

Substituting it into p_2 above, we can get:

$$p_2(z_1, z_2; \tau) = \sum_{k=0}^{\infty} \frac{r^k}{k!} \cdot \frac{1}{2\pi} H_k(z_1) \mathrm{e}^{-\frac{z_1^2}{2}} H_k(z_2) \mathrm{e}^{-\frac{z_2^2}{2}} \tag{4.69}$$

Then substitute the previous equation into eq. (4.59) and the integral variables z_1, z_2 are rewritten as z to get:

$$R_Y(\tau) = \sum_{k=0}^{\infty} \frac{r^k}{k!} \cdot \left[\frac{1}{\sqrt{2\pi}} \int_{-\infty}^{\infty} f(\sigma z) H_k(z) \mathrm{e}^{-\frac{z^2}{2}} \, \mathrm{d}z \right]^2 \tag{4.70}$$

Make

$$C_k = \frac{1}{\sqrt{2\pi}} \int_{-\infty}^{\infty} f(\sigma z) H_k(z) \mathrm{e}^{-\frac{z^2}{2}} \, \mathrm{d}z \tag{4.71}$$

The general expression of the output correlation function is:

$$R_Y(\tau) = \sum_{k=0}^{\infty} \frac{r^k(\tau)}{k!} \cdot C_k^2 \tag{4.72}$$

Taking the Fourier transform of the above formula, the general expression of the output power spectral density can be obtained as follows:

$$G_Y(\omega) = G_{Y_0}(\omega) + G_{Y_1}(\omega) + G_{Y_2}(\omega) + \cdots \tag{4.73}$$

Each item on the right-hand side of the eq. (4.73) respectively represents the power spectral density of the components, like low-frequency (including DC), high-frequency fundamental waves, high-frequency second harmonic, etc.

Example 4.2.4. It is known that input $X(t)$ is stationary normal noise. Its mean is zero and its variance is σ^2. The correlation coefficient is $r(\tau)$. Find the output correlation function $R_Y(\tau)$ through a half-wave linearity device. The characteristics of the half-wave linear device are:

$$y = f(x) = \begin{cases} bx\,, & x \geq 0 \\ 0\,, & x < 0 \end{cases} \tag{4.74}$$

Solution: Use a nonlinear function to make variable substitution $x = \sigma z$ to obtain:

$$y = f(\sigma z) = \begin{cases} b\sigma z\,, & z \geq 0 \\ 0\,, & z < 0 \end{cases} \tag{4.75}$$

From eq. (4.71), C_k can be obtained:

$$C_0 = \frac{1}{\sqrt{2\pi}} \int_{-\infty}^{\infty} b\sigma z \cdot 1 \cdot \mathrm{e}^{-\frac{z^2}{2}}\,\mathrm{d}z = \frac{b\sigma}{\sqrt{2\pi}} \tag{4.76}$$

$$C_1 = \frac{1}{\sqrt{2\pi}} \int_{-\infty}^{\infty} b\sigma z \cdot z \cdot \mathrm{e}^{-\frac{z^2}{2}}\,\mathrm{d}z = \frac{b\sigma}{2} \tag{4.77}$$

$$C_2 = \frac{1}{\sqrt{2\pi}} \int_{-\infty}^{\infty} b\sigma z \cdot (z^2 - 1) \cdot \mathrm{e}^{-\frac{z^2}{2}}\,\mathrm{d}z = \frac{b\sigma}{\sqrt{2\pi}} \tag{4.78}$$

$$C_3 = \frac{1}{\sqrt{2\pi}} \int_{-\infty}^{\infty} b\sigma z \cdot (z^3 - 3z) \cdot \mathrm{e}^{-\frac{z^2}{2}}\,\mathrm{d}z = 0 \tag{4.79}$$

$$C_4 = \frac{1}{\sqrt{2\pi}} \int_{-\infty}^{\infty} b\sigma z \cdot (z^4 - 6z^2 + 3) \cdot \mathrm{e}^{-\frac{z^2}{2}}\,\mathrm{d}z = -\frac{b\sigma}{\sqrt{2\pi}} \tag{4.80}$$

Substitute them into eq. (4.72) to obtain:

$$R_Y(\tau) = \frac{b^2\sigma^2}{2\pi}\left[1 + \frac{\pi}{2}r(\tau) + \frac{1}{2}r^2(\tau) + \frac{1}{24}r^4(\tau) + \cdots\right] \tag{4.81}$$

4.2.5 Stationary normal noise through half-wave linear devices

The above example obtained the expression of the output correlation function, which can be divided into two parts:

$$R_Y(\tau) = R_{Y=}(\tau) + R_{Y\sim}(\tau) \tag{4.82}$$

where $R_{Y=}(\tau) = \frac{b^2\sigma^2}{2\pi}$ represents the DC component, and

$$R_{Y\sim}(\tau) = \frac{b^2\sigma^2}{2\pi}\left[\frac{\pi}{2}r(\tau) + \frac{1}{2}r^2(\tau) + \frac{1}{24}r^4(\tau) + \cdots\right] \tag{4.83}$$

represents the undulating component.

Use $E\{[Y(t)]^n\} = \int_{-\infty}^{\infty}[f(x)]^n p(x,t)\,dx$ to get the mean square value of the output process $Y(t)$:

$$R_Y(0) = \overline{Y^2} = \int_{-\infty}^{\infty} f^2(x)p(x)\,dx = \frac{b^2}{\sqrt{2\pi}\sigma}\int_0^{\infty} x^2 e^{\frac{x^2}{2\sigma^2}}\,dx = \frac{b^2\sigma^2}{2} \tag{4.84}$$

Thus, the variance of the process $Y(t)$ is:

$$\begin{aligned}\sigma_Y^2 &= \overline{Y^2} - m_Y^2 = R_Y(0) - R_{Y=}(\tau)\\ &= \frac{b^2\sigma^2}{2} - \frac{b^2\sigma^2}{2\pi} = \frac{b^2\sigma^2}{2}\left(1 - \frac{1}{\pi}\right)\end{aligned} \tag{4.85}$$

or

$$\frac{b^2\sigma^2}{2\pi} = \frac{\sigma_Y^2}{\pi - 1} \tag{4.86}$$

Substituting into the above expression $R_{Y\sim}(\tau)$, we obtain:

$$\begin{aligned}R_{Y\sim}(\tau) &= \frac{\sigma_Y^2}{\pi - 1}\left[\frac{\pi}{2}r(\tau) + \frac{1}{2}r^2(\tau) + \frac{1}{24}r^4(\tau) + \cdots\right]\\ &= \sigma_Y^2[0.734r(\tau) + 0.233r^2(\tau) + 0.0194r^4(\tau) + \cdots]\end{aligned} \tag{4.87}$$

The infinite series above converges quickly. For example, when $\tau = 0$, we have $r(0) = 1$. It can be worked out that the first three items of eq. (4.87) account for 98.6% of the output fluctuation intensity σ_Y^2. So, generally, only the first three items are needed to obtain:

$$R_{Y\sim}(\tau) \approx \frac{\sigma_Y^2}{\pi - 1}\left[\frac{\pi}{2}r(\tau) + \frac{1}{2}r^2(\tau) + \frac{1}{24}r^4(\tau)\right] \tag{4.88}$$

It should be noted that the above analysis results apply to both narrowband processes and broadband processes.

Let us analyze the output power spectrum of the narrowband process through the linear envelope detector. We assume the input is stationary narrowband noise with a symmetrical power spectrum (relative to the central frequency ω_0). From the previous analysis it can be shown the correlation coefficient is:

$$r(\tau) = r_0(\tau)\cos\omega_0\tau \tag{4.89}$$

where $r_0(\tau)$ is the envelope of the correlation coefficient.

Substituting into eq. (4.87), we obtain:

$$R_{Y\sim}(\tau) \approx \frac{b^2\sigma^2}{2\pi}\left[\frac{\pi}{2}r_0(\tau)\cos\omega_0\tau + \frac{1}{2}r_0^2(\tau)\cos^2\omega_0\tau + \frac{1}{24}r_0^4(\tau)\cos^4\omega_0\tau\right] \tag{4.90}$$

Using trigonometric function expansion:

$$\begin{aligned} \cos^2\omega_0\tau &= \frac{1}{2}\left(1+\cos^2\omega_0\tau\right) \\ \cos^4\omega_0\tau &= \frac{1}{8}\left(\cos^4\omega_0\tau + 4\cos^2\omega_0\tau + 3\right) \end{aligned} \tag{4.91}$$

Ignoring the item $\cos^4\omega\tau$, whose amplitude is smaller and substituting into eq. (4.91) we obtain:

$$\begin{aligned} R_{Y\sim}(\tau) \approx \frac{b^2\sigma^2}{2\pi}\left\{\frac{1}{4}\left[r_0^2(\tau) + \frac{1}{16}r_0^4(\tau)\right] + \frac{\pi}{2}r_0(\tau)\cos\omega_0\tau\right. \\ \left. + \frac{1}{4}\left[r_0^2(\tau) + \frac{1}{12}r_0^4(\tau)\right]\cos 2\omega_0\tau\right\} \end{aligned} \tag{4.92}$$

The envelope detector output is the low-frequency components and the first item on the right-hand side of eq. (4.92):

$$R_Z(\tau) \approx \frac{b^2\sigma^2}{8\pi}\left[r_0^2(\tau) + \frac{1}{16}r_0^4(\tau)\right] \tag{4.93}$$

where $r_0(\tau)$ is a variable with the shape of the corresponding frequency characteristics of the narrowband system (narrowband intermediate frequency amplifier). As was obtained above, when the rectangular frequency characteristic is ideal, the correlation coefficient is:

$$r(\tau) = \frac{\sin\left(\frac{\Delta\omega\tau}{2}\right)}{\frac{\Delta\omega\tau}{2}}\cos\omega_0\tau \tag{4.94}$$

We have:

$$r_0(\tau) = \frac{\sin\left(\frac{\Delta\omega\tau}{2}\right)}{\frac{\Delta\omega\tau}{2}} = Sa\left(\frac{\Delta\omega\tau}{2}\right) \tag{4.95}$$

Substitute into eq. (4.93) to obtain:

$$R_Z(\tau) \approx \frac{b^2\sigma^2}{8\pi}\left[Sa^2\left(\frac{\Delta\omega\tau}{2}\right) + \frac{1}{16}Sa^4\left(\frac{\Delta\omega\tau}{2}\right)\right] \tag{4.96}$$

If we ignore the high-order item $Sa^4(\Delta\omega\tau/2)$ of the previous equation and apply the Fourier transform, since the sample function $Sa(t)$ and the rectangular spectrum function $S(\omega)$ are a pair of Fourier transforms, we obtain eq. (4.93) by the convolution theorem in the frequency domain:

$$Sa(t)\cdot Sa(t) \leftrightarrow \frac{1}{2\pi}S(\omega)\otimes S(\omega) \tag{4.97}$$

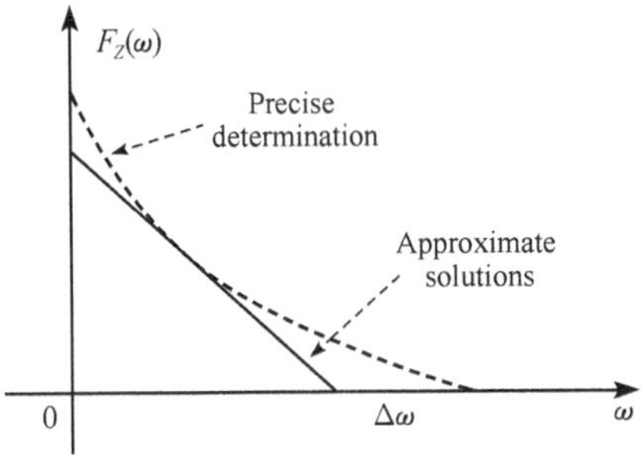

Fig. 4.8: Narrowband noise of a rectangular power spectrum measured by the low-frequency power spectral density of a linear detector.

We can see that the output low-frequency power spectral density $F_X(\omega)$ obtained by the convolution of two rectangular spectrums is the triangle, which is shown by the solid line in Fig. 4.8. The dotted lines in the figure show the exact solution when the item $Sa^4(\Delta\omega\tau/2)$ is included.

Example 4.2.5. It is known that the nonlinear function relationship of the half-wave linear rate device is eq. (4.74). Input $X(t)$ is the stationary normal noise and its power spectral density is:

$$G_X(f) = \begin{cases} c\,, & f_0 - \frac{\Delta f}{2} < |f| < f_0 + \frac{\Delta f}{2} \\ 0\,, & \text{others} \end{cases} \tag{4.98}$$

as shown in Fig. 4.9 (a). Find the correlation function and the power spectral density of the output process $Y(t)$.

Solution: Taking the first three items of eq. (4.81), the correlation function of the process $Y(t)$ is:

$$R_Y(\tau) \approx \frac{b^2\sigma^2}{2\pi}\left[1 + \frac{\pi}{2}r(\tau) + \frac{1}{2}r^2(\tau)\right] = \frac{b^2\sigma^2}{2\pi} + \frac{b^2}{4}R_X(\tau) + \frac{b^2}{4\pi\sigma^2}R_X^2(\tau) \tag{4.99}$$

Use Fourier transform on eq. (4.99), the power spectral density of the process $Y(t)$ is:

$$G_Y(f) = \frac{b^2\sigma^2}{2\pi}\delta(f) + \frac{b^2}{4}G_X(f) + \frac{b^2}{4\pi\sigma^2}\int_{-\infty}^{\infty} G_X(f')G_X(f-f')\,\mathrm{d}f' \tag{4.100}$$

$$\begin{aligned} &= \frac{b^2\sigma^2}{2\pi}\delta(f) + \begin{cases} \frac{b^2c}{4}\,, & f_0 - \frac{\Delta f}{2} < |f| < f_0 + \frac{\Delta f}{2} \\ 0\,, & \text{otherwise} \end{cases} \\ &+ \begin{cases} \frac{b^2c}{4\pi}\left(1 - \frac{|f|}{\Delta f}\right), & 0 < |f| < \Delta f \\ \frac{b^2c}{8\pi}\left(1 - \frac{1}{\Delta f}\big||f| - 2f_0\big|\right), & 2f_0 - \Delta f < |f| < 2f_0 + \Delta f \\ 0\,, & \text{elsewhere} \end{cases} \end{aligned} \tag{4.101}$$

as shown in Fig. 4.9 (b).

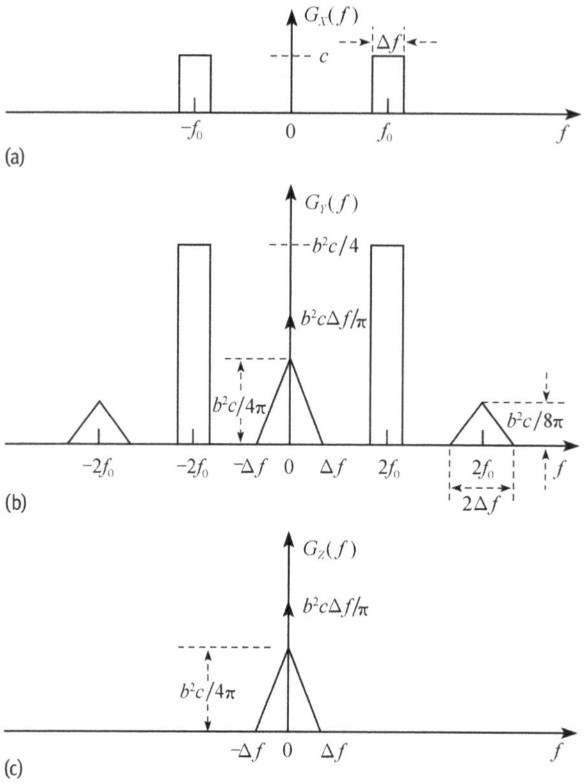

Fig. 4.9: Power density of stationary narrowband noise through a half-wave linear law device.

Comparing this with Fig. 4.4 (e), we can see that with full-wave square-law devices, the output exists only in the vicinity of zero-frequency and high-frequency second harmonics. With half-wave linear devices, the output also exists near the high-frequency fundamental waves [if the higher-order terms in the expression $R(t)$ are included, the output also exists near all the high-frequency harmonics]. However, if the ideal lowpass filter is used to form the envelope detector, the low-frequency power spectral density images are similar and they are triangular, as shown in Fig. 4.9 (c).

4.3 Transformation method of random process nonlinear transformation

When the nonlinear function is complicated, the transformation function can be transformed into the transfer function by using the Fourier transform or the Laplace transform; the probability density function is transformed into the characteristic function, and we can calculate after the integral form is changed. In this section, the transfer function is introduced first. Then the moment function and the power spectrum of the output of the nonlinear device are solved. Finally, a special case of the transformation method – the price method – is introduced.

4.3.1 Transfer function

We assume that the transmission characteristics between the nonlinear system output value y and input value x is:

$$y = f(x) \tag{4.102}$$

If this nonlinear function $f(x)$ and its derivative is continuous by segment, and $f(x)$ satisfies absolute integral conditions:

$$\int_{-\infty}^{\infty} |f(x)|\,\mathrm{d}x < \infty \tag{4.103}$$

then the Fourier transform $F(\lambda)$ of $f(x)$ exists:

$$F(\lambda) = \int_{-\infty}^{\infty} f(x)\mathrm{e}^{-\mathrm{j}\lambda x}\,\mathrm{d}x \tag{4.104}$$

Then $F(\lambda)$ is called a transfer function of a nonlinear system. Thus, the output value y of the nonlinear system can be expressed by the Fourier transform of the transfer function $F(\lambda)$:

$$y = f(x) = \frac{1}{2\pi}\int_{-\infty}^{\infty} F(\lambda)\mathrm{e}^{\mathrm{j}\lambda x}\,\mathrm{d}\lambda \tag{4.105}$$

However, some common transmission characteristics (such as half-wave linear law devices, etc.) are not absolutely integral, so the Fourier transform does not exist, and we cannot use the above formula for calculation. Nevertheless, the definition of the transfer function can be generalized from the Fourier transform to the Laplace transform, as described below in conjunction with the actual transmission characteristics.

Figure 4.10 shows the unidirectional hard limiting characteristics for when the transmission characteristic $f(x)$ in the positive infinite interval is not absolutely integral and is zero in the negative infinite interval, for example.

$$f(x) = \begin{cases} 1\,, & x > 0 \\ 0\,, & x \le 0 \end{cases} \tag{4.106}$$

Although the function $f(x)$ does not satisfy the absolute integral condition, it will satisfy the absolute integral conditions if the function $f(x)$ multiplied by the attenuation

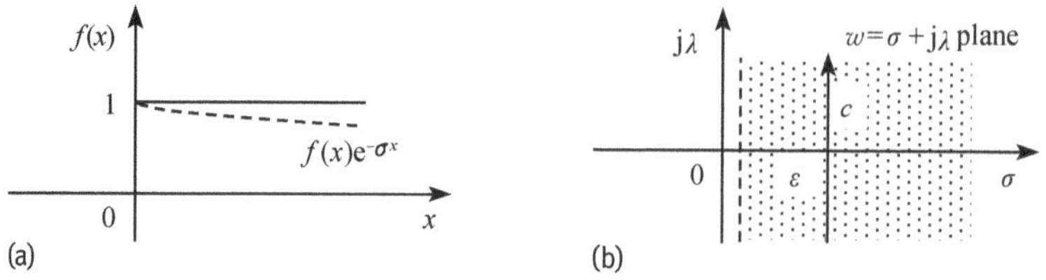

Fig. 4.10: One-way hard limiting properties and convergence domains of the single Laplace transform.

factor $e^{-\sigma x}$ (where $\sigma > 0$), and the Fourier transform expression is:

$$F(\lambda) = \int_{-\infty}^{\infty} f(x) e^{-\sigma x} e^{-j\lambda x} \, dx = \int_{-\infty}^{\infty} f(x) e^{-(\sigma + j\lambda)x} \, dx \tag{4.107}$$

Using complex variable $w = \sigma + j\lambda$, the transfer function is generalized:

$$F(w) = \int_{-\infty}^{\infty} f(x) e^{-wx} \, dx \tag{4.108}$$

Finally, letting $\sigma \to 0$, we can obtain the transfer function of the original function:

$$F(\lambda) = \lim_{\sigma \to 0} F(w) \tag{4.109}$$

As can be seen from eq. (4.108), $F(w)$ is the unilateral Laplace transform of transmission characteristics $f(x)$. For the characteristics shown in Fig. 4.10 (a), we can use eq. (4.108) to obtain:

$$F(w) = \int_{0}^{\infty} e^{-wx} \, dx = \frac{1}{w} \tag{4.110}$$

So, once the transfer function $F(w)$ is known, transmission characteristics $f(x)$ can be expressed by the Laplace inverse transform, that is,

$$f(x) = \frac{1}{2\pi j} \int_{c} F(w) e^{wx} \, dw \tag{4.111}$$

where c is the integral path and must be selected in the convergence domain of $f(x)$. For the characteristics shown in Fig. 4.11 (a), the convergence domain is the shaded area shown in Fig. 4.11 (b), which is the right half of the plane of the imaginary axis $w = j$ (excluding the virtual axis), so the optional integral path c is a straight line $w = \varepsilon + j\lambda$ (where $\varepsilon > 0$, $-\infty < \lambda < \infty$).

Figure 4.1 (a) shows the bidirectional hard limiting characteristics in the case when some of the transmission characteristics in the semi-infinite interval are not

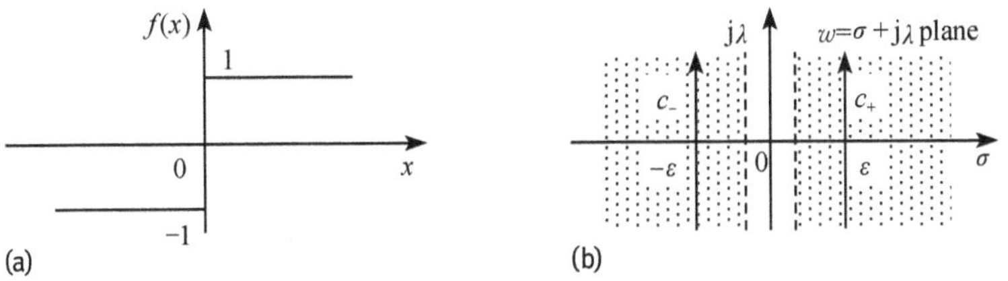

(a) (b)

Fig. 4.11: One-way hard limiting properties and convergence domains of the single Laplace transform.

zero:

$$f(x) = \begin{cases} 1\,, & x > 0 \\ 0\,, & x = 0 \\ -1\,, & x < 0 \end{cases} \tag{4.112}$$

In this case, the half-wave transmission characteristic can be defined as:

$$f_+(x) = \begin{cases} f(x)\,, & x > 0 \\ 0\,, & x \le 0 \end{cases} \tag{4.113}$$

$$f_-(x) = \begin{cases} 0\,, & x \ge 0 \\ f(x)\,, & x < 0 \end{cases} \tag{4.114}$$

Thus, the entire transmission characteristic is

$$f(x) = f_+(x) + f_-(x) \tag{4.115}$$

Since both $f_+(x)$ and $f_-(x)$ satisfy the conditions above, the Laplace transforms exist and are set to $F_+(w)$ and $F_-(w)$, respectively:

$$F_+(w) = \int_0^{\infty} f_+(x)\mathrm{e}^{-wx}\,\mathrm{d}x \tag{4.116}$$

$$F_-(w) = \int_{-\infty}^{0} f_-(x)\mathrm{e}^{-wx}\,\mathrm{d}x \tag{4.117}$$

For the characteristics shown in Fig. 4.11 (a), we can work out:

$$F_+(w) = \frac{1}{w} \tag{4.118}$$

$$F_-(w) = -F_+(-w) = \frac{1}{w} \tag{4.119}$$

So, after the transfer functions $F_+(w)$ and $F_-(w)$ are known, the entire transmission characteristics can be obtained from eq. (4.115) as:

$$f(x) = \frac{1}{2\pi j}\int_{c+} F_+(w)e^{wx}\,dw + \frac{1}{2\pi j}\int_{c-} F_-(w)e^{wx}\,dw \tag{4.120}$$

where integral paths $c+$ and $c-$ must be selected in the convergence of the respective functions. For the characteristics shown in Fig. 4.11 (a), the origin is a discontinuous point, which is the first-order pole in the plane $w = \sigma + j\lambda$. The convergence domain of $f_+(x)$ is in the right half plane of the imaginary axis (excluding the imaginary axis), and the convergence domain of $f_-(x)$ is in the left half of the plane (excluding the imaginary axis). These are, respectively, shown in the shaded area in Fig. 4.11 (b), so the integral path $c+$ is a straight line $w = \varepsilon + j\lambda$ and $c-$ is a straight line $w = -\varepsilon + j\lambda$, where $\varepsilon > 0$, $-\infty < \lambda < \infty$.

If the transmission characteristics $f(x)$ are not the same in the functions within the two half-infinite domains, as shown in Fig. 4.12 (a),

$$f(x) = \begin{cases} f_1(x) = 1\,, & x > 0 \\ f_2(x) = e^x\,, & x < 0 \end{cases} \tag{4.121}$$

$f(x)$ will be multiplied by the attenuation factor $e^{-\sigma x}$ and x integrated from $-\infty$ to $+\infty$:

$$\int_{-\infty}^{\infty} f(x)e^{-\sigma x}\,dx = \int_{0}^{\infty} e^{-\sigma x}\,dx + \int_{-\infty}^{0} e^{(1-\sigma)x}\,dx \tag{4.122}$$

When $\sigma > 0$, the first integral on the right-hand side converges, and when $\sigma < 1$, the second integral converges. So, in the range from 0 to 1, function $f(x)e^{-\sigma x}$ satisfies the absolute integral condition and the bilateral Laplace transform exists, while there is no bilateral Laplace transform for the other values of σ.

When the bilateral Laplace transform of function $f(x)$ exists, the transfer functions $F_1(w)$ and $F_2(w)$, which correspond to functions $f_1(x)$ and $f_2(x)$, have the overlapping convergent domain in the w-plane as shown in Fig. 4.12 (b). At this time, the transfer function of the transmission characteristics $f(x)$ can be defined by a bilateral Laplace

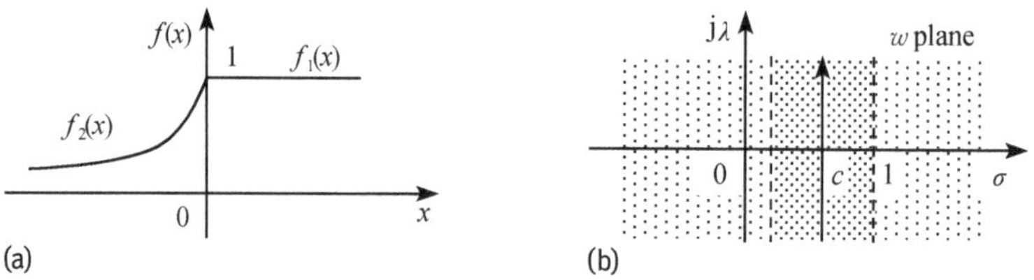

Fig. 4.12: Nonlinear function and its convergent domain of the unilateral Laplace transform.

transform as follows:

$$F(w) = F_1(w) + F_2(w) = \int_0^{\infty} f_1(x)e^{-wx}\,dx + \int_{-\infty}^{0} f_2(x)e^{-wx}\,dx = \int_{-\infty}^{\infty} f(x)e^{-wx}\,dx \quad (4.123)$$

Knowing transfer function $F(w)$, the entire transmission characteristic is:

$$f(x) = \frac{1}{2\pi j}\int_c F(w)e^{wx}\,dw \quad (4.124)$$

where the integral path c is optional in the overlapping convergence domain, and we can use the same integral path as shown in Fig. 4.12 (b).

From the above analysis, we can see that when solving the transfer function, it is necessary to use the Fourier transform, or the unilateral or bilateral Laplace transform according to the specific situation of the transmission characteristics $f(x)$. The Laplace transform is the Fourier transform when there is an overlapping convergence domain and the convergence domain contains an imaginary axis and is integrated along the imaginary axis. In the Laplace inverse transformation operation, the integral path in eqs. (4.111) or (4.124) ($\varepsilon - j\infty$ to $\varepsilon + j\infty$) constitutes a closed loop surrounding all the poles for the closed-circuit integral and it can be solved by the residue theorem.

4.3.2 Moment functions of nonlinear device output processes

When the transfer functions $F(\lambda)$ or $F(w)$ of nonlinear devices is known, the mathematical expectation of the output process $Y(t)$ can be obtained:

$$\begin{aligned} E[Y] = E[f(X)] = E\left[\frac{1}{2\pi}\int_{-\infty}^{\infty} F(\lambda)e^{j\lambda X}\,d\lambda\right] &= \frac{1}{2\pi}\int_{-\infty}^{\infty} F(\lambda)E\left[e^{j\lambda X}\right]d\lambda \\ &= \frac{1}{2\pi}\int_{-\infty}^{\infty} F(\lambda)\Phi_X(\lambda)\,d\lambda \end{aligned} \quad (4.125)$$

where $\Phi_X(\lambda) = E[e^{j\lambda X}]$ is the one-dimensional characteristic function of the input process $X(t)$, or:

$$\begin{aligned} E[Y] = E\left[\frac{1}{2\pi j}\int_c F(w)e^{wX}\,dw\right] &= \frac{1}{2\pi j}\int_c F(w)E[e^{wX}]\,dw \\ &= \frac{1}{2\pi j}\int_c F(w)\Phi_X(\lambda)\,dw \end{aligned} \quad (4.126)$$

where $\Phi_X(w) = E[e^{jwX}]$ is also the one-dimensional characteristic function of the input process $X(t)$, but its variable is $w = \sigma + j\lambda$.

When input $X(t)$ is the stationary process, the correlation function of the output process is:

$$R_Y(\tau) = E[Y(t)Y(t+\tau)] \tag{4.127}$$

Now make $Y(t) = Y_1$, $Y(t+\tau) = Y_2$ and then:

$$\begin{aligned} R_Y(\tau) &= E[Y_1 Y_2] = E[f(X_1)f(X_2)] \\ &= E\left[\frac{1}{2\pi}\int_{-\infty}^{\infty} F(\lambda_1)e^{j\lambda_1 X_1}\,d\lambda_1 \frac{1}{2\pi}\int_{-\infty}^{\infty} F(\lambda_2)e^{j\lambda_2 X_2}\,d\lambda_2\right] \end{aligned} \tag{4.128}$$

Exchanging the order of operations of the integral and the mean, we have:

$$R_Y(\tau) = \frac{1}{(2\pi)^2}\int_{-\infty}^{\infty} F(\lambda_1)\int_{-\infty}^{\infty} F(\lambda_2)\Phi_X(\lambda_1,\lambda_2;\tau)\,d\lambda_1\,d\lambda_2 \tag{4.129}$$

where $\Phi_X(\lambda_1,\lambda_2;\tau) = E[e^{j\lambda_1 X_1 + j\lambda_2 X_2}]$.

Similarly, we can obtain:

$$R_Y(\tau) = \frac{1}{(2\pi j)^2}\int_C F(w_1)\int_C F(w_2)\Phi_X(w_1,w_2;\tau)\,dw_1\,dw_2 \tag{4.130}$$

where $\Phi_X(\lambda_1,\lambda_2;\tau) = E[e^{jw_1 X_1 + jw_2 X_2}]$.

The two equations above are the general formula of the correlation function of the output of the nonlinear system (this is a stationary process but the transformation method can also be used for nonstationary processes). Because the equation contains the characteristic function, it is also called the characteristic function method. Because the equation contains the circuit integral, sometimes it is called the loop integral method.

The conversion method is used more widely than the direct method, but if the input process $X(t)$ is non-normal distribution, the calculation is still quite difficult. So, the following only discusses two kinds of typical input: (1) stationary normal noise, and (2) stationary normal noise and sinusoidal signal input.

4.3.2.1 Stationary normal noise input

We assume input stationary normal noise $n(t)$ whose mean is zero, variance is σ^2 and correlation coefficient is $r(\tau)$. So, the correlation function is:

$$R_n(\tau) = \sigma^2 r(\tau) \tag{4.131}$$

The two-dimensional characteristic function of normal process $n(t)$ is:

$$\Phi_n(\lambda_1,\lambda_2;\tau) = e^{-\frac{\sigma^2}{2}[\lambda_1^2+\lambda_2^2+2\lambda_1\lambda_2 r(\tau)]} = e^{-\frac{\sigma^2}{2}(\lambda_1^2+\lambda_2^2)}e^{-\lambda_1\lambda_2 R_n(\tau)} \tag{4.132}$$

Replacing $j\lambda$ by w, we obtain the characteristic function:

$$\Phi_n(w_1,w_2;\tau) = e^{-\frac{\sigma^2}{2}(w_1^2+w_2^2)}e^{w_1 w_2 R_n(\tau)} \tag{4.133}$$

Expanding $e^{w_1 w_2 R_n(\tau)}$ into Mark Lauryn series:

$$e^{w_1 w_2 R_n(\tau)} = \sum_{k=0}^{\infty} \frac{R_n^k(\tau)}{k!}(w_1 w_1)^k \tag{4.134}$$

So:

$$\Phi_n(w_1, w_2; \tau) = e^{-\frac{\sigma^2}{2}(w_1^2 + w_2^2)} \sum_{k=0}^{\infty} \frac{R_n^k(\tau)}{k!}(w_1 w_1)^k \tag{4.135}$$

Substituting into eq. (4.130), the correlation function of the output process $Y(t)$ is:

$$\begin{aligned} R_Y(\tau) &= \sum_{k=0}^{\infty} \frac{R_n^k(\tau)}{k!}\left[\frac{1}{(2\pi j)^2}\int_C F(w_1) w_1^k e^{\sigma^2 w_1^2/2}\, dw_1 \int_C F(w_2) w_2^k e^{\sigma^2 w_2^2/2}\, dw_2\right] \\ &= \sum_{k=0}^{\infty} \frac{R_n^k(\tau)}{k!} h_{0k}^2 \end{aligned} \tag{4.136}$$

where:

$$h_{0k} = \frac{1}{2\pi j}\int_C F(w) w^k e^{\sigma^2 w^2/2}\, dw \tag{4.137}$$

From the above, we can see that as long as a nonlinear function $y = f(x)$ is given to obtain its transfer function $F(w)$, then h_{0k} can be solved from eq. (4.127), and we can solve the correlation function $R_Y(\tau)$.

4.3.2.2 Stationary normal noise and sine-type signal common input

Let the input process be:

$$X(t) = s(t) + n(t) \tag{4.138}$$

where noise $n(t)$ is as above and $s(t)$ is the random initial phase signal:

$$s(t) = A\cos(\omega_0 t + \theta) \tag{4.139}$$

where θ is uniform distributed in $[0, 2\pi]$. Signal and noise are statistically independent.

Here, $X(t)$ is still a stationary process, and its two-dimensional characteristic function is:

$$\Phi_X(w_1, w_2; \tau) = \Phi_s(w_1, w_2; \tau)\, \Phi_n(w_1, w_2; \tau) \tag{4.140}$$

According to the definition of the characteristic function, the two-dimensional characteristic function of the signal is:

$$\Phi_s(w_1, w_2; \tau) = E\left[e^{w_1 s_1 + w_2 s_2}\right] \tag{4.141}$$

where:

$$s_1 = s_1(t) = A\cos(\omega_0 t_1 + \theta) = A\cos\varphi_1 \tag{4.142}$$

$$s_2 = s_2(t) = A\cos(\omega_0 t_2 + \theta) = A\cos\varphi_2 \tag{4.143}$$

$e^{z\cos\varphi}$ can be expanded into the series by using the following Jacobi–Anger formula:

$$e^{z\cos\varphi} = \sum_{m=0}^{\infty} \varepsilon_m I_m(z)\cos m\varphi \tag{4.144}$$

where $I_m(z)$ is the first class of m-order Bessel modified functions, ε_m is called the Nyman factor, and:

$$\varepsilon_m = \begin{cases} 1, & m = 0 \\ 2, & m = 1, 2, \ldots \end{cases} \tag{4.145}$$

Thus, we can obtain:

$$\Phi_s(w_1, w_2; \tau) = \sum_{m=0}^{\infty}\sum_{n=0}^{\infty} \varepsilon_m \varepsilon_n I_m(w_1 A) I_n(w_2 A)\overline{\cos m\varphi_1 \cos n\varphi_2} \tag{4.146}$$

Due to:

$$\overline{\cos m\varphi_1 \cos n\varphi_2} = \begin{cases} 0\,, & n \neq m \\ (1/\varepsilon_m)\cos m\omega_0\tau\,, & n = m \end{cases} \tag{4.147}$$

and so, we have:

$$\Phi_s(w_1, w_2; \tau) = \sum_{m=0}^{\infty} \varepsilon_m I_m(w_1 A)\cos m\omega_0\tau \tag{4.148}$$

We substitute eqs. (4.148) and (4.135) into eq. (4.140) and then substitute the results into eq. (4.130) to obtain the general expression of the correlation function of the output process $Y(t)$ as:

$$\begin{aligned} R_Y(\tau) &= \frac{1}{(2\pi j)^2}\int_C F(w_1)\int_C F(w_2)\Phi_X(w_1, w_2; \tau) \\ &= \Phi_s(w_1, w_2; \tau)\,\Phi_n(w_1, w_2; \tau)\,\mathrm{d}w_1\,\mathrm{d}w_2 \\ &= \sum_{k=0}^{\infty}\frac{R_n^k(\tau)}{k!}\sum_{m=0}^{\infty}\varepsilon_m\cos m\omega_0\tau \\ &\quad \left[\frac{1}{(2\pi j)^2}\int_C F(w_1)w_1^k I_m(w_1 A)e^{\sigma^2 w_1^2/2}\,\mathrm{d}w_1 \int_C F(w_2)w_2^k I_m(w_2 A)e^{\sigma^2 w_2^2/2}\,\mathrm{d}w_2\right] \\ &= \sum_{m=0}^{\infty}\sum_{k=0}^{\infty}\frac{\varepsilon_m h_{mk}^2}{k!}R_n^k(\tau)\cos m\omega_0\tau \end{aligned} \tag{4.149}$$

where:

$$h_{mk} = \frac{1}{2\pi j}\int_C F(w)w^k I_m(wA)e^{\sigma^2 w^2/2}\,\mathrm{d}w \tag{4.150}$$

Analyzing eq. (4.149), m represents the various components of the input signal and k represents the various components of the input noise, which will produce DC components, various high-frequency harmonic components, and kinds of beat frequency components after a nonlinear transformation. Although the number of items is large, it can always be divided into the following four components: the DC component h_{00}^2, the signal–signal component $R_{ss}(\tau)$, the noise–noise component $R_{nn}(\tau)$, and the signal–noise component $R_{sn}(\tau)$. Thus, the formula can be rewritten as:

$$R_Y(\tau) = h_{00}^2 + R_{ss}(\tau) + R_{nn}(\tau) + R_{sn}(\tau) \tag{4.151}$$

where:

$$R_{ss}(\tau) = 2\sum_{m=0}^{\infty} h_{m0}^2 \cos m\omega_0\tau \tag{4.152}$$

$$R_{nn}(\tau) = \sum_{k=1}^{\infty} \frac{h_{0k}^2}{k!} R_n^k(\tau) \tag{4.153}$$

$$R_{sn}(\tau) = 2\sum_{m=1}^{\infty}\sum_{k=1}^{\infty} \frac{h_{mk}^2}{k!} R_n^k(\tau)\cos m\omega_0\tau \tag{4.154}$$

Applying the Fourier transform on eq. (4.151) to obtain the general expression of the power spectral density output process $Y(t)$, we have:

$$G_Y(f) = h_{00}^2\delta(f) + G_{ss}(f) + G_{nn}(f) + G_{sn}(f) \tag{4.155}$$

The first term on the right-hand side is the DC component spectrum, and the other three items are the corresponding alternating component spectrums, where:

$$G_{ss}(f) = \sum_{m=0}^{\infty} h_{m0}^2\,[\delta(f + mf_0) + \delta(f - mf_0)] \tag{4.156}$$

$$G_{nn}(f) = \sum_{k=1}^{\infty} \frac{h_{0k}^2}{k!}\, {}_kG_n(f) \tag{4.157}$$

$$G_{sn}(f) = \sum_{m=1}^{\infty}\sum_{k=1}^{\infty} \frac{h_{mk}^2}{k!}\,[{}_kG_n(f + mf_0) + {}_kG_n(f - mf_0)] \tag{4.158}$$

where ${}_kG_n(f)$ is the Fourier transform of $R_n^k(\tau)$ and the k-times convolution integral of $G_n(f)$ with itself:

$${}_kG_n(f) = \underbrace{G_n(f) \otimes G_n(f) \otimes \cdots \otimes G_n(f)}_{k\text{ times}} \tag{4.159}$$

For example, when input noise $n(t)$ has the Gaussian narrowband power spectrum [as shown in Fig. 4.13 (a)], then ${}_1G_n(f) = G_n(f)$, ${}_2G_n(f) = G_n(f) \otimes G_n(f)$, and ${}_3G_n(f) = {}_2G_n(f) \otimes G_n(f)$, as shown in Fig. 4.13 (b) and (c). Since the convolution result between the Gaussian function and itself is still a Gaussian function, the shape of each component spectrum in the graph is still a Gaussian shape, and the peak value may be different.

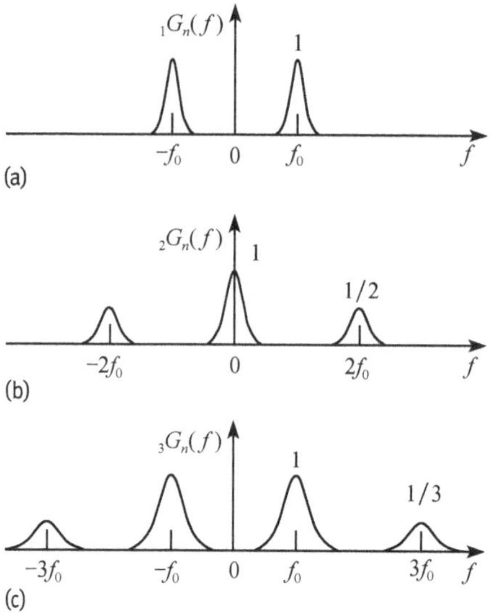

Fig. 4.13: Convolution of Gaussian-shaped narrowband noise.

It should be pointed out that eqs. (4.151) and (4.155) are the results obtained when the input signal is an equal amplitude sinusoidal signal. The same method can also be used to derive the result when the input signal is an amplitude modulated sinusoidal signal. It should also be pointed out that the two equations above are the correlation function and power spectral density of nonlinear device outputs. According to the use of nonlinear devices, one generally need to add a filter behind the nonlinear devices to select the required component. For example, if the nonlinear device is an envelope detector, then the addition of a lowpass filter is needed. At this time, the output only contains DC and low-frequency components. Now, because the input is an equal amplitude signal, the output signal only contains h_{00}^2, while the output noise only contains the low-frequency components of $R_{nn}(\tau)$ and $R_{sn}(\tau)$, which can be obtained from the low-frequency component spectrum after convolution in eqs. (4.157) and (4.158). For example, if the nonlinear device is a limiting IF amplifier, we need to add a high-frequency narrowband filter whose central frequency is w_0, and then the output only contains the frequency components near the high-frequency fundamental wave ($m = 1$). So, the output signal only contains the item $2h_{10}^2 \cos \omega_0 \tau$, and the output noise only contains the high-frequency fundamental components of $R_{nn}(\tau)$ and $R_{sn}(\tau)$, which can be obtained from the high-frequency fundamental components' spectrum after convergence in eqs. (4.157) and (4.158).

4.3.3 The price method

This method is a special case of the transformation method and applies for the certain condition where the input is a stationary normal process and nonlinear function can change to the δ-function after k-times derivation. This method can greatly simplify the integral operation since it skillfully uses the integral characteristic of the δ-function.

We assume that the input $X(t)$ is a stationary normal process, whose mean is zero, variance is σ^2 and correlation coefficient is $r(\tau)$. It is known that the two-dimensional probability density is:

$$p_2(x_1, x_2; \tau) = \frac{1}{2\pi\sigma^2\sqrt{1-r^2}} \exp\left[-\frac{x_1^2 + x_1^2 - 2rx_1x_2}{2\sigma^2(1-r^2)}\right] \tag{4.160}$$

where $x_1 = X(t_1)$, $x_2 = X(t_2)$, $r = r(\tau)$.

The corresponding two-dimensional characteristic function is:

$$\Phi_X(\lambda_1, \lambda_2; \tau) = \exp\left[-\frac{\sigma^2}{2}\left(\lambda_1^2 + \lambda_2^2 - 2r\lambda_1\lambda_2\right)\right] \tag{4.161}$$

At the two ends of the equation we take the k-order derivative for r:

$$\frac{\partial^k \Phi_X}{\partial r^k} = (-\sigma^2\lambda_1\lambda_2)^k \Phi_X(\lambda_1, \lambda_2; \tau) = (-\sigma^2\lambda_1\lambda_2)^k E[\mathrm{e}^{\mathrm{j}\lambda_1 X_1 + \mathrm{j}\lambda_2 X_2}] \tag{4.162}$$

and at the two ends of eq. (4.129) we take the k-order derivative for r to obtain:

$$\frac{\partial^k R_Y}{\partial r^k} = \frac{1}{(2\pi)^2} \int_{-\infty}^{\infty} F(\lambda_1) \int_{-\infty}^{\infty} F(\lambda_2) \frac{\partial^k \Phi_X}{\partial r^k} \, \mathrm{d}\lambda_1 \, \mathrm{d}\lambda_2 \tag{4.163}$$

Substituting eq. (4.162) into the formula, we obtain:

$$\frac{\partial^k R_Y}{\partial r^k} = \sigma^{2k} E\left[\frac{1}{(2\pi)^2} \int_{-\infty}^{\infty} (\mathrm{j}\lambda_1)^k F(\lambda_1) \mathrm{e}^{\mathrm{j}\lambda_1 X_1} \, \mathrm{d}\lambda_1 \int_{-\infty}^{\infty} (\mathrm{j}\lambda_2)^k F(\lambda_2) \mathrm{e}^{\mathrm{j}\lambda_2 X_2} \, \mathrm{d}\lambda_2\right] \tag{4.164}$$

The nonlinear function $f(X)$ and the transfer function $F(\lambda)$ are a pair of Fourier transforms, and the relationship is:

$$f(X) = \frac{1}{2\pi} \int_{-\infty}^{\infty} F(\lambda) \mathrm{e}^{\mathrm{j}\lambda X} \, \mathrm{d}\lambda \tag{4.165}$$

For this equation, we take the k-order derivative for X to obtain:

$$f^{(k)}(X) = \frac{1}{2\pi} \int_{-\infty}^{\infty} (\mathrm{j}\lambda)^k F(\lambda) \mathrm{e}^{\mathrm{j}\lambda X} \, \mathrm{d}\lambda \tag{4.166}$$

Thus, eq. (4.164) can be rewritten as:

$$\begin{aligned} \frac{\partial^k R_Y}{\partial r^k} &= \sigma^{2k} E\left[f^{(k)}(X_1) f^{(k)}(X_2)\right] \\ &= \sigma^{2k} \int_{-\infty}^{\infty} \int_{-\infty}^{\infty} f^{(k)}(x_1) f^{(k)}(x_2) p_2(x_1, x_2; \tau) \, \mathrm{d}x_1 \, \mathrm{d}x_2 \end{aligned} \tag{4.167}$$

Substitute eq. (4.160) into eq. (4.168) to obtain:

$$\frac{\partial^k R_Y}{\partial r^k} = \frac{\sigma^{2k-2}}{2\pi\sqrt{1-r^2}} \int_{-\infty}^{\infty}\int_{-\infty}^{\infty} f^{(k)}(x_1) f^{(k)}(x_2) \exp\left[-\frac{x_1^2 + x_1^2 - 2rx_1x_2}{2\sigma^2(1-r^2)}\right] \mathrm{d}x_1\,\mathrm{d}x_2 \quad (4.168)$$

If, after taking the k-times derivative, the function $f(x)$ can become a δ-function, then the following integral property of δ-function is used:

$$\int_{-\infty}^{\infty} f(x)\delta(x - x_0)\,\mathrm{d}x = f(x_0) \quad (4.169)$$

The double integral in eq. (4.168) can be greatly simplified so as to obtain differential equations. Solving the differential equations, and we obtain the correlation function $R_Y(\tau)$.

Example 4.3.1. Suppose that the nonlinear device has a half-wave linearity law characteristic:

$$y = f(x) = \begin{cases} bx\,, & x \geq 0 \\ 0\,, & x < 0 \end{cases} \quad (4.170)$$

Input stationary normal noise $X(t)$ is described as above. Solve the correlation function of the output process $Y(t)$.

Solution: The above characteristics are shown in Fig. 4.14 (a). It can become a δ-function after derivation twice:

$$f'(x) = b\,, \quad f''(x) = b\delta(x)\,, \quad x \geq 0 \quad (4.171)$$

As is shown in Figs. 4.14 (b) and (c). So, the price method can be used for the solution.

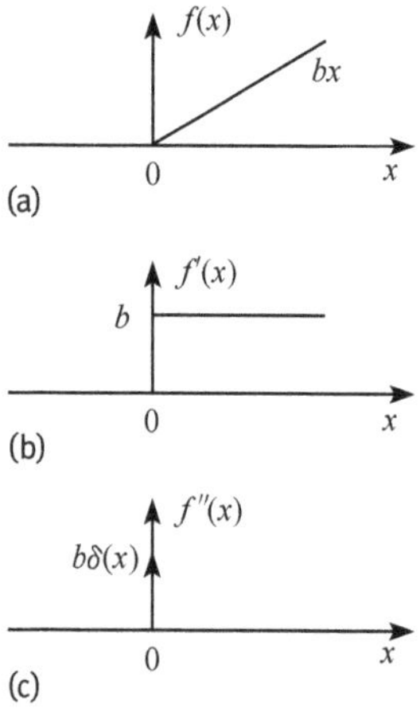

Fig. 4.14: Characteristics of the half wave linear law and their derivatives.

Letting $k = 2$, eq. (4.168) we have:

$$\frac{\partial^2 R_Y(\tau)}{\partial r^2} = \frac{\sigma^2}{2\pi\sqrt{1-r^2}} \int_{-\infty}^{\infty}\int_{-\infty}^{\infty} b^2 \delta(x_1)\delta(x_2) \exp\left[-\frac{x_1^2 + x_1^2 - 2rx_1x_2}{2\sigma^2(1-r^2)}\right] \mathrm{d}x_1\,\mathrm{d}x_2 \tag{4.172}$$

We use eq. (4.169) to obtain:

$$\frac{\partial^2 R_Y(\tau)}{\partial r^2} = \frac{b^2\sigma^2}{2\pi\sqrt{1-r^2}} \tag{4.173}$$

We solve the second-order differential equation to obtain:

$$\frac{\partial R_Y(\tau)}{\partial r} = \int \frac{b^2\sigma^2}{2\pi\sqrt{1-r^2}} \mathrm{d}r = \frac{b^2\sigma^2}{2\pi} \sin^{-1} r + C_1 \tag{4.174}$$

We use the boundary condition when $\tau \to \infty$, $r(\infty) = 0$ and from eq. (4.168) we can obtain:

$$\left.\frac{\partial R_Y(\tau)}{\partial r}\right|_{r=0} = \frac{1}{2\pi} \int_0^{\infty}\int_0^{\infty} b^2 \exp\left[-\frac{x_1^2 + x_1^2}{2\sigma^2}\right] \mathrm{d}x_1\,\mathrm{d}x_2 = \frac{b^2\sigma^2}{4} \tag{4.175}$$

Substituting into the above formula we obtain the determined constant:

$$C_1 = \left.\frac{\partial R_Y(\tau)}{\partial r}\right|_{r=0} = \frac{b^2\sigma^2}{4} \tag{4.176}$$

So, the first-order differential equation:

$$\frac{\partial R_Y(\tau)}{\partial r} = \frac{b^2\sigma^2}{4} + \frac{b^2\sigma^2}{2\pi} \sin^{-1} r \tag{4.177}$$

Thus:

$$\begin{aligned} R_Y(\tau) &= \int \left(\frac{b^2\sigma^2}{4} + \frac{b^2\sigma^2}{2\pi} \sin^{-1} r\right) \mathrm{d}r \\ &= \frac{b^2\sigma^2}{4} r + \frac{b^2\sigma^2}{2\pi}\left(r \sin^{-1} r + \sqrt{1-r^2}\right) + C_2 \end{aligned} \tag{4.178}$$

We use the boundary condition when $\tau \to \infty$, $r(\infty) = 0$. At this time, by eq. (4.168) we can obtain:

$$R_Y(\tau)|_{r=0} = \frac{1}{2\pi\sigma^2} \int_0^{\infty}\int_0^{\infty} b^2 x_1 x_2 \exp\left[-\frac{x_1^2 + x_1^2}{2\sigma^2}\right] \mathrm{d}x_1\,\mathrm{d}x_2 = \frac{b^2\sigma^2}{2\pi} \tag{4.179}$$

Substituting into eq. (4.179), the determined constant can be obtained:

$$C_2 = R_Y(\tau)|_{r=0} - \frac{b^2\sigma^2}{2\pi} = 0 \tag{4.180}$$

The correlation function of output process $Y(t)$ is:

$$R_Y(\tau) = \frac{b^2\sigma^2}{4} r(\tau) + \frac{b^2\sigma^2}{2\pi}\left(r(\tau)\sin^{-1} r(\tau) + \sqrt{1 - r^2(\tau)}\right) \quad (4.181)$$

It should be pointed out that this equation is different from eq. (4.81), but the actual results are consistent. We use the following series:

$$(1 - r^2)^{1/2} = 1 - \frac{r^2}{2} - \frac{r^4}{2\cdot 4} - \frac{1\cdot 3r^6}{2\cdot 4\cdot 6} - \frac{1\cdot 3\cdot 5r^8}{2\cdot 4\cdot 6\cdot 8} - \cdots, \quad -1 \le r \le 1 \quad (4.182)$$

$$\sin^{-1} r = r + \frac{r3}{6} + \frac{1\cdot 3}{2\cdot 4}\cdot\frac{r^5}{5} + \frac{1\cdot 3\cdot 5}{2\cdot 4\cdot 6}\cdot\frac{r7}{7} + \cdots, \quad r^2 < 1 \quad (4.183)$$

Equation (4.181) can be changed into eq. (4.81).

4.4 Slowly changing envelopment method for random process nonlinear transformation

As was mentioned above, the high-frequency narrowband process can be approximated as the quasi-sinusoidal oscillation of a random amplitude for slow modulation, the expression is:

$$X(t) = A(t)\cos\left[\omega_0 t + \varphi(t)\right] \quad (4.184)$$

where relative to ω_0, envelope $A(t)$ and phase $\varphi(t)$ are random slow-changing time functions.

For this high-frequency narrowband process with envelope detection, we only consider the nonlinear transformation of envelope $A(t)$ to greatly simplify the calculation, which is the outstanding advantages of the slowly changing envelope method. However, it should be noticed that the use of this method is limited. The nonlinear system input must be a high-frequency narrowband process.

Depending on whether or not the load reaction of a nonlinear system is noted, the slowly changing envelope method can be divided into two cases: no load reaction and load reaction.

4.4.1 Slowly changing envelope method without load reaction

The transmission characteristic of the nonlinear device is:

$$y = f(x) \quad (4.185)$$

The input is a high-frequency narrowband process:

$$X(t) = A(t)\cos[\omega_0 t + \varphi(t)] \quad (4.186)$$

Thus, the output random process of the nonlinear device is:

$$Y(t) = f\{X(t)\} = f\{A(t)\cos[\omega_0 t + \varphi(t)]\} \tag{4.187}$$

Due to nonlinear transformation, the output process $Y(t)$ will be different from the input process $X(t)$ and will produce waveform distortion of the high-frequency signal and many new spectral components.

When input $X(t)$ is the high-frequency narrowband process ($\Delta\omega/\omega_0 \ll 1$), because it can be approximated as quasi-sinusoidal oscillation, that is, in a short period of time (but there are many cycles of sinusoidal oscillation), that is, $1/\omega_0 \ll T \ll 1/\Delta\omega$, phase $\varphi(t)$ can be seen as a constant and $\varphi(t) \approx \varphi$; similarly, the amplitude $A(t) \approx A$. Thus, during this time, the random process $Y(t)$ is an approximately distorted sine wave, which can be treated as a periodic process, i.e.,

$$Y(t) = f\left\{A\cos\left[\omega_0 t + \varphi\right]\right\} \tag{4.188}$$

With $\varphi = \omega_0 t + \varphi$, the value of the random process at the moment t is:

$$Y(t) = f\left\{A\cos\varphi\right\} \tag{4.189}$$

This equation indicates that the random process Y is the periodic function of nonrandom variable ϕ, so it can be expanded into the following Fourier series:

$$Y(t) = f_0(A) + f_1(A)\cos\varphi + f_2(A)\cos 2\varphi + \cdots + f_n(A)\cos n\varphi \tag{4.190}$$

where

$$f_0(A) = \frac{1}{\pi}\int_0^{\pi} f(A\cos\varphi)\,\mathrm{d}\varphi$$

$$f_n(A) = \frac{2}{\pi}\int_0^{\pi} f(A\cos\varphi)\cos n\varphi\,\mathrm{d}\varphi\,, \quad n \neq 0$$

The formula above will be recorded as:

$$Y = I_0 + I_1 + I_2 + \cdots + I_n \tag{4.191}$$

where $I_0 = f_0(A)$, indicating the low-frequency component of the output (including DC), and $I_n = f_n(A)\cos n(\omega_0 t + \varphi)$, indicating the respective high-frequency harmonic components of the output.

In fact, A and φ are slowly changing and can be replaced by $A(t)$ and $\varphi(t)$, that is, we can approximately obtain the output random process $Y(t)$.

Depending on the analysis of the harmonic components of the narrowband random process, the power spectrum of the output random process $Y(t)$ is intermittent, which distributes the narrowband range near $\omega = n\omega_0 (n = 0, 1, 2, \ldots)$, as shown in Fig. 4.15 below.

In the following we discuss the process of solving the moment function and the power spectrum of the output process $Y(t)$.

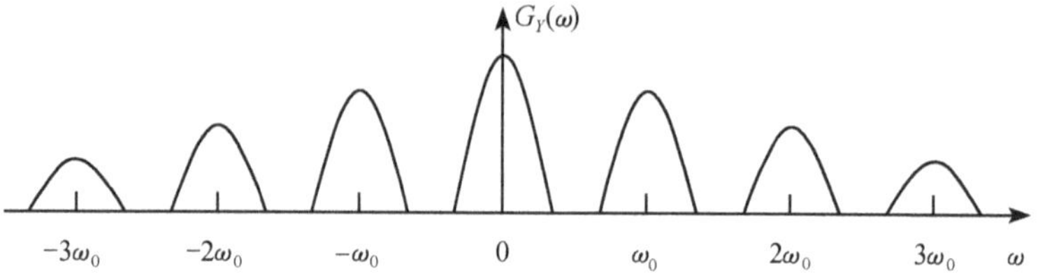

Fig. 4.15: Power spectrum of an output random process with high-frequency narrowband process input.

4.4.1.1 Statistical mean and variance

Calculate the statistical average for eq. (4.191) to obtain the statistical mean:

$$\overline{Y} = \overline{I_0} + \overline{I_1} + \overline{I_2} + \cdots + \overline{I_n} \tag{4.192}$$

because when $n > 0$, we have:

$$\overline{I_n} = \overline{f_n(A)\cos n\varphi} = \overline{f_n(A)} \cdot \overline{\cos n\varphi} = 0 \tag{4.193}$$

This formula uses the previous conclusion, that is, at some point t the envelope value A and the phase value φ of the narrowband process are statistically independent, so their functions $f_n(A)$ and $\cos n\varphi$ are also statistically independent. So:

$$\overline{Y} = \overline{I_0} = \overline{f_0(A)} = \int_{-\infty}^{\infty} f_0(A)p(A)\,\mathrm{d}A \tag{4.194}$$

The low-frequency fluctuation component of the output process is:

$$I_L = I_0 - \overline{I_0} \tag{4.195}$$

So, the variance of the low-frequency component of the output process is:

$$\sigma_L^2 = \overline{I_0^2} - \left(\overline{I_0}\right)^2 \tag{4.196}$$

where $\overline{I_0^2}$ is the mean square value of the low-frequency component of the output process. Its value is:

$$\overline{I_0^2} = \overline{f_0^2(A)} = \int_{-\infty}^{\infty} f_0^2(A)p(A)\,\mathrm{d}A \tag{4.197}$$

From the above, we can see that when we calculate the statistical mean and the variance of the low-frequency component of the output process, we only need to know the one-dimensional distribution of the envelope value A of the input narrowband process, and it is unrelated to the phase value φ. (This value has been obtained previously. When there is no signal, it is Rayleigh distribution. Otherwise, it is generalized Rayleigh distribution.)

4.4.1.2 Correlation function and power spectrum

For the moment t, the output process is a random variable:

$$Y = I_0 + I_1 + I_2 + \cdots + I_n \tag{4.198}$$

For the moment $t + \tau$, the output process is also a random variable:

$$Y_\tau = I_{0\tau} + I_{1\tau} + I_{2\tau} + \cdots + I_{n\tau} \tag{4.199}$$

When the input is a stationary narrowband process, according to the definition, the correlation function of the output process is:

$$R_Y(\tau) = \overline{YY_\tau} = \overline{(I_0 + I_1 + I_2 + \cdots + I_n)(I_{0\tau} + I_{1\tau} + I_{2\tau} + \cdots + I_{n\tau})} \tag{4.200}$$

where

$$\overline{I_m I_{n\tau}} = \overline{f_m(A) f_n(A_\tau) \cos n(\omega_0 t + \varphi) \cos n[\omega_0(t+\tau) + \varphi_\tau]}$$

Because the envelope process $A(t)$ and phase process $\varphi(t)$ are not statistically independent, when solving the statistical average value by eq. (4.200), the following fourth-order integral should be used:

$$\overline{I_m I_{n\tau}} = \int_0^\infty \int_0^\infty \int_0^{2\pi} \int_0^{2\pi} [\bullet] p_4(A, A_\tau; \varphi, \varphi_\tau) \, \mathrm{d}A \, \mathrm{d}A_\tau \, \mathrm{d}\varphi \, \mathrm{d}\varphi_\tau$$
$$[\bullet] = f_m(A) f_n(A_\tau) \cos n(\omega_0 t + \varphi) \cos n[\omega_0(t+\tau) + \varphi_\tau] \tag{4.201}$$

where:

$$\begin{aligned} p_4(A, A_\tau; \varphi, \varphi_\tau) &= \frac{AA_\tau}{4\pi^2\sigma^2(1-r^2)} \exp\left\{-\frac{1}{2\sigma^4(1-r^2)}\left[\sigma^2\left(A^2 + A_\tau^2\right) - 2A(\tau)AA_\tau \cos(\varphi_\tau - \varphi)\right]\right\} \\ &\qquad A, A_\tau \geq 0\,, \quad 0 \leq \varphi, \varphi_\tau \leq 2\pi \end{aligned} \tag{4.202}$$

We expand $\exp[\cos(\varphi_\tau - \varphi)]$ to the power series of $\cos\varphi_\tau$ and $\cos\varphi$ and use the orthogonal characteristics of the cosine function, which can prove:

$$\begin{cases} \overline{I_m I_{n\tau}} = 0\,, & m \neq n \\ \overline{I_m I_{n\tau}} \neq 0\,, & m = n \end{cases} \tag{4.203}$$

It can be obtained that:

$$\begin{aligned} R_Y(\tau) &= \overline{I_0 I_{0\tau}} + \overline{I_1 I_{1\tau}} + \overline{I_2 I_{2\tau}} + \cdots + \overline{I_n I_{n\tau}} \\ &= R_0(\tau) + R_1(\tau) + R_2(\tau) + \cdots + R_n(\tau) \end{aligned} \tag{4.204}$$

where:

$$R_0(\tau) = \overline{I_0 I_{0\tau}} = \overline{f_0(A) f_0(A_\tau)} = \int_{-\infty}^{\infty}\int_{-\infty}^{\infty} f_0(A) f_0(A_\tau) p_2(A, A_\tau)\,\mathrm{d}A\,\mathrm{d}A_\tau \qquad (4.205)$$

$$p_2(A, A_\tau) = \begin{cases} \frac{AA_\tau}{\sigma^4(1-r^2)} I_0\left[\frac{AA_\tau r}{\sigma^2(1-r^2)}\right] \exp\left\{-\frac{A^2+A_\tau^2}{2\sigma^2(1-r^2)}\right\}, & A, A_\tau \geq 0 \\ 0, & \text{otherwise} \end{cases} \qquad (4.206)$$

$$R_n(\tau) = \overline{I_n I_{n\tau}} = \overline{f_n(A) f_n(A_\tau) \cos n(\omega_0 t + \varphi) \cos n[\omega_0(t+\tau) + \varphi_\tau]} \qquad (4.207)$$

Applying the Fourier transform to eq. (4.204) we obtain the power spectrum of the output process as:

$$G_Y(\omega) = G_0(\omega) + G_1(\omega) + G_2(\omega) + \cdots + G_n(\omega) \qquad (4.208)$$

In summary, solving the correlation function and power spectrum of high-frequency components of the output by using the slowly changing envelope method, we must know the four-dimensional distribution, whose operation is difficult and not practical. However, when this method is used for the envelope detector, it is only necessary to solve the low-frequency components, and the covariance function of the low-frequency component of the output process is:

$$R_L(\tau) = \overline{I_L I_{L\tau}} = \overline{I_0 I_{0\tau}} - \left(\overline{I_0}\right)^2 = \overline{f_0(A) f_0(A_\tau)} - \left[\overline{f_0(A)}\right]^2 \qquad (4.209)$$

It can be seen that if the two-dimensional distribution of the input process is known, the operation is relatively easy. Although the actual operation is still quite difficult, the main purpose of solving correlation function and power spectrum of the output process is to solve the DC and low-frequency power of the output process. At this time, we only need to know the one-dimensional distribution of the input process, which can simplify the operation. If we want to know the power spectrum of the output process, we can use the difference beat method described above.

Example 4.4.1. Stationary normal narrowband noise $X(t)$ passes through a half-wave linear detector. It is known that the nonlinear function relationship is described as eq. (4.74), $X(t) = A(t)\cos[\omega_0 t + \varphi(t)]$. Solve the DC power and low-frequency fluctuation power of the output process $Y(t)$.

Solution: The output values corresponding to the input value $X = A\cos\varphi$ at the moment t are:

$$Y = f(A\cos\varphi) = \begin{cases} bA\cos\varphi, & -\frac{\pi}{2} \leq \varphi \leq \frac{\pi}{2} \\ 0, & -\pi < \varphi < -\frac{\pi}{2},\ -\frac{\pi}{2} < \varphi < \pi \end{cases} \qquad (4.210)$$

as shown in Fig. 4.16.

The low-frequency component (including DC) of the output process is:

$$I_0 = f_0(A) = \frac{1}{\pi}\int_0^{\pi} f(A\cos\varphi)\,\mathrm{d}\varphi = \frac{1}{\pi}\int_0^{\pi} bA\cos\varphi\,\mathrm{d}\varphi = \frac{bA}{\pi} \qquad (4.211)$$

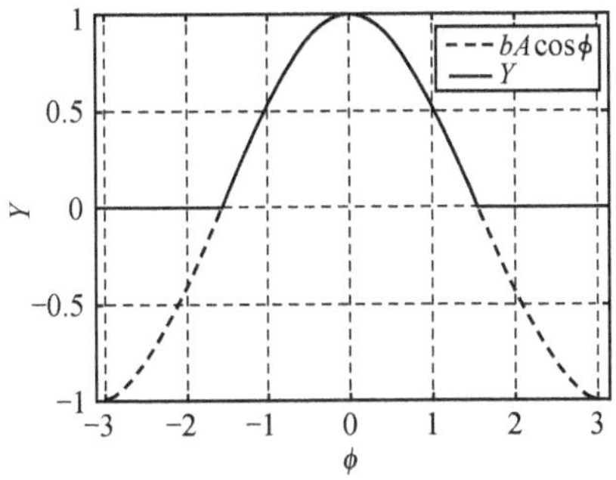

Fig. 4.16: The range of variation of $Y = bA\cos\varphi$.

The DC component of the output process is:

$$\overline{I_0} = \overline{f_0(A)} = \int_{-\infty}^{\infty} f_0(A)p(A)\,\mathrm{d}A = \int_{0}^{\infty} \frac{bA}{\pi} \cdot \frac{A}{\sigma^2}\mathrm{e}^{-\frac{A}{2\sigma^2}}\,\mathrm{d}A = \frac{b\sigma}{\sqrt{2\pi}} \tag{4.212}$$

So, the DC power of the output process is:

$$\left(\overline{I_0}\right)^2 = \frac{b^2\sigma^2}{2\pi} \tag{4.213}$$

The low-frequency fluctuation power of the output process is the variance:

$$\sigma_L^2 = \overline{I_0^2} - \left(\overline{I_0}\right)^2 \tag{4.214}$$

where

$$\overline{I_0^2} = \overline{f_0^2(A)} = \int_{-\infty}^{\infty} f_0^2(A)p(A)\,\mathrm{d}A = \int_{0}^{\infty} \left(\frac{bA}{\pi}\right)^2 \cdot \frac{A}{\sigma^2}\mathrm{e}^{-\frac{A}{2\sigma^2}}\,\mathrm{d}A = 2\left(\frac{b\sigma}{\pi}\right)^2 \tag{4.215}$$

So that:

$$\sigma_L^2 = 2\left(\frac{b\sigma}{\pi}\right)^2 - \frac{b^2\sigma^2}{2\pi} = b^2\sigma^2\frac{4-\pi}{2\pi^2} \tag{4.216}$$

Example 4.4.2. Stationary normal narrowband noise $X(t)$ passes through a full-wave square-law detector. It is known that the nonlinear function relationship is $y = bx^2$, $X(t) = A(t)\cos[\omega_0 t + \varphi(t)]$. Solve the DC power and low-frequency fluctuation power of the output process $Y(t)$.

Solution: The output values corresponding to the input value $X = A\cos\varphi$ at the moment t are:

$$Y = bX^2 = b(A\cos\varphi)^2 = \frac{bA^2}{2} + \frac{bA^2}{2}\cos 2\varphi \tag{4.217}$$

Comparing this with eq. (4.190), we can see:

$$f_0(A) = \frac{bA^2}{2}, \quad f_2(A) = \frac{bA^2}{2} \tag{4.218}$$

The low-frequency component (including DC) of the output process is:

$$I_0 = f_0(A) = \frac{bA^2}{2} \tag{4.219}$$

So, the DC component is:

$$\overline{I_0} = \overline{f_0(A)} = \int_{-\infty}^{\infty} f_0(A)p(A)\,\mathrm{d}A = \int_0^{\infty} \frac{bA^2}{2} \cdot \frac{A}{\sigma^2}\mathrm{e}^{-\frac{A}{2\sigma^2}}\,\mathrm{d}A = b\sigma^2 \tag{4.220}$$

The DC power of the output process is:

$$\left(\overline{I_0}\right)^2 = b^2\sigma^4 \tag{4.221}$$

The low-frequency component of the output process is:

$$\overline{I_0^2} = \overline{f_0^2(A)} = \int_{-\infty}^{\infty} f_0^2(A)p(A)\,\mathrm{d}A = \int_0^{\infty} \left(\frac{bA^2}{2}\right)^2 \cdot \frac{A}{\sigma^2}\mathrm{e}^{-\frac{A}{2\sigma^2}}\,\mathrm{d}A = 2b^2\sigma^4 \tag{4.222}$$

So, the low-frequency fluctuation power of the output process is:

$$\sigma_L^2 = \overline{I_0^2} - \left(\overline{I_0}\right)^2 = 2b^2\sigma^4 - b^2\sigma^4 = b^2\sigma^4 \tag{4.223}$$

Comparing the above two cases with the cases of Examples 4.2.1 and 4.2.2, which were solved with the direct method, it can be seen that the results are the same, but obviously the slowly changing envelope method is much simpler for calculation. So, the slowly changing envelope method is an important method to solve the stationary narrowband process through the envelope detector.

4.4.2 Slowly changing envelope method with load reaction

In the actual envelope detector, the voltage on the detector load resistor reacts to the nonlinear device. When this reaction is to be taken into account in the nonlinear transformation, the reaction effect can be reflected in the detector's voltage transmission coefficient, which is the same as the case where the known signal is through the envelope detector.

The typical circuit of the envelope detector is shown in Fig. 4.17. The voltage transmission coefficient is defined as the ratio of the output signal envelope to the input signal envelope $R(t)$, that is,

$$K_d = \frac{Y(t)}{R(t)} \tag{4.224}$$

where K_d is related to load R_L, C_L and the inner product r_1 of the nonlinear device, and while square-law detecting, it is also related to the size of the input signal envelope $R(t)$. When these values are known, K_d is determined.

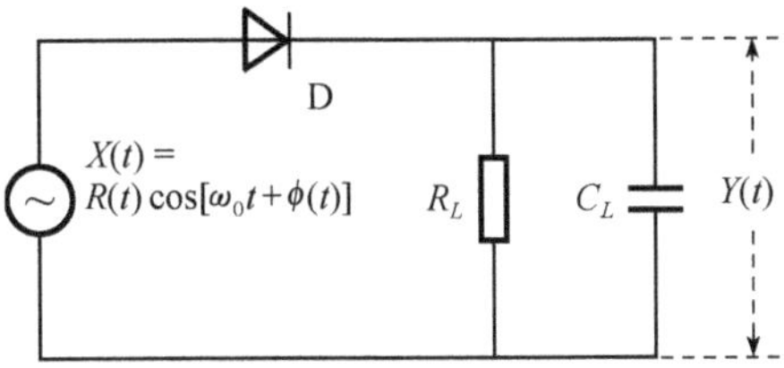

Fig. 4.17: Typical circuit of the envelope detector.

The effect of the envelope detector is to demodulate the amplitude modulation high-frequency oscillation and remove the envelope carrying information, which should satisfy the narrowband conditions $\Delta\omega/\omega_0 \ll 1$. So, the load of the envelope detector should satisfy the following two conditions:

(1) Time constant $\tau(= RC)$ should be larger than the high-frequency cycle $T_0(= \omega_0/2\pi)$, that is,

$$\tau \gg T_0 \tag{4.225}$$

to filter out the high-frequency components in the output process.

(2) The voltage changes on the load should keep up with the envelope changes of the input process so as to not produce cutting distortion of the waveform, so the needed time constant τ should be much smaller than the relevant time τ_0 of the input process envelope, that is,

$$\tau \ll \tau_0 \tag{4.226}$$

A correctly designed envelope detector can satisfy these two conditions, so the output process of general envelope detector is:

$$Y(t) = K_d R(t) \tag{4.227}$$

Depending on the type of envelope detector (linear or square law) and the component numbers of the circuits, the corresponding voltage transmission factor K_d is checked from the graphs shown in Figs. 4.18 or 4.19. We use the above formula to calculate the statistical properties of the output process. The derivation theory of the curves shown in Figs. 4.18 and 4.19 is given in Appendix A.3. In Fig. 4.17, R_L is the resistance of the detector load, R_i is the internal resistor of detection diode, $b = 1/R_i$ and A is the voltage of the input envelope.

We assume that the input signal $s(t)$ is a random initial phase signal:

$$s(t) = A\cos(\omega_0 t + \theta) \tag{4.228}$$

Input noise $n(t)$ is a stationary normal narrowband process whose mean is zero and variance is σ^2:

$$n(t) = n_c(t)\cos\omega_0 t - n_s(t)\sin\omega_0 t \tag{4.229}$$

So, the input synthesis process is still a stationary narrowband process, that is:

$$X(t) = s(t) + n(t) = R(t)\cos\left[\omega_0 t + \varphi(t)\right] \tag{4.230}$$

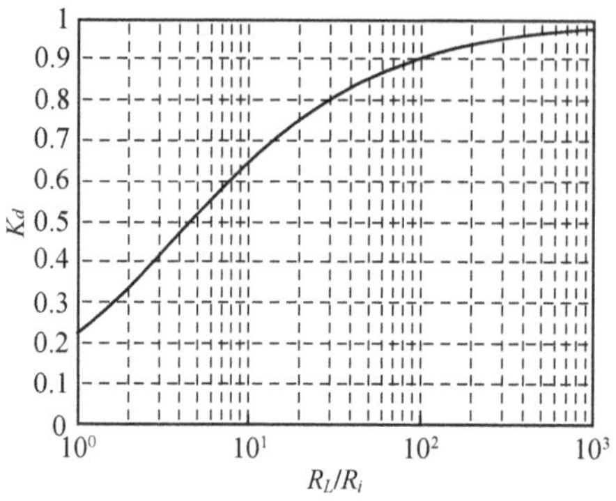

Fig. 4.18: The relationship curve of K_d and R_L/R_i of the linear detector.

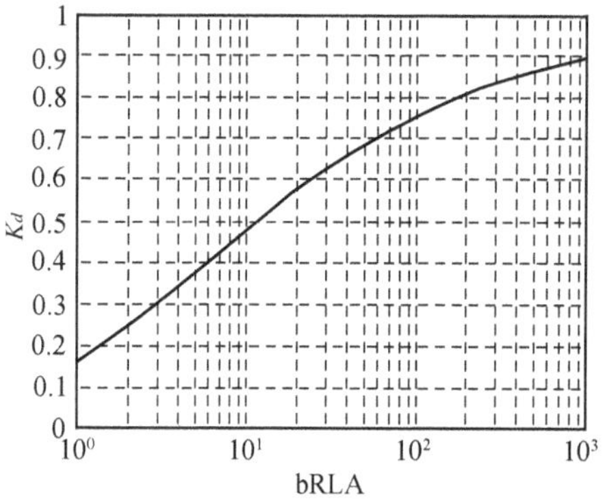

Fig. 4.19: The relationship curve of K_d and R_L/R_i of the linear detector.

From the previous analysis we can see that the envelope $R(t)$ obeys the generalized Rayleigh distribution; the probability density is:

$$p(R) = \frac{R}{\sigma^2} \exp\left[-\frac{R^2 + A^2}{2\sigma^2}\right] I_0\left(\frac{RA}{\sigma^2}\right) \tag{4.231}$$

4.4.2.1 DC voltage of the detector output

When there is no signal but only noise, the output voltage is:

$$U_n = K_d A' \tag{4.232}$$

where A' is the envelope of the input noise voltage, and K_d is the transmission coefficient of the detector voltage.

We calculate the average on the both sides of the formula:

$$\overline{U_n} = K_d \overline{A'} \tag{4.233}$$

When signal and noise are input at the same time, the output voltage is:

$$U_{sn} = K_d R \tag{4.234}$$

where R is the composite voltage envelope of the input.

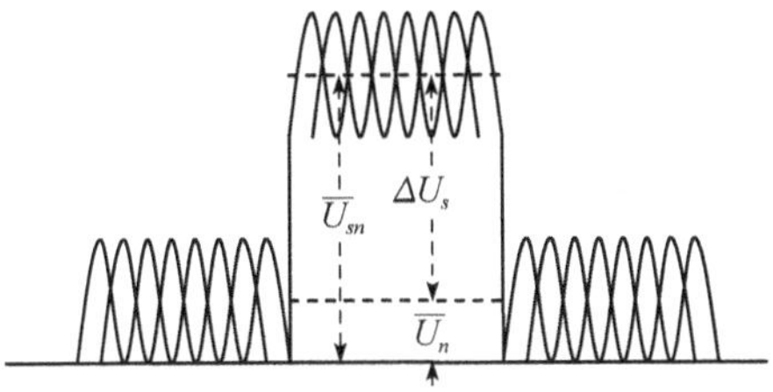

Fig. 4.20: Output voltage of narrowband noise and a constant amplitude sine signal passing a detector.

We calculate the average on the both sides of the formula:

$$\overline{U_{sn}} = K_d\overline{R} \tag{4.235}$$

The output voltage increment caused by the input signal is:

$$\Delta U_s = \overline{U_{sn}} - \overline{U_n} \tag{4.236}$$

When detecting a radar signal, it represents the signal component in the output voltage. The arguments above are shown in Fig. 4.20. The following is the solution of the detection characteristic and the ratio of the signal power to noise power at the output side in this detector.

$$\overline{R} = \int_{-\infty}^{\infty} R \cdot p(R)\,\mathrm{d}R = \int_{-\infty}^{\infty} \frac{R^2}{\sigma^2} \exp\left[-\frac{R^2 + A^2}{2\sigma^2}\right] I_0\left(\frac{RA}{\sigma^2}\right) \mathrm{d}R \tag{4.237}$$

After the integral operation for this formula (see Appendix 4), we can obtain:

$$\overline{R} = \sqrt{\frac{\pi}{2}}\sigma \cdot \mathrm{e}^{-\sigma^2/2}\left[\left(1 + Q^2\right) I_0\left(\frac{Q^2}{2}\right) + Q^2 I_1\left(\frac{Q^2}{2}\right)\right] \tag{4.238}$$

where $Q = \frac{A}{\sqrt{2}\sigma}$ is the ratio of signal voltage to noise voltage at the input; $I_0(\bullet)$ and $I_1(\bullet)$ are, respectively, the zero-order and the first-order modified Bessel function.

If there is no signal but only noise, $Q = 0$. Because $I_0(0) = 1$ and $I_1(0) = 0$, from the above formula, we can obtain:

$$\overline{R} = \overline{A'} = \sqrt{\frac{\pi}{2}}\sigma \tag{4.239}$$

Therefore, when in these two cases the input is where there is signal or not, the ratio of detector load voltage is:

$$\frac{\overline{U_{sn}}}{\overline{U_n}} = \frac{K_d\overline{R}}{K_d\overline{A'}} = \sigma \cdot \mathrm{e}^{-\sigma^2/2}\left[\left(1 + Q^2\right) I_0\left(\frac{Q^2}{2}\right) + Q^2 I_1\left(\frac{Q^2}{2}\right)\right] \tag{4.240}$$

The above relation of the linear detector with noise is shown in the graph (4) in Fig. 4.21. The figure shows the relationship of the linear detector with no noise as the curve (2) and the relationship of the square law detector with noise as the curve (3).

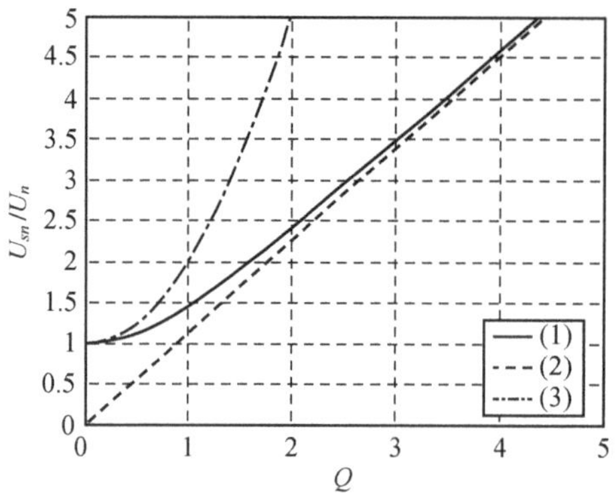

Fig. 4.21: Lines of the detector $\overline{U_{sn}}/\overline{U_n} \sim Q$ relationship.

It is not difficult to prove that the curve (1) has the following two asymptotic properties. When the ratio of signal to noise is strong ($Q^2/2 \gg 1$):

$$\frac{\overline{U_{sn}}}{\overline{U_n}} \approx \frac{2}{\sqrt{\pi}}Q \tag{4.241}$$

When $Q \gg \sqrt{2}$, the upper section of the curve (1) approaches the line (2), which indicates that at this time the impact of noise can be almost negligible, that is, the linear detector with noise is equivalent to that without noise.

When the ratio of signal to noise is $Q^2/2 \ll 1$:

$$\frac{\overline{U_{sn}}}{\overline{U_n}} \approx 1 + \frac{Q^2}{2} \tag{4.242}$$

When $Q \ll \sqrt{2}$ the lower section of curve (1) approaches the curve (3), which indicates that at this time due to the impact of noise, $\overline{U_{sn}}/\overline{U_n}$ and Q have a square relationship, that is, a linear detector with noise is equivalent to the square factor detector with noise.

Because:

$$\frac{\Delta U_s}{\overline{U_n}} = \frac{\overline{U_{sn}} - \overline{U_n}}{\overline{U_n}} = \frac{\overline{U_{sn}}}{\overline{U_n}} - 1 \tag{4.243}$$

the relationship curve of $\Delta U_s/\overline{U_n} \sim Q$ is as shown as Fig. 4.22; it is actually a detector characteristic curve. From Fig. 4.22, we can see that the detection characteristics are linear when the ratio of signal to noise is strong and that they are square when the ratio of signal to noise is weak.

The above two asymptotic properties of curve (1) can be physically explained by the "strong and weak" phenomena in the nonlinear system. When the inputs of the nonlinear system are the strong signal and the weak signal together, the load voltage depends on the input strong signal. If the voltage can be added to the nonlinear device, it will produce a reaction so that the work point changes, which is bad for weak signal transmission and results in the weak signal of the output being less than the strong signal, which is the so-called "strong pressure weak" phenomenon.

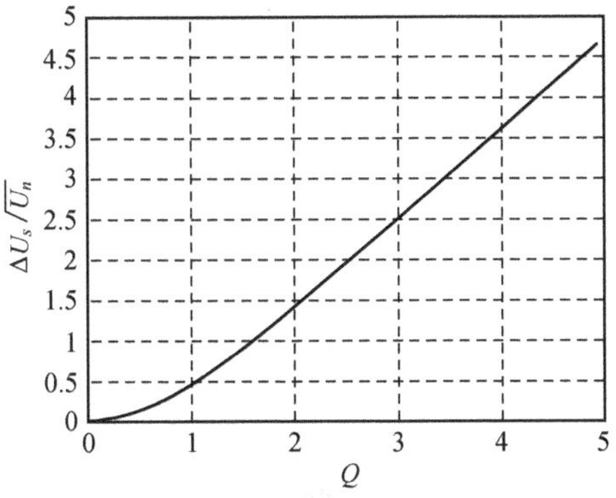

Fig. 4.22: Linear detector $\Delta U_s/\overline{U_n} \sim Q$ relationship lines.

It should be pointed out that when a weak signal is suppressed by a strong signal and results in the relative smaller value of the output, it is not that the weak signal is submerged by the strong signal and unclear. When the input signal is much stronger than the noise, the noise is suppressed by the strong signal. So, noise has almost no effect and the detection characteristic is a linear relationship. When the input signal is weaker than the noise, the signal is suppressed by the relatively strong noise. So, the effect of the signal is very small, and the detection characteristic is a square relationship. When the input signal and noise intensity is basically the same, this "strong and weak" phenomenon is not very obvious.

It should be pointed out that in the general nonlinear system, the "strong and weak" phenomenon will exist, so the initial section of the detection characteristics of kinds of detectors is square due to the impact of noise.

When a radar detects a signal, the signal-to-noise power ratio at the output of the envelope detector is usually defined as:

$$\left(\frac{S}{N}\right)_0 = \frac{(\Delta U_S)^2}{\sigma_n^2} \tag{4.244}$$

where $\Delta U_S = \overline{U_{sn}} - \overline{U_n}$ is the output DC voltage increment caused by the signal; σ_n^2 is the output noise variance when there is no signal.

It is not difficult to prove that for linear detectors, when SNR is strong ($Q^2/2 \gg 1$):

$$\left(\frac{S}{N}\right)_0 \approx \frac{1}{1-\frac{\pi}{4}}\left(\frac{S}{N}\right)_i \tag{4.245}$$

when SNR is weak ($Q^2/2 \ll 1$):

$$\left(\frac{S}{N}\right)_0 \approx \frac{1}{4(\frac{4}{\pi}-1)}\left(\frac{S}{N}\right)_i^2 \tag{4.246}$$

Therefore, the relationship between the output signal-to-noise ratio and the input signal-to-noise ratio of the envelope detector has the characteristics shown in Fig. 4.23: the characteristic curve of the weak SNR is square and the characteristic curve of strong SNR (Signal Noise Ratio) is linear relationship.

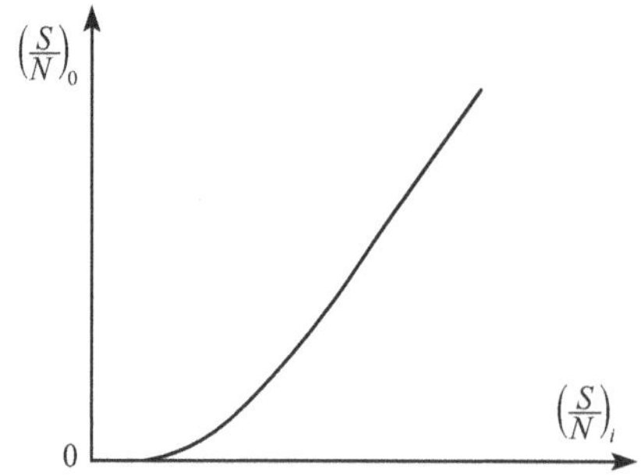

Fig. 4.23: Relationship curve of the ratio of the output SNR ratio of the envelope detector.

4.4.2.2 Low-frequency fluctuation power at the detector output

According to eq. (4.196), the low-frequency fluctuation power is the variance:

$$\sigma_L^2 = \overline{U_0^2} - (\overline{U_0})^2 = K_d^2\overline{R^2} - (K_d\overline{R})^2 = K_d^2[\overline{R^2} - (\overline{R})^2] \tag{4.247}$$

where:

$$\overline{R} = \sqrt{\frac{\pi}{2}}\sigma \cdot e^{-\sigma^2/2}\left[\left(1 + Q^2\right)I_0\left(\frac{Q^2}{2}\right) + Q^2 I_1\left(\frac{Q^2}{2}\right)\right] \triangleq \sigma f\left(\frac{Q^2}{2}\right) \tag{4.248}$$

$$\overline{R^2} = \int_{-\infty}^{\infty} R^2 \cdot p(R)\,\mathrm{d}R = \int_{-\infty}^{\infty} R^2 \frac{R}{\sigma^2}\exp\left[-\frac{R^2 + A^2}{2\sigma^2}\right] I_0\left(\frac{RA}{\sigma^2}\right)\mathrm{d}R \triangleq \sigma^2 g\left(\frac{Q^2}{2}\right) \tag{4.249}$$

So:

$$\sigma_L^2 = K_d^2\sigma^2\left[g\left(\frac{Q^2}{2}\right) - f^2\left(\frac{Q^2}{2}\right)\right] \triangleq K_d^2\sigma^2\varphi\left(\frac{Q^2}{2}\right) \tag{4.250}$$

where the relationship curve of functions $\varphi(Q^2/2)$ and $Q^2/2$ is as shown in Fig. 4.24.

Figure 4.24 shows that when input variance σ^2 is certain, undulating strength σ_L^2 increases with the ratio Q^2 of the input signal to noise increasing, and then gradually tends to a certain value.

When the input is a weak signal $[(S/N)_i \ll 1]$, since the input signal amplitude is very small, the weak signal is suppressed by the relatively strong noise, so the output signal-noise component is very small, and the output noise is mainly noise–noise component, which is the amplitude of the top of pulse shown in Fig. 4.25 (a). The

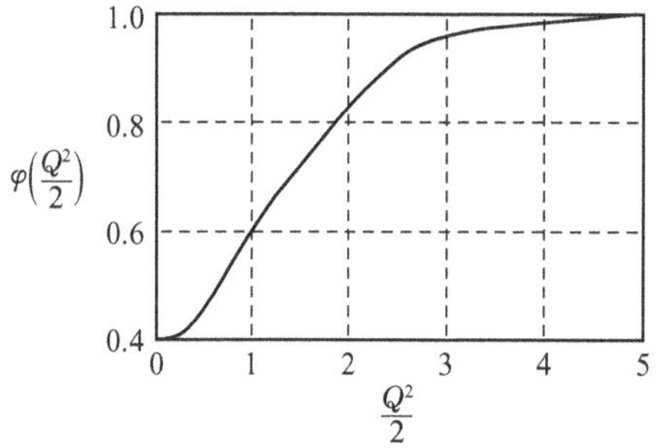

Fig. 4.24: Relationship curves of linear detector $\varphi(Q^2/2) \sim Q^2/2$.

height of the amplitude is basically equal to the height of the amplitude on both sides of the pulse.

When the input is a medium intensity signal $[(S/N)_i \approx 1]$, the signal amplitude is no longer small, so the output noise has signal–noise components in addition to noise–noise components, and it increases as signal amplitude increases, which causes the height of the amplitude at the top of the pulse to be greater than the height of the amplitude on both sides of the pulse, as is shown in Fig. 4.25 (b).

When the input is a strong signal $[(S/N)_i \gg 1]$, the noise is suppressed by the strong signal, and the signal–noise component does not increase as much as the signal amplitude increases, so the height of the amplitude at the top of the pulse does not increase as much, as is shown in Fig. 4.25 (c).

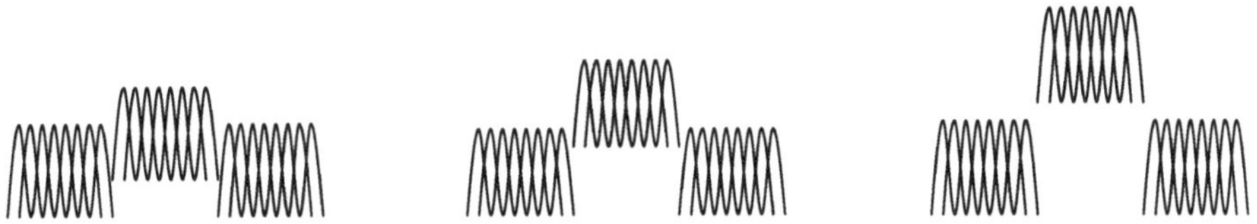

(a) Weak signal to noise ratio (b) Medium intensity signal to noise ratio (c) Strong signal to noise ratio

Fig. 4.25: Output SNR of a detector at different intensity SNR input.

4.5 Analysis of random processes through a limiter

The limiter is also a nonlinear circuit, which is used more frequently in radar and jammers. It also changes the probability distribution and power spectrum of random signals, like the detector.

4.5.1 Effect of limiting on probability distribution

The two typical properties of the limiter are shown in Fig. 4.26.

Figure 4.26 (a) is the two-way fold line limit, and the limiting characteristics are:

$$y = f(x) = \begin{cases} a_0\,, & x > U_L \\ sx\,, & |x| < U_L \\ -a_0\,, & x < -U_L \end{cases} \tag{4.251}$$

where s is the slope of the linear segment of the limiting property, U_L is the input limiting level, and $|a_0|$ is the output limiting level.

Figure 4.26 (b) shows the smooth limit and its limiting feature can be approximately represented by using an error function or a hyperbolic tangent function.

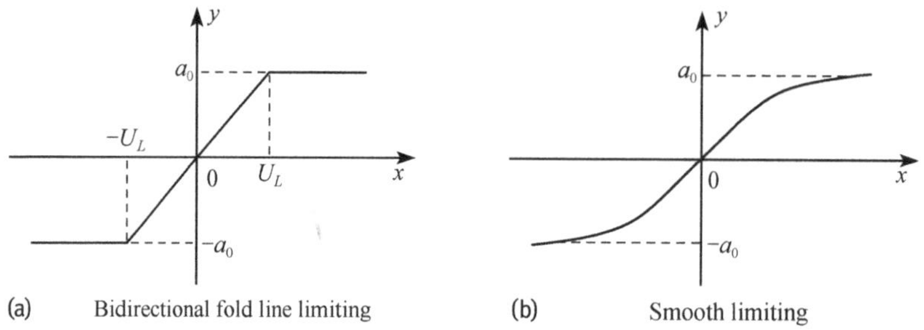

Fig. 4.26: Limiting characteristic curve.

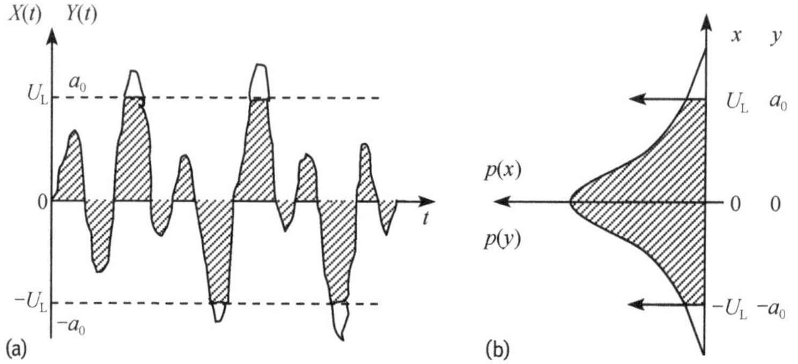

Fig. 4.27: Limiting waveform and probability distribution of noise.

In the case of normal noise passing the bidirectional fold line limiter shown in Fig. 4.26 (a), the variation of the noise waveform and the probability distribution is shown in Fig. 4.27 (a).

The shaded portion of Fig. 4.27 (a) represents the output waveform of the limiter, and the shaded portion in Fig. 4.27 (b) represents the output probability distribution after limiting. Figure 4.27 (b) shows that although the input probability density is normal, the output probability density has become a limiting normal distribution. Because $-a_0 \sim a_0$ is the linear transformation, the probability density $p(y)$ should still be kept for the input normal distribution curve shape. As $y > a_0$ and $y < -a_0$, $p(y)$ become zero, the area below the distribution curve must be equal to 1. So, the δ-function will appear separately at a_0 and $-a_0$, whose intensity is, respectively, equal to the integral value on the interval of $U_L \sim +\infty$ and $-\infty \sim -U_L$.

It can be seen that, if the input limit level U_L is smaller, more parts are removed because of limiting, and the structure of the noise becomes bigger.

For example, if $U_L = \sigma$, then the probability of the limiting part is $1 - P[|X| \le \sigma] = 1 - 0.683 = 0.317$. At this time the strength of the two δ functions are $0.317/2$.

For example, if $U_L = 2\sigma$, then the probability of limiting part is $1 - P[|X| \le 2\sigma] = 1 - 0.955 = 0.045$. At this time the strength of the two δ-functions is $0.045/2$.

In jamming emitters, limiters are often used to change the probability distribution of the noise modulation jamming. When producing noise FM jamming, in order to make the jamming evenly cover a wide band, it is usually hoped that the power spectrum of FM jamming is a rectangular uniform distribution, which makes the noise FM jamming change from normal distribution to rectangular uniform distribution. This change in the distribution rate is usually done with a smooth limiter.

We assume that the limiting characteristic of the smooth limiter is the error function, and the expression is:

$$y = \frac{1}{K\sqrt{2\pi}\sigma_L}\int_0^x \exp\left[-\frac{t^2}{2\sigma_L^2}\right]\mathrm{d}t\,, \quad -\infty < x < \infty \tag{4.252}$$

where K is a constant; σ_L represents a parameter that the limiting characteristic deviates from the bidirectional ideal limiting characteristic.

Now, the input is normal noise and its probability density is:

$$p(x) = \frac{1}{\sqrt{2\pi}\sigma}\exp\left[-\frac{x^2}{2\sigma^2}\right], \quad -\infty < x < \infty \tag{4.253}$$

The probability density of the output noise is:

$$\begin{aligned} p(y) = p(x)\left|\frac{\mathrm{d}x}{\mathrm{d}y}\right| &= \frac{1}{\sqrt{2\pi}\sigma}\exp\left[-\frac{x^2}{2\sigma^2}\right]\cdot K\sqrt{2\pi}\sigma_L\exp\left[\frac{x^2}{2\sigma_L^2}\right] \\ &= K\frac{\sigma_L}{\sigma}\exp\left[-\frac{x^2}{2\sigma^2}\left(1-\frac{\sigma^2}{\sigma_L^2}\right)\right] \end{aligned} \tag{4.254}$$

We set $a = \frac{\sigma_L^2}{\sigma^2}$ to obtain:

$$p(y) = K\sqrt{a}\exp\left[-\frac{x^2}{2\sigma^2}\left(1-\frac{1}{a}\right)\right], \quad -\frac{1}{2K} \le y \le \frac{1}{2K} \tag{4.255}$$

This formula indicates that corresponding to different values of a, the distribution of the output probability density is different.

(1) When $a = 1$:

$$p(y) = K\,, \quad -\frac{1}{2K} \le y \le \frac{1}{2K} \tag{4.256}$$

It can be seen that the probability density is uniform distribution in the limiting interval, as is shown in Fig. 4.28.

(2) When $a \approx \infty$:

$$p(y) \approx K\sqrt{a}\exp\left[-\frac{x^2}{2\sigma^2}\right], \quad -\frac{1}{2K} \le y \le \frac{1}{2K} \tag{4.257}$$

Here, the output probability density is still normally distributed. This is not difficult to understand because $a \approx \infty$ is equivalent to a_L. The limiting characteristic curve

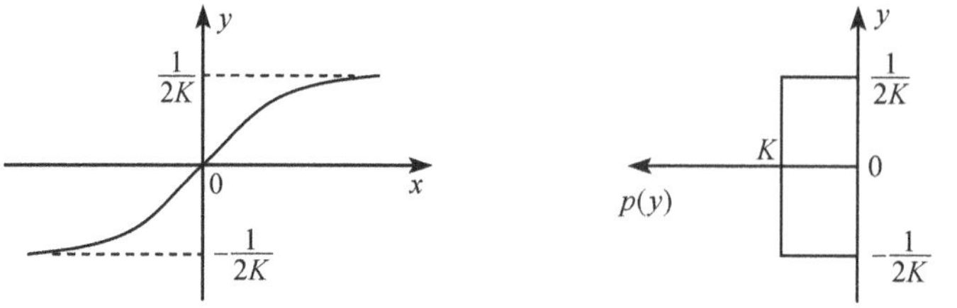

Fig. 4.28: The limiting characteristic and output distribution curve of $a = 1$.

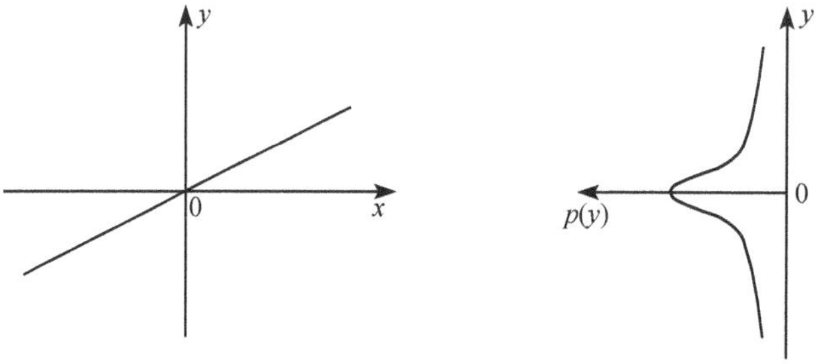

Fig. 4.29: The limiting characteristic and output distribution curve of $a = \infty$.

becomes straight and in linear amplification in the limiting interval, so the shape of the distribution curve will not change, as is shown in Fig. 4.29.
(3) When $a = 0$:

$$p(y) = \begin{cases} \infty\,, & y = -\frac{1}{2K}, \frac{1}{2K} \\ 0\,, & -\frac{1}{2K} < y < \frac{1}{2K} \end{cases} \tag{4.258}$$

Since the area under the distribution curve should be equal to 1, the probability density at the two limiter levels is a δ-function, and its intensity is $1/2$, as is shown in Fig. 4.30.

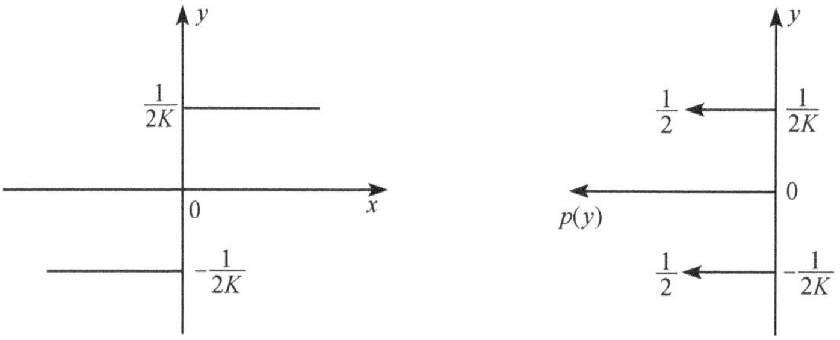

Fig. 4.30: The limiting characteristic and output distribution curve of $a = 0$.

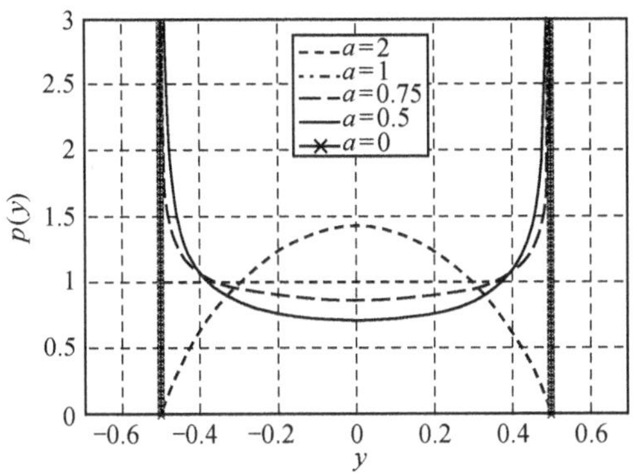

Fig. 4.31: Output distribution curve of a smooth limiter at different values of a.

Figure 4.31 shows the output distribution curve of the smooth limiter corresponding to different values of a. The appropriate value of a can be chosen so that the output distribution curve becomes the desired shape. In order to make the output distribution curve rectangular, in the noise FM jammer limiter, it is usually selected that $a = 1$.

4.5.2 Effect of limiting on the power spectrum

Given the statistical parameters of the limiting characteristics and the input random signal, the nonlinear transformation method can be used to obtain the output correlation function and then apply the Fourier transform to obtain output power spectrum. However, the specific calculation is very complicated, so the following is only a brief introduction.

Suppose that the limiting feature is a one-way ideal upper limit, as shown in Fig. 4.32. Its expression is:

$$y = f(x) = \begin{cases} a_0 , & x \geq 0 \\ 0 , & x < 0 \end{cases} \tag{4.259}$$

The input is a stationary narrowband normal noise, whose mean is zero and variance is σ^2.

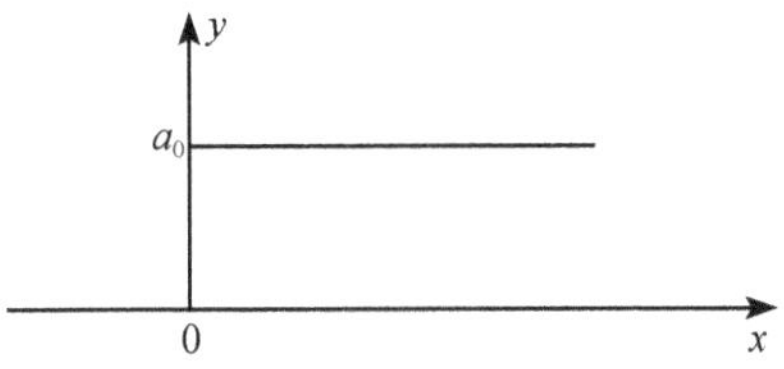

Fig. 4.32: One-way ideal upper limit characteristic.

We can use the price method or the Hermitian polynomial method to obtain the output correlation function:

$$R_Y(\tau) = \frac{\sigma_0^2}{4}\left[1 + \frac{2}{\pi}\sin^{-1} r(\tau)\right] \tag{4.260}$$

where $\sin^{-1} r(\tau)$ can be expanded into a MacLaurin series as follows:

$$\sin^{-1} r(\tau) = r(\tau) + \sum_{n=1}^{\infty} \frac{(2n-1)!!}{(2n)!!} \cdot \frac{r^{2n+1}(\tau)}{2n+1} \tag{4.261}$$

If the input smooth narrowband normal noise has a symmetrical power spectrum (symmetrical to the central frequency ω_0), then the correlation coefficient is:

$$r(\tau) = r_0(\tau)\cos\omega_0\tau \tag{4.262}$$

where $r_0(\tau)$ is the envelope of the correlation coefficient.

From this equation, we obtain:

$$r^{2n+1}(\tau) = r_0^{2n+1}(\tau)\cos^{2n+1}\omega_0\tau \tag{4.263}$$

where the cosine function can also be expanded into a MacLaurin series:

$$\cos^{2n+1}\omega_0\tau = 2^{-2n}\sum_{k=0}^{n}\binom{2n+1}{k}\cos(2n+1-2k)\omega_0\tau \tag{4.264}$$

Now there is a limiting IF amplifier. The narrowband filter characteristics only allow the high-frequency fundamental components near the carrier frequency ω_0 to pass, and the rest of the frequency components (DC, low-frequency and high-frequency harmonics) are removed, so the reserved items of the equation above are only the items of $k = n$. Now, the output function is:

$$R_Y(\tau)\big|_{\omega=\omega_0} = \frac{\sigma_0^2}{2\pi}\left[r_0(\tau) + \sum_{n=1}^{\infty}\frac{(2n-1)!!}{(2n)!!}\cdot\frac{2^{-2n}}{2n+1}\binom{2n+1}{k} r_0^{2n+1}(\tau)\right]\cos\omega_0\tau \tag{4.265}$$

This formula shows that the envelope of $R_Y(\tau)$ only contains the odd items of $r_0(\tau)$: $r_0(\tau), r_0^3(\tau), r_0^5(\tau), \ldots$ and when n is larger, the coefficient is smaller and the corresponding power spectral density is smaller. However, the envelope of the correlation function of the input narrowband noise is $A(\tau) = \sigma^2 r_0(\tau)$, and the correlation function and the power spectral density have the relationship of the Fourier transform, so:

$$A_0^2(\tau) \leftrightarrow G_{A0}(f) \otimes G_{A0}(f) \otimes G_{A0}(f) \tag{4.266}$$

because the number of convolutions is larger when the spectrum is wider, and the power spectral density corresponding to the high-power items of $r_0(\tau)$ will have a lower intensity and a bigger width with increasing power.

If the limiting level $U_L \neq 0$, the limiting characteristic is as shown in Fig. 4.33, and the formula is:

$$y = f(x) = \begin{cases} a_0\,, & x \geq U_L \\ 0\,, & x < U_L \end{cases} \tag{4.267}$$

Now, the amplitude of the output noise is limited at a_0, the output noise becomes a series of pulses, and the pulse interval varies randomly (related to the characteristics of the input noise). The starting and ending times of the pulses are consistent with the input noise, as is shown in Fig. 4.33.

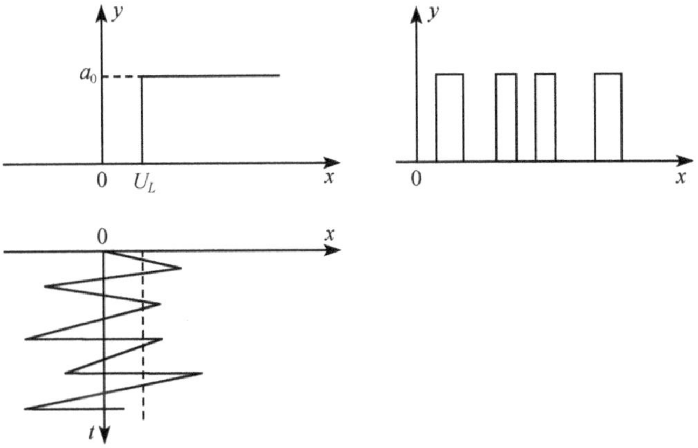

Fig. 4.33: Limiting characteristics and waveforms of input and output.

The method above can be followed to calculate the output correlation function, and then the result is different from the one above. There are homogeneous power terms of the $r_0(\tau)$ and even power terms, which is cumbersome. Levin (Левин) [16] analyzed the ideal limiter by using the transformation method, and the following only lists the results.

The output correlation function is:

$$R_Y(\tau) = a_0^2\left[1 - \Phi\left(\frac{U_L}{\sigma}\right)\right]^2 + \frac{a_0^2}{2\pi}\mathrm{e}^{-\frac{U_L^2}{\sigma^2}}\sum_{k=1}^{\infty} H_{k-1}^2\left(\frac{U_L}{\sigma}\right)\frac{r^k(\tau)}{k!} \tag{4.268}$$

where σ^2 is the variance of the input noise envelope.

Substituting eq. (4.262) into eq. (4.268):

$$\begin{aligned}R_Y(\tau) = a_0^2\left[1-\Phi\left(\frac{U_L}{\sigma}\right)\right]^2 + \frac{a_0^2}{2\pi}\mathrm{e}^{-\frac{U_L^2}{\sigma^2}}\Bigg\{\sum_{k=1}^{\infty} H_{2k-1}^2\left(\frac{U_L}{\sigma}\right)\frac{\binom{2k}{k}}{(2k)!2^{2k}}r_0^{2k}(\tau)\\
+\left[\sum_{k=1}^{\infty} H_{2k-2}^2\left(\frac{U_L}{\sigma}\right)\frac{\binom{2k-1}{k-1}}{(2k-1)!2^{2k-2}}r_0^{2k-1}(\tau)\right]\cos\omega_0\tau\\
+\sum_{r=2}^{\infty}\left[\sum_{k=r}^{\infty} H_{2k-2}^2\left(\frac{U_L}{\sigma}\right)\frac{\binom{2k-1}{k-r}}{(2k-1)!2^{2k-2}}r_0^{2k-1}(\tau)\right]\cos(2r-1)\omega_0\tau\\
+\sum_{r=1}^{\infty}\left[\sum_{k=r}^{\infty} H_{2k-1}^2\left(\frac{U_L}{\sigma}\right)\frac{\binom{2k}{k-r}}{(2k)!2^{2k-1}}r_0^{2k}(\tau)\right]\cos 2r\omega_0\tau\Bigg\}\end{aligned} \tag{4.269}$$

The Fourier transform is used to obtain the output power spectrum. This equation shows that the output power spectrum contains DC and low-frequency and high-frequency fundamentals, and high-frequency harmonic components, which are all related to the limiter level U_L. In the following, we know the influence of the limiter level on the power spectrum only from the power spectrum curve drawn by the analysis above of the results.

Assuming that the input narrowband noise power spectrum $F_X(f)$ shown in Fig. 4.34 (a) and the output power spectrum $F_Y(f)$ at different limiter levels can now be obtained.

(1) At the time of a low limiting level ($U_L \ll \sigma$), the output power spectrum is as shown in Fig. 4.34 (b) below. The dashed box in the figure represents the component corresponding to the input power spectrum. Figure 4.34 shows that the output noise power is mainly concentrated near the carrier frequency f_0, but about 20% is converted to DC and low-frequency and high-frequency harmonic components. Because the output noise mostly turned flat-top pulse, the component of the side frequency increases, and the power spectrum at f_0 is broadened by about 5 ~ 15%.
(2) At the time of the high limiting level ($U_L = \sigma$), the output power spectrum is as shown in Fig. 4.34 (c). The figure shows that about 50% is converted to low-frequency (including DC) and high-frequency harmonic components, and each one accounts for about 25%. Since the number of flat-top pulses that the output noise converts to is small, the power spectrum at f_0 is broadened a little.

If the limiting level is very high ($U_L \gg \sigma$), from the figure of limiting characteristics and the waveform of the input and the output, we can see that only a very small amount of high-amplitude noise can pass the limiter, so the output power spectrum is basically the same as the input power spectrum.

It should be pointed out that the expression $R_Y(\tau)$ above and Fig. 4.34 include all the frequency components. In fact, due to the narrowband filter characteristics of the narrow band, all the low-frequency and high-frequency harmonics are removed, so

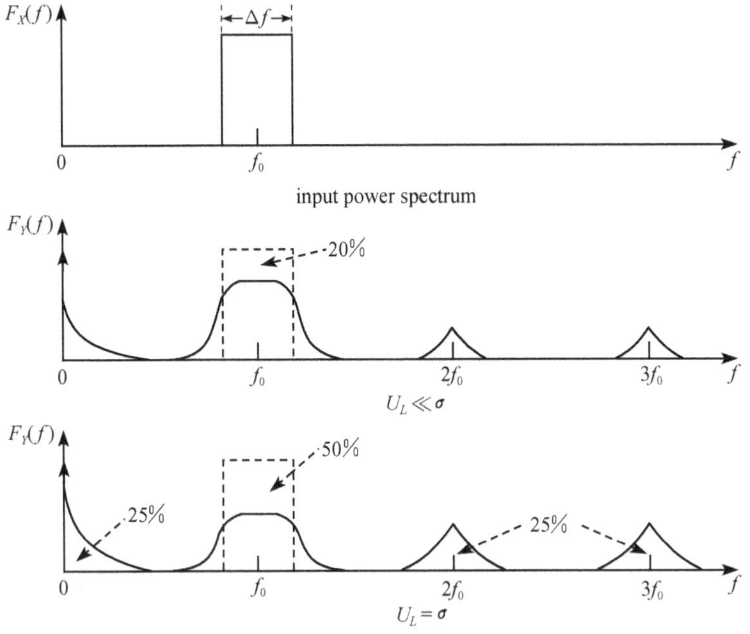

Fig. 4.34: The output power spectrum of different limiting levels.

the output is only the components near the high-frequency fundamental f_0. If it is a narrowband ideal bandpass filter, it can be restored to the input rectangular power spectrum, but the energy in other components cannot be recovered.

4.5.3 Noise and sinusoidal signals together through limiting IF amplifier

Davenport [16] analyzes the output signal–noise ratio of the stationary narrowband normal noise and the equal-amplitude sinusoidal signal through the limiting IF amplifier by using the transformation method. The nonlinear function of the limiter is shown in Fig. 4.35 (a) and is the two-way fold line limit. The limiting IF amplifier has ideal bandpass characteristics, whose center is at the high-frequency fundamental wave w_0. According to the analysis of Section 4.3, the output signal is determined by the signal–signal component $2h_{10}^2 \cos \omega_0 \tau$, and the output noise is determined by the noise–noise component and the signal–noise component.

When the ratio of the input signal–noise power is small, the decrease of the ratio of the output signal–noise power will not exceed $\pi/4$ (approximately -1dB) times the ratio of the input signal–noise power. That is, when the SNR of the input is weak, although the weak signal is suppressed by noise, the loss of the ratio of the signal–noise power is not significant. When the ratio of the input signal–noise power is large, if the conductivity s of the limiter is so large that any input signal is limited, the output signal to the noise power ratio can be improved twice due to suppression from strong signal to weak noise. However, it should be pointed out that the noise power spectrum is also widened.

Gardner [16] conducted a theoretical analysis and obtained the following relationship:

$$\left(\frac{S}{N}\right)_0 = \left(\frac{S}{N}\right)_i \frac{1 + 2\left(\frac{S}{N}\right)_i}{\frac{4}{\pi} + \left(\frac{S}{N}\right)_i} \tag{4.270}$$

So, when $(S/N)_i \ll 1$:

$$\left(\frac{S}{N}\right)_0 = \frac{\pi}{4}\left(\frac{S}{N}\right)_i \tag{4.271}$$

And when $(S/N)_i \gg 1$:

$$\left(\frac{S}{N}\right)_0 = 2\left(\frac{S}{N}\right)_i \tag{4.272}$$

The curve is shown in Fig. 4.35.

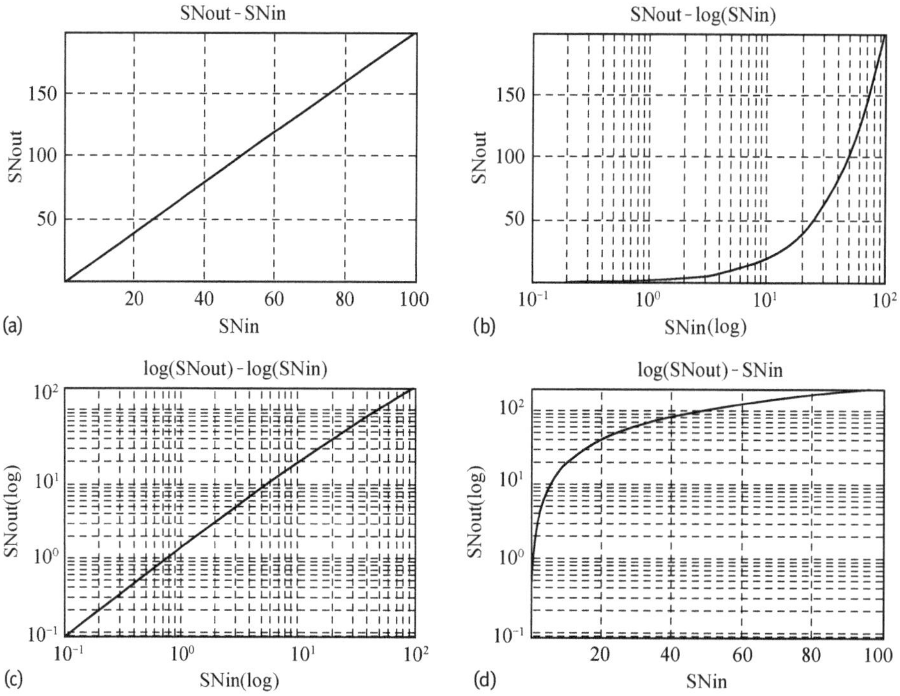

Fig. 4.35: The input and output signal ratio relationship curve when signal and noise pass through the limiting amplifier together.

4.6 Calculation of SNR at the output of a radio system

When a radio system transmits a signal, it is unavoidably accompanied by noise (internal noise or external), which can affect the detection of the signal and the estimation of the signal parameters. In engineering, the signal–noise power ratio (referred to as the signal–noise ratio) is usual and is written as (S/N) to measure its impact. For example, for linear circuits, the common noise coefficient F is used to measure the size of the impact of internal noise:

$$F = \frac{\left(\frac{S}{N}\right)_i}{\left(\frac{S}{N}\right)_O} \tag{4.273}$$

where $(S/N)_i$, $(S/N)_O$ are, respectively, the SNR at the input and output in the linear system. The noise coefficient does not apply to nonlinear circuits, and the radio system usually also contains nonlinear circuits, so when estimating the impact of noise on the entire radio system or characterizing the system's antijamming performance, the signal–noise ratio at the output of the system is often regarded as a standard.

A radio system usually contains many unit circuits, either as a linear circuit or as a nonlinear circuit. For example, superheterodyne receivers can usually be divided into three parts as shown in Fig. 4.36, where the linear circuit I contains A high-frequency amplifier, A mixer (this is a nonlinear circuit, but can be regarded as quasi-linear circuit), and an IF amplifier; the nonlinear circuit is the envelope detector. The linear circuit II is a low-frequency or video amplifier. To calculate the signal–noise ratio of the output of this system, it is only necessary to obtain the signal power and the noise power from before to after by stages.

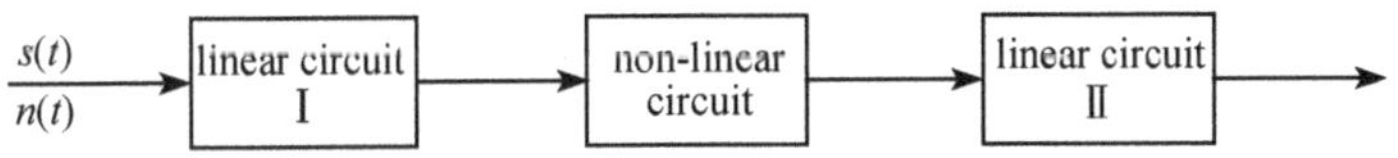

Fig. 4.36: The components of a superheterodyne receiver.

Linear circuits, because of their superposition and the fact that signal and noise are statistically independent, allow us to consider the signal and noise separately and pass the linear circuit. If they are stationary processes, according to the previous conclusion, we have:

$$G_Y(\omega) = G_X(\omega)\,|H(\omega)|^2 \tag{4.274}$$

We solve the power spectral density of the output signal or noise, and then calculate their power.

Because they do not have the superposition characteristic and the signal and noise in the nonlinear circuit interact, nonlinear circuits produce new spectral components and the noise to be considered cannot be separated from the signal, so the calculation is more complex. However, the alternating component power spectrums of the output can be divided into the following three parts:

$G_{ss}(\omega)$ is the component spectrum generated by the interaction between the components of the signal;

$G_{nn}(\omega)$ is the component spectrum generated by the interaction between the components of the noise;

$G_{sn}(\omega)$ is the component spectrum generated by the interaction between the signal and the various components of the noise.

The three component spectrums above contain a variety of sum frequency, difference frequency and harmonic components, but the nonlinear circuit, in addition to nonlinear devices, is usually followed by a filter network (which consists of inert components and can be merged into the linear circuit II), which can filter the unnecessary frequency components and only select the desired frequency components. For example, the envelope detector has a lowpass filter with the IF bandpass filter in the IF limiting amplifier.

In the three component spectrums above, the signal–signal component $G_{ss}(\omega)$ belongs to the signal at the output, and the noise–noise component $G_{nn}(\omega)$ belongs to the noise at the output. The problem should be can the signal–noise components $G_{sn}(\omega)$, which contain both the noise and the signal, be regarded as the signal or noise of the output? It is necessary to make a decision based on the using condition of the radio system, which can be divided into the two cases below:

(1) $G_{sn}(\omega)$ is regarded as noise. Communication systems generally use this principle, because the signal–noise component is an irregular signal, which affects the quality of communication (for listening-detection equipment, it will affect the clarity of the signal sound; for the observation equipment, it will affect the fidelity of signal waveforms).
Similarly, the principle is used to estimate radar signal parameters because errors in the measured value will be caused by signal–noise components.
(2) $G_{sn}(\omega)$ is regarded as a signal. This principle is used when a radar system detects a signal because the amount of information is contained in the signal–noise component, which helps to find the target signal from the noise background.

When the respective component spectra of the nonlinear circuit of the output are calculated and the frequency response characteristic $H_{II}(\omega)$ of the linear circuit II is known, the SNR at the output of the radio system can be calculated as follows:

(1) $G_{sn}(\omega)$ is regarded as noise.
Output noise is dependent on $G_{nn}(\omega)$ and $G_{sn}(\omega)$, and the output signal is dependent on $G_{ss}(\omega)$. So, the SNR of the system output is:

$$\left(\frac{S}{N}\right)_0 = \frac{\int_0^\infty |H_{II}(\omega)|^2 \, G_{ss}(\omega)\,\mathrm{d}\omega}{\int_0^\infty |H_{II}(\omega)|^2 \,[G_{nn}(\omega) + G_{sn}(\omega)]\,\mathrm{d}\omega} \tag{4.275}$$

(2) $G_{sn}(\omega)$ is regarded as a signal.
Output noise is dependent on $G_{nn}(\omega)$, and the output signal is dependent on $G_{ss}(\omega)$ and $G_{sn}(\omega)$. So, the SNR of the system output is:

$$\left(\frac{S}{N}\right)_0 = \frac{\int_0^{\infty} |H_{II}(\omega)|^2 \left[G_{ss}(\omega) + G_{sn}(\omega)\right] \mathrm{d}\omega}{\int_0^{\infty} |H_{II}(\omega)|^2 \, G_{nn}(\omega) \, \mathrm{d}\omega} \tag{4.276}$$

Example 4.6.1. Noise $n(t)$ and the signal $s(t)$ pass commonly through the full-wave square-law detector, and the relationship of the nonlinear function is $y = bx^2$; $n(t)$ is stationary normal noise, whose mean is zero and variance is σ^2. Its power spectral density is:

$$G_n(f) = \begin{cases} c\,, & f_0 - \frac{\Delta f}{2} < |f| < f_0 + \frac{\Delta f}{2} \\ 0\,, & \text{otherwise} \end{cases} \tag{4.277}$$

The input signal is a random amplitude and random initial phase signal $s(t) = A(t)\cos(\omega_0 t + \theta)$, where phase θ is uniformly distributed in $(0, 2\pi)$ and $A(t)$ is a low-frequency random amplitude modulation wave subject to Rayleigh distribution and whose width is less than Δf. Random variables $A(t)$ are irrelevant with θ. $s(t)$ and $n(t)$ statistically independent.

It is known that this detector is used in communication receivers where the amplitude-frequency characteristics of lowpass filters are:

$$G_n(f) = \begin{cases} 1\,, & 0 < |f| < \Delta f \\ 0\,, & \text{otherwise} \end{cases} \tag{4.278}$$

Solve the relationship between the SNR at this detector output $(S/N)_0$ and input SNR $(S/N)_\mathrm{i}$.

Solution: Because $A(t)$ and θ are not related, the autocorrelation function of the input signal $s(t)$ is:

$$R_s(\tau) = E\left[A(t)A(t+\tau)\right] \cdot E\left\{\cos(\omega_0 t + \theta) \cdot \cos\left[\omega_0(t+\tau) + \theta\right]\right\} = R_A(\tau)\frac{1}{2}\cos\omega_0\tau \tag{4.279}$$

where $R_A(\tau) = E[A(t)A(t+\tau)]$ is the correlation function of the input signal envelope.

According to the previous analysis, we can obtain the three components of the autocorrelation function of the nonlinear device output as follows:

Signal–signal component:

$$R_{ss}(\tau) = b^2\left[\frac{R_{A^2}(\tau)}{4} + \frac{R_{A^2}(\tau)}{8}\cos 2\omega_0\tau\right] \tag{4.280}$$

Noise–noise component:

$$R_{nn}(\tau) = b^2\left[\sigma^4 + 2R_n^2(\tau)\right] \tag{4.281}$$

Signal–noise component:

$$R_{sn}(\tau) = b^2 \left[R_A(0)\sigma^2 + 2R_A(\tau)R_n(\tau)\cos\omega_0\tau\right] \tag{4.282}$$

We use the Fourier transform for the four correlation functions to obtain the corresponding power spectral density whose graphs are similar to the ones in Example 4.2.2. The difference is only that the lines should be changed into a narrowband continuous spectrum in Fig. 4.4 (a) and (b) in Example 4.2.2.

With $\tau = 0$ in the above expression $R_s(\tau)$ above, the signal power of input is:

$$S_i = R_s(0) = \frac{1}{2}R_A(0) = \frac{1}{2}\int_0^\infty \frac{A^3}{\sigma_A^2} e^{-\frac{A^2}{\sigma_A^2}}\, dA = \sigma_A^2 \tag{4.283}$$

and the input noise power is $N_i = A^2$.

This detector is used in communication receivers where the noise-to-noise component should be counted as noise. However, when determining the output signal and output noise of the detector, the amplitude-frequency characteristics of the filter must be taken into account. From the expression of the amplitude-frequency characteristics of the lowpass filter we can see that a lowpass filter only lets all the low-frequency components pass, but no DC and high-frequency component can be output. So, the detector output (low-frequency) signal power can be obtained by the above the expression $R_{ss}(\tau)$:

$$S_o = b^2 \frac{R_{A^2}(0)}{4} = \frac{b^2}{4}\int_0^\infty \frac{A^5}{\sigma_A^2} e^{-\frac{A^2}{\sigma_A^2}}\, dA = 2b^2\sigma_A^2 = 2b^2 S_i^2 \tag{4.284}$$

The detector output (low-frequency) noise power can be obtained by the above expression R_{nn} and R_{sn}:

$$N_o = \frac{1}{2}\left[2b^2 R_n^2(0) + 2b^2 R_A(0)R_n(0)\right] = b^2\left[\sigma^4 + \sigma^2 R_A(0)\right] = b^2 N_i^2\left[1 + 2\frac{S_i}{N_i}\right] \tag{4.285}$$

The coefficients $1/2$ in the formula are because half of the output noise power of the square-law device is concentrated in the vicinity of zero frequency, while the other half is concentrated in the vicinity of the double carrier frequency.

The SNR at the output of the detector is obtained from the above two equations:

$$\left(\frac{S}{N}\right)_0 = \frac{2\left(\frac{S}{N}\right)_i^2}{1 + 2\left(\frac{S}{N}\right)_i} \tag{4.286}$$

If $(S/N)_i \gg 1$:

$$\left(\frac{S}{N}\right)_0 \approx \left(\frac{S}{N}\right)_i \tag{4.287}$$

It is in linear relationship.

If $(S/N)_i \ll 1$:

$$\left(\frac{S}{N}\right)_0 \approx 2\left(\frac{S}{N}\right)_i^2 \tag{4.288}$$

It is in square relationship.

The result of this example is compared with the last result in Section 4.5. It can be seen that the relationship of the square-law detector is similar to that of the linear detector, and the difference is only the proportionality constant. For other types of detectors, the same characteristics are also present.

It should be pointed out that because the noise is a random process, and the radar echo signal is the random ups and downs caused by the target flicker and other reasons, the so-called "discovery target" is a random event and we need to use "detection probability" to express it, and it is not enough to only know the SNR at the output of the radar system. Similarly, in order to accurately estimate the antijamming performance of the radio system, it should also be based on the different conditions of use of the radio system to take into account the corresponding probability criteria. These questions are left for the next chapter.

According to the probability criterion to evaluate the specific composition of the radio system, it is usually necessary to obtain the probability distribution of the output of the system. Therefore, the probability distribution of its output should be analyzed step by step according to the specific composition of the radio system. For example, for the receiver shown in the schematic diagram of the super heterodyne receiver components in Section 4.6, according to the analysis of the third chapter, we can see that the system output probability distribution is as follows:

If the input of the system is a sinusoidal signal with noise, the IF instantaneous value of the output process is normally distributed through the narrow band, and its envelope value is the generalized Rayleigh distribution. If the envelope detector has a characteristic of half-wave linear law, the probability distribution of the output of the detector is a generalized Rayleigh distribution. If the video amplifier is a broadband circuit, the probability distribution of the output of the system is constant, and it is still the generalized Rayleigh distribution.

If the envelope detector has a square-law characteristic, we use independent sampling by the interval (signal cycle) for the video process of the output of detection in the video circuit, and use video accumulation, so the probability distribution of the cumulative value is χ^2 distribution (no signal input) or noncentral χ^2 distribution (with sinusoidal signal input).

Exercises

4.1 Assuming that the relationship of the non-inert nonlinear function is $y = e^{ax}$, enter a stationary normal random process whose mean is zero, variance is σ^2 and correlation coefficient is $r(\tau)$. Try to prove the equation with the direct method of nonlinear transformation: $R_Y(\tau) = \exp\{a^2\sigma^2[1 + r(\tau)]\}$.

4.2 For the zero-mean normal real random variables X_1, X_2, X_3, X_4, $\overline{X_1X_2X_3X_4} = \overline{X_1X_2} \cdot \overline{X_3X_4} + \overline{X_1X_3} \cdot \overline{X_2X_4} + \overline{X_1X_4} \cdot \overline{X_2X_3}$. Assume that the nonlinear function of the full-wave square-law device is $y = bx^2$, enter a smooth normal noise whose mean is zero, variance is σ^2 and correlation coefficient is $r(\tau)$. Prove that the correlation coefficient of the output stochastic process is $r_Y(\tau) = r^2(\tau)$.

4.3 Assume a narrowband stochastic process $Z(t) = s(t) + n(t)$, where $s(t) = A\cos(\omega_0 t + \theta)$ is a known signal, and $n(t) = X(t)\cos\omega_0 t - Y(t)\sin\omega_0 t$ is the stationary normal noise, whose mean is zero and variance is σ^2. Try to use the known analysis results to prove that the correlation function of the square of the envelope of $Z(t)$ is $R_Z(\tau) = A^4 + 4A^2\sigma^2 + 4\sigma^4 + 4[A^2 R_X(\tau) + R_X^2(\tau) + R_{XY}^2(\tau)]$.

4.4 Assuming that the stationary normal noise $X(t)$, whose mean is zero and variance is σ^2 is through the full-wave square-law device, the nonlinear function relation of which is $y = bx^2$. Use the Hermitian polynomial method to calculate the correlation function R_Y of the output noise Y_t.

4.5 Assume that the nonlinear function of the full-wave square-law detector and the half-wave linear detector is:

$$y_1 = x^2 \tag{4.289}$$

$$y_2 = \begin{cases} x\,, & x > 0 \\ 0\,, & x \le 0 \end{cases} \tag{4.290}$$

The input is a stationary narrowband normal noise (ideal rectangular symmetry spectrum), whose mean is zero and variance is σ^2. Calculate the DC power, low-frequency fluctuation power and high-frequency fluctuation power (ignoring the load influence of the detector) of each detector using the known analysis results.

4.6 Assume that the smooth normal noise $X(t)$, whose mean is zero and variance is σ^2 is through the ideal limiter, and the limiting characteristic is:

$$y = f(x) = \begin{cases} a_0\,, & x > 0 \\ 0\,, & x \le 0 \end{cases} \tag{4.291}$$

(1) Try to use the price method to solve the correlation function $R_Y(t)$ of the output noise $Y(t)$;
(2) Use the Hermitian polynomial method to solve the correlation function $R_Y(\tau)$ of the output noise $Y(t)$.

4.7 Assume that the nonlinear function of the detection element is $y = bx^4$, and the input is symmetrical high-frequency narrowband stationary normal noise, whose mean is zero and variance is σ^2. Try to use the envelope method to solve the power of DC and low-frequency fluctuant power in the detecting device.

4.8 Assume that the limiting characteristics of the bidirectional fold line limiter are as shown in Fig. 4.37. The input is normal noise (the mean is zero and the variance is 1).

(1) Try to write the expression of the limiting characteristics.
(2) Try to solve the probability density of the output noise of the limiter $p(y)$, and draw the curve.
(3) Calculate the variance of the output noise of the limiter.

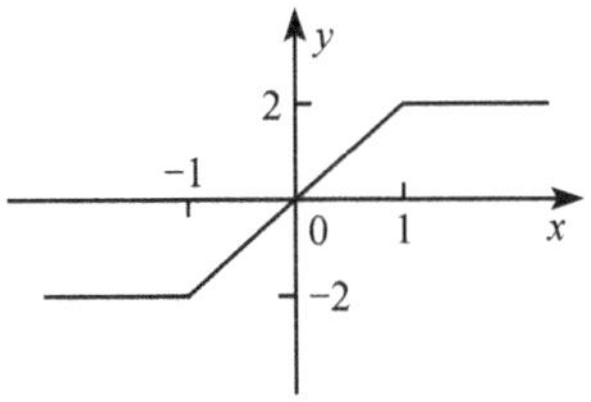

Fig. 4.37: Bidirectional fold line limiter.

4.9 Assume that the limiting characteristic of the limiter is:

$$y = a + \frac{1}{K\sqrt{2\pi}\sigma_L} \int_0 e^{-\frac{t^2}{2\sigma_L^2}} dt \tag{4.292}$$

Try to solve the probability density $p(y)$ of the normal noise (the mean is zero, and the variance is σ^2) after passing the limiter.

4.10 Assume that the composition of the system is as shown in Fig. 4.37; $n(t)$ is the real stationary normal signal, whose mean is zero and spectral density is uniform, and $G_n(\omega) = N_0/2$. The limiting characteristic of the lowpass filter is:

$$|H_3(\omega)| = \begin{cases} 1, & |\omega| \le \frac{2}{RC} \\ 0, & |\omega| > \frac{2}{RC} \end{cases} \tag{4.293}$$

where RC is the parameter of the integrating circuit.
(1) Which distribution are $n_1(t)$ and $n_2(t)$?
(2) Solve the autocorrelation function and the power spectral density of $n_2(t)$.
(3) Draw the graph of the autocorrelation function and the power spectral density of $n(t)$, $n_1(t)$ and $n_2(t)$.
(4) Calculate the DC power and the fluctuation power at the output of the system.

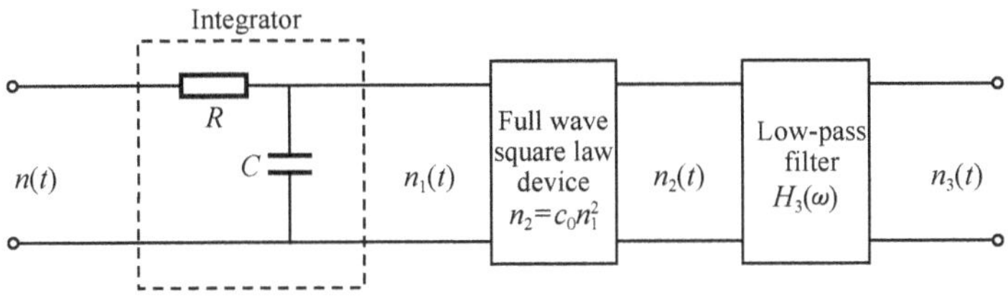

Fig. 4.38: Given nonlinear system.

4.11 Assume that a nonlinear system transmission characteristic is: $y = g(x) = \beta\alpha^x(\alpha > 0)$. The input $X(t)$ is stationary Gaussian noise, whose mean is zero and variance is σ_X^2, and the correlation function is $R_X(\tau)$. Try to use the polynomial moment function method to calculate autocorrelation function of the output [$g(x)$ only take three items].

4.12 On the machine: use the MATLAB program to design a sine-type signal, Gaussian white noise signal. Analyze, respectively, the power spectrum and the amplitude distribution change of the sinusoidal signal, the Gaussian white noise signal and the composite signals after they pass the following four nonlinear devices (1) the full-wave square-law device, (2) the half-wave linearizer, (3) the half-one-way ideal limiter; and (4) the stationary limiter.

5 Nonstationary random processes

5.1 Statistical description of nonstationary random signals

5.1.1 Probability density of nonstationary random signals

The probability density of a nonstationary random signal is a function of time; when time $t = t_i$, the probability density function is defined as:

$$p(x, t_i) = \lim_{\Delta x \to 0} \frac{P[x < X(t_i) < x + \Delta x]}{\Delta x} \tag{5.1}$$

where P denotes probability. Moreover:

$$\int_{-\infty}^{\infty} p(x, t_i)\, \mathrm{d}x = 1 \tag{5.2}$$

5.1.2 Digital characteristics of nonstationary random signals

Based on the probability density function, similarly to stationary random signals, the numerical characteristics of the nonstationary random signals can be defined as follows:

Mean:

$$m_X(t) \overset{\text{def}}{=} E[X(t)] = \int_{-\infty}^{\infty} x p(x, t)\, \mathrm{d}x \tag{5.3}$$

Mean square value:

$$D_X(t) \overset{\text{def}}{=} E[X^2(t)] = \int_{-\infty}^{\infty} x^2 p(x, t)\, \mathrm{d}x \tag{5.4}$$

Variance

$$\sigma_X^2(t) \overset{\text{def}}{=} D_X(t) - m_X^2(t) \tag{5.5}$$

They are all functions of time. It should be noted that nonstationary random signals are only statistically significant in the aggregate sense, and there is no statistical significance in the sense of time. These digital features can also be estimated by using methods similar to those in a stationary random case. Setting $x_i(t) (i = 1, 2, \ldots, N)$ are N samples of the signal $X(t)$. The mean is estimated as follows:

$$\widehat{m_X}(t) = \frac{1}{N} \sum_{i=1}^{N} x_i(t) \tag{5.6}$$

https://doi.org/10.1515/9783110593808-006

It can be shown that when the samples are independent of each other, it is an unbiased, consistent estimate, i.e.:

$$E[\widehat{m_X}(t)] = m_X(t)$$
$$\mathrm{Var}[\widehat{m_X}(t)] = \frac{1}{N}\sigma_X^2(t) \tag{5.7}$$

The mean square value is estimated as:

$$\widehat{D_X}(t) = \frac{1}{N}\sum_{i=1}^{N} x_i^2(t) \tag{5.8}$$

It can be proved that it is also unbiased to estimate it as:

$$E[\widehat{D_X}(t)] = D_X(t) \tag{5.9}$$

It is difficult to do a general analysis for its variance characteristics, but the nonstationary random signal $x(t)$ obeys a Gaussian distribution at any time t whose mean is $m_X(t)$ and variance is $G_X^2(t)$. It can be proved that:

$$\mathrm{Var}[\widehat{D_X}(\mathrm{t})] = \frac{2}{N}\left[D_X^2(t) - m_X^4(t)\right] \tag{5.10}$$

When $N \to \infty$, $\mathrm{Var}[\widehat{D_X}(\mathrm{t})] \to 0$, so $D_X(t)$ is consistent estimate in this situation.

Setting the second-order joint probability density of the nonstationary random signals $X_1(t)$ and $X_2(t)$ as $p(x_1, x_2; t_1, t_2)$, its autocorrelation function is defined as:

$$\begin{aligned} R_{XX}(t_1, t_2) &= E[X(t_1)X(t_2)] \\ &= \int_{-\infty}^{\infty}\int_{-\infty}^{\infty} x_1 x_2 p(x_1, x_2; t_1, t_2)\,\mathrm{d}x_1\,\mathrm{d}x_2 \end{aligned} \tag{5.11}$$

The cross-correlation function of nonstationary random signals $X(t)$, $Y(t)$ is defined as:

$$\begin{aligned} R_{XY}(t_1, t_2) &= E[X(t_1)Y(t_2)] \\ &= \int_{-\infty}^{\infty}\int_{-\infty}^{\infty} xyp(x, y; t_1, t_2)\,\mathrm{d}x\,\mathrm{d}y \end{aligned} \tag{5.12}$$

where $p(x, y; t_1, t_2)$ is the second-order joint probability density function of $X(t)$ and $Y(t)$.

Similarly, the autocovariance function and the cross-covariance function are defined as:

$$\begin{cases} C_{XX}(t_1, t_2) = E\{[X(t_1) - m_X(t_1)][X(t_2) - m_X(t_2)]\} \\ C_{XY}(t_1, t_2) = E\{[X(t_1) - m_X(t_1)][Y(t_2) - m_Y(t_2)]\} \end{cases} \tag{5.13}$$

The relationship between $C_{XX}(t_1, t_2)$ with $R_{XX}(t_1, t_2)$ is:

$$C_{XX}(t_1, t_2) = R_{XX}(t_1, t_2) - m_X(t_1)m_X(t_2) \tag{5.14}$$

The relationship between $C_{Xy}(t_1, t_2)$ with $R_{Xy}(t_1, t_2)$ is:

$$C_{XY}(t_1, t_2) = R_{XY}(t_1, t_2) - m_X(t_1)m_Y(t_2) \tag{5.15}$$

By the original definition, the correlation function has the following symmetry:

$$\begin{aligned} R_{XX}(t_2, t_1) &= R_{XX}(t_1, t_2) \\ R_{XY}(t_2, t_1) &= R_{YX}(t_1, t_2) \end{aligned} \tag{5.16}$$

Similarly, use N record samples $x_i(t)(i = 1, 2, \ldots, N)$ to estimate $R_{XX}(t_1, t_2)$, which is:

$$\widehat{R_{XX}}(t_1, t_2) = \frac{1}{N}\sum_{i=1}^{N} x_i(t_1)x_i(t_2) \tag{5.17}$$

If t_1 is fixed, change $t_2 = t_1 + \tau$, τ is the delay and then

$$R_{XX}(t_1, t_1 + \tau) = R_{XX}(t_1, \tau) = E[X(t_1)X(t_1 + \tau)] \tag{5.18}$$

It is not only the function of $t = t_1$, but also the function of delay τ. However, it is not an even function due to $R_{XX}(t_1, t_1 + \tau) \neq R_{XX}(t_1, t_1 - \tau)$ for the nonstationary random signals. In the same way that the Fourier transform of $R_{XX}(t_1, t_1 + \tau)$ is not a real function. To this end, the autocorrelation function for nonstationary random signals is given by another form of definition as follows:

$$R_{XX}(t, \tau) \overset{\text{def}}{=} E[X(t + \tau/2)X(t - \tau/2)] \tag{5.19}$$

Obviously, this defined autocorrelation function has the characteristic of an even function. This is

$$R_{XX}(t, \tau) = R_{XX}(t, -\tau) \tag{5.20}$$

When $X(t)$ is real signal, $R_{XX}(t, \tau)$ is also a real function, so that its Fourier transform can maintain the desired real-even characteristic.

As shown in Fig. 5.1, the t in $R_{XX}(t, \tau)$ is the mean in the position t_1, t_2 of the time axis of X_1, X_2, $t = (t_1 + t_2)/2$ and τ is the time interval of X_1, X_2. Thus, this autocorrelation function is called a symmetric autocorrelation function and is often used in some analysis and processing of nonstationary random signals.

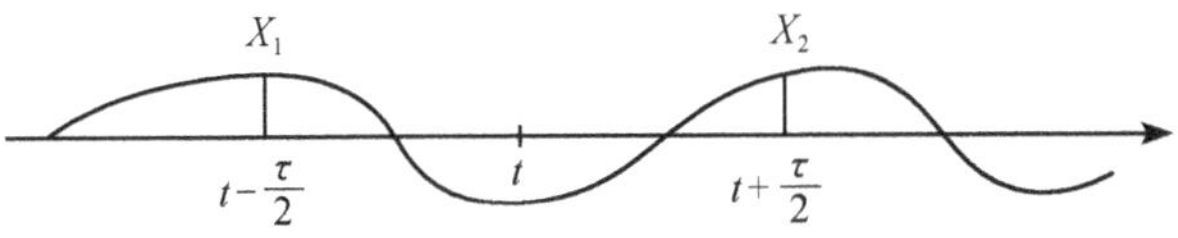

Fig. 5.1: Symmetric autocorrelation function.

5.1.3 The time-varying power spectrum and the short-time power spectrum

Autocorrelation function $R_{XX}(t_1, t_2)$ of nonstationary random signals is the function of t_1 and t_2, and if two-dimensional Fourier transform is used for it, we have:

$$S_{XX}(\omega_1, \omega_2) = \int_{-\infty}^{\infty}\int_{-\infty}^{\infty} R_{XX}(t_1, t_2)\mathrm{e}^{-\mathrm{j}(\omega_2 t_2 - \omega_1 t_1)}\,\mathrm{d}t_1\,\mathrm{d}t_2 \tag{5.21}$$

$$R_{XX}(t_1, t_2) = \frac{1}{4\pi^2}\int_{-\infty}^{\infty}\int_{-\infty}^{\infty} S_{XX}(\omega_1, \omega_2)\mathrm{e}^{\mathrm{j}(\omega_2 t_2 - \omega_1 t_1)}\,\mathrm{d}\omega_1\,\mathrm{d}\omega_2 \tag{5.22}$$

We define $S_{XX}(\omega_1, \omega_2)$ as the spectral density of nonstationary random signals, and this spectral density definition can be the same for the stationary random case mathematically, because $S_{XX}(\omega_1, \omega_2)$ has a value only at $\omega_1 = \omega_2$, and it equals to 0 when $\omega_1 \neq \omega_2$, i.e.,

$$S_{XX}(\omega_1, \omega_2) = S_{XX}(\omega_1)\delta(\omega_2 - \omega_1) \tag{5.23}$$

It degenerates into the power spectrum of the stationary random signal. However, this definition has obvious flaws in the physical concept; $R_{XX}(t_1, t_2)$ is not an even function, so $S_{XX}(\omega_1, \omega_2)$ is not a real function. To overcome this difficulty, we use the time-varying power spectrum defined as the following to characterize the spectral characteristics in practice, namely:

$$S_{XX}(t, \omega) \stackrel{\text{def}}{=} \int_{-\infty}^{\infty} R_{XX}(t, \tau)\mathrm{e}^{-\mathrm{j}\omega\tau}\,\mathrm{d}\tau \tag{5.24}$$

In the formula, $R_{XX}(t, \tau)$ is the symmetric autocorrelation function. From the above equation it can be seen that the time-varying power spectrum $S_{XX}(t, \omega)$ is a partial Fourier transform of delay τ variable for the symmetric autocorrelation function $R_{XX}(t, \tau) = E[X(t + \tau/2)X^*(t - \tau/2)]$.

The short-term power spectrum was proposed by Fano [1] and is defined as the square of the amplitude of the short Fourier transform of the signal $X(t)$:

$$ST_X(t, \omega) = \left|\int_{-\infty}^{\infty} X(\tau)\mathrm{e}^{-\mathrm{j}\omega\tau} w(t - \tau)\,\mathrm{d}\tau\right|^2 \tag{5.25}$$

In the formula, $w(t-\tau)$ is the window function or weight function that slides over time, and its function is to weight the data, so that the weight of the past data in the current spectral estimation is weakened over time. The most commonly used window function is the negative exponential function:

$$w(t - \tau) = \mathrm{e}^{-\omega(t-\tau)}u(t - \tau) \tag{5.26}$$

As in Fig. 5.2, in the study of the sum signal of multiple signals, the short-time Fourier transform is linear, but its short-term power spectrum is quadratic, so there is cross-term interference affecting the resolution of the spectrum, which is its drawbacks.

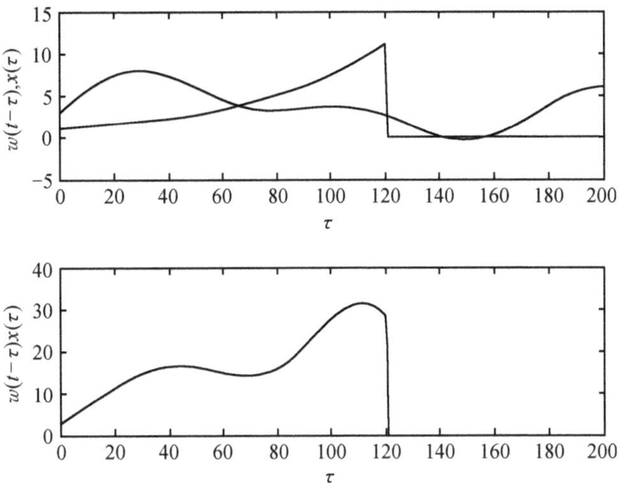

Fig. 5.2: Short-term power spectrum.

5.1.4 The Wiener process

The Wiener process is a nonstationary normal random process. If we use $X(t)$ to show it, we have

$$X(0) = 0 \tag{5.27}$$

The mean and autocorrelation functions are:

$$\begin{cases} E[X(t)] = 0 \\ R_{XX}(t_1, t_2) = \begin{cases} \alpha t_2\,, & t_1 \geq t_2 \\ \alpha t_1\,, & t_1 < t_2 \end{cases} \end{cases} \tag{5.28}$$

The Wiener process $X(t)$ can be seen as the result of the white noise $W(t)$ after the integrator, as shown in Fig. 5.3. If $W(t)$ is stationary normal white noise, with zero mean and variance σ_W^2, the Wiener process $X(t)$ is:

$$X(t) = \int_0^t W(\tau)\,\mathrm{d}\tau \tag{5.29}$$

Since the integral is a linear operation, $X(t)$ is a normal distribution process. The following shows the mean of the Wiener process and the character of the autocorrelation function.

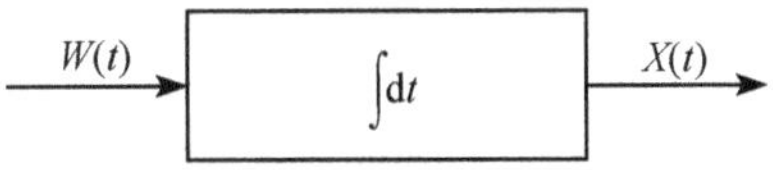

Fig. 5.3: Generation of Wiener processes.

Test:

(1) Because $E[W(t)] = 0$, so $E[X(t)] = 0$.

(2) When $0 < t_1 < t_2$:

$$R_{XX}(t_1, t_2) = E[X(t_1)X(t_2)] = E\left[\int_0^{t_1} W(u)\,\mathrm{d}u \int_0^{t_2} W(v)\,\mathrm{d}v\right]$$

$$= E\left[\int_0^{t_1}\int_0^{t_2} W(u)W(v)\,\mathrm{d}u\,\mathrm{d}v\right] = \int_0^{t_1}\int_0^{t_2} \sigma_W^2\delta(u-v)\,\mathrm{d}u\,\mathrm{d}v$$

$$= \int_0^{t_1} \sigma_W^2\,\mathrm{d}u = \sigma_W^2 t_1$$

When $0 < t_2 < t_1$:

$$R_{XX}(t_1, t_2) = E\left[\int_0^{t_1} W(u)\,\mathrm{d}u \int_0^{t_2} W(v)\,\mathrm{d}v\right]$$

$$\int_0^{t_2}\int_0^{t_1} \sigma_W^2\delta(u-v) = \int_0^{t_2} \sigma_W^2\,\mathrm{d}u = \sigma_W^2 t_1$$

With $\alpha = \sigma_W^2$, by the above proofs (1) and (2), $R_{XX}(t_1, t_2)$ can be expressed as:

$$R_{XX}(t_1, t_2) = \alpha \min(t_1, t_2)\,.$$

If the white noise power spectrum $S_{WW}(\omega) = \alpha$, because of the nature of the function δ: $F^{-1}[1] = \delta(x)$, we have

$$R_{WW}(\tau) = \frac{1}{2\pi}\cdot\int_{-\pi}^{\pi} S_{WW}(\omega)\mathrm{e}^{\mathrm{j}\omega\tau}\,\mathrm{d}\omega = \alpha\frac{1}{2\pi}\int_{-\pi}^{\pi} \mathrm{e}^{\mathrm{j}\omega\tau}\,\mathrm{d}\omega = \alpha\delta(\tau)$$

Then, $R_{WW}(0) = \sigma_W^2 = \alpha$, namely α is the power spectrum of white noise $W(t)$.

5.2 Linear time-varying systems and nonstationary random signals

5.2.1 Description of linear time-varying discrete system characteristics

It is well known that the characteristics of linear invariant discrete systems can be described by the impulse response or by a transfer function in the Z-domain. Corresponding to this, the characteristics of a linear time-varying discrete system can be described in the time domain using a time-varying impulse response or Green's function, whereas it can be described by a generalized transfer function in the Z-domain.

5.2.1.1 Time-varying impulse response and Green's function

The input and output relationships of linear time-varying discrete systems are described in the time domain by Green's function, $g(n, m)$, or a time-varying impulse response, $h(n, m)$. Green's function $g(n, m)$ is defined as the output of the resulting system at the moment n after the unit pulse entered at the moment m. The time-varying impulse response $h(n, m)$ is defined as the output of the system obtained at time n by inputting the unit pulse at time $n - m$. Obviously, the relationship of time varying pulses $h(n, m)$ with the Green function $g(n, m)$ is

$$\begin{aligned} h(n, m) &= g(n, n - m) \\ g(n, m) &= h(n, n - m) \end{aligned} \tag{5.30}$$

If $g(n, m)$ is the relationship of input and output for the linear and invariant discrete system, then $g(n, m)$ is only related with the input and output time difference $n - m$, i.e.:

$$g(n, m) = g(n - m) \tag{5.31}$$

Due to the type of substitution, $h(n, m)$, available:

$$h(n, m) = g(n, n - m) = g[n - (n - m)] = g(m) = h(m) \tag{5.32}$$

Visibly, at this time $h(n, m)$ has nothing to do with n, as the impulse response $h(m)$ of a general linear time-invariant discrete system.

If there is a linear time-varying system with a Green function $g(n, m)$ or time-varying impulse response $h(n, m)$, when the input is $x(n)$, the output is:

$$y(n) = \sum_{m=-\infty}^{\infty} g(n, m)x(m) \tag{5.33}$$

Or:

$$y(n) = \sum_{m=-\infty}^{\infty} h(n, m)x(n - m) = \sum_{m=-\infty}^{\infty} h(n, n - m)x(m) \tag{5.34}$$

If the linear time-varying discrete system is causal, then:

$$g(n, m) = 0, n < m \tag{5.35}$$

5.2.1.2 Generalized transfer function

The generalized transfer function of linear time-varying discrete systems can be defined by Green's function:

$$G(z, n) = \sum_{m=-\infty}^{\infty} g(n, m)z^{-(n-m)} \tag{5.36}$$

It can also be defined by the time-varying impulse response:

$$H(z, n) = \sum_{m=-\infty}^{\infty} h(n, m)z^{-m} \tag{5.37}$$

These two definitions are the same, i.e.,

$$G(z, n) = H(z, n) \tag{5.38}$$

We calculate $G(z, n)$ on the unit circle:

$$G(\omega, n) = G(z, n)\big|_{z=e^{j\omega}} = \sum_{m=-\infty}^{\infty} g(n, m)e^{-j(n-m)\omega} \tag{5.39}$$

$G(\omega, n)$ is usually plural and can be denoted by:

$$G(\omega, n) = |G(\omega, n)|e^{j\arg|G(\omega,n)|} \tag{5.40}$$

Visibly, the amplitude characteristics and phase characteristics of $G(\omega, n)$ are both functions of ω and n, so $G(\omega, n)$ is called a generalized frequency response or time-varying frequency response.

If the input signal sequence is $x(m) = e^{jm\omega}(-\infty < m < \infty)$, then the output of the linear time-varying discrete system with Green's function $g(n, m)$ is:

$$y(n) = \sum_{m=-\infty}^{\infty} g(n, m)e^{jm\omega} = e^{jn\omega} \sum_{m=-\infty}^{\infty} g(n, m)e^{-j(n-m)\omega} = e^{jn\omega}G(\omega, n) \tag{5.41}$$

Thus, the physical explanation of the generalized transfer function $G(\omega, n)$ can be derived from the above equation:

$$G(\omega, n) = \frac{\exp(j\omega n) \text{ the resulting system response } y(n)}{\exp(j\omega n)} \tag{5.42}$$

$H(z, n)$ is calculated in the unit circle as:

$$H(\omega, n) = \sum_{m=-\infty}^{\infty} h(n, m)e^{-jm\omega} \tag{5.43}$$

Because $H(\omega, n) = G(\omega, n)$, the generalized frequency response or time-varying frequency response $H(\omega, n)$ is a partial Fourier transform of the time-varying impulse response $h(n, m)$ on m.

Example 5.2.1. If the linear discrete system is a causal system, its Green function is

$$g(n, m) = \begin{cases} a^{n-m} \sin[b(n-2m)], & n \geq m \\ 0, & n < m \end{cases}$$

Try to solve its generalized transfer function.

Solution: Order $l = n - m$, then:

$$h(n, l) = g(n, n-l) = \begin{cases} a^{l} \sin[b(2l-n)], & l \geq 0 \\ 0, & l < 0 \end{cases}$$

Therefore, the generalized transfer function is:

$$
\begin{aligned}
H(z,n) &= \sum_{l=0}^{\infty} a^l \sin[b(2l-n)]z^{-1} \\
&= \sum_{l=0}^{\infty} [\sin(2bl)\cos(bn) - \cos(2bl)\sin(bn)]z^{-1} \\
&= \cos(bn)\sum_{l=0}^{\infty} a^l \sin(2bl)z^{-1} - \sin(bn)\sum_{l=0}^{\infty} a^l \cos(2bl)z^{-1} \\
&= \cos(bn)\frac{z^{-1}a\sin(2b)}{1-2a\cos(2b)z^{-1}+a^2z^{-2}} - \sin(bn)\frac{z^{-1}a\cos(2b)}{1-2a\cos(2b)z^{-1}+a^2z^{-2}} \\
&= \frac{-\sin(bn)+a\sin[b(n+2)]z^{-1}}{1-2a\cos(2b)z^{-1}+a^2z^{-2}}
\end{aligned}
$$

Example 5.2.2. If the linear time-varying discrete system is a causal system, its Green function is:

$$
g(n,m) = \begin{cases} (\mathrm{e}^{-an})^{n-m-1}\sin(bn)\,, & n \geq m+1 \\ 0\,, & n < m+1 \end{cases}
$$

Try to solve its generalized transfer function.

Solution: Order $l = n - m$, then:

$$
h(n,l) = g(n,n-l) = \begin{cases} (\mathrm{e}^{-an})^{l-1}\sin(bn)\,, & l \geq 1 \\ 0\,, & l < 1 \end{cases}
$$

Therefore, the generalized transfer function is:

$$
H(z,n) = \sum_{l=1}^{\infty} (\mathrm{e}^{-an})^{l-1}\sin(bn)z^{-1} = \frac{\sin(bn)z^{-1}}{1-\mathrm{e}^{-an}z^{-1}}
$$

5.2.2 Characterization of linear time-varying continuous systems

In this paper, we introduce the time-varying impulse response and generalized transfer function of linear time-varying discrete systems, and there are similar definitions for continuous systems.

Time-varying impulse response $h(t,t')$ of linear time-varying continuous systems is defined as the output of the system obtained at t by inputting the unit pulse at time $t-t'$.

The linear time-varying system with a time-varying impulse response $h(t,t')$, whose output $y(t)$ is related with input $x(t)$ is as follows:

$$
y(t) = \int_{-\infty}^{\infty} h(t,t')x(t')\,\mathrm{d}t' \tag{5.44}
$$

The generalized transfer function of linear time-varying continuous systems is:

$$H(t,\omega) = \int_{-\infty}^{\infty} h(t,\xi)\mathrm{e}^{\mathrm{j}\omega(t-\xi)}\,\mathrm{d}\xi \tag{5.45}$$

If the Fourier spectrum of $x(t)$ is $X(\omega)$, then the output $y(t)$ can be made by the inverse Fourier transform for the product of $H(t,\omega)$ and $X(\omega)$, namely:

$$y(t) = \frac{1}{2\pi}\int_{-\infty}^{\infty} H(t,\omega)X(\omega)\mathrm{e}^{\mathrm{j}\omega t}\,\mathrm{d}\omega \tag{5.46}$$

5.2.3 Time-varying parameters of AR, MA and ARMA models nonstationary random signals

As mentioned earlier, for some types of stationary random signal, it can be described by AR, MA and ARMA models. This kind of stationary random signal can be generated by the output of a causal linear time-invariant discrete system with white noise excitation, where the ARMA model is generally:

$$\begin{aligned} X(n) + a_1X(n-1) + a_2X(n-2) + \cdots + a_pX(n-p) \\ = W(n) + b_1W(n-1) + b_2W(n-2) + \cdots + b_qW(n-q) \end{aligned} \tag{5.47}$$

Similarly, there is also a certain type of nonstationary random signal, which can be described by time-varying AR, MA and ARMA models.

For general nonstationary random signals $X(n)$, it can be made by the output of a causal linear time-varying discrete system with white noise $W(n)$ to excite:

$$X(n) = \sum_{m=0}^{\infty} h(n,m)W(n-m) \tag{5.48}$$

In the formula, $h(n,m)$ is the system time-varying impulse response. The above formula can also use the system's Green function $g(n,m)$ to display it, as shown below:

$$X(n) = \sum_{m=0}^{\infty} g(n,m)W(n) \tag{5.49}$$

It can be seen that for any nonstationary random signal, it corresponds to a time-varying discrete system with the system function $g(n,m)$. It is easy to prove when the Green function $g(n,m)$ is the following p-order degradation sequence, i.e.,

$$g(n,m) = \sum_{i=1}^{p} u_i(n)\varphi_i(m) \tag{5.50}$$

The corresponding nonstationary random signals can be expressed as the following time-varying parameter ARMA (p, q) model:

$$X(n) + a_1(n)X(n-1) + a_2(n)X(n-2) + \cdots + a_p(n)X(n-p)$$
$$= W(n) + b_1(n)W(n-1) + b_2(n)W(n-2) + \cdots + b_q(n)W(n-q) \quad (5.51)$$

Therefore, any causal linear time-varying discrete system with white noise excitation can produce nonstationary random signals, but the resulting nonstationary random signals can be described by the time-varying parameter ARMA model; the Green function of the linear time-varying discrete system must be a degraded sequence.

5.3 Wigner–Ville spectrum of nonstationary random signals

5.3.1 Overview of time-frequency analysis

Many of the parameters of the signal in the practical application are time varying, so the spectral structure also changes with time. The general spectral density can only give the energy distribution of the signal throughout the duration in the frequency, but it cannot give energy distribution at a certain time or for a certain period of time in the frequency. Therefore, people began to study signal time-frequency analysis technology. It is in the time-frequency plane to study the distribution of signal energy to make a time-domain signal map a two-dimensional signal in the time-frequency plane. People are interested in its nature and application. Now, time-frequency analysis plays an increasingly important role in signal processing.

The development of time-frequency analysis has a considerable history but also gives a wealth of results, from the short-time Fourier transform to Wigner distribution and wavelet analysis, and then to nuclear-based general time-frequency distribution; from deterministic signal analysis to random signal analysis, and so on. The initial time-frequency analysis technique is a short-time Fourier transform, which is a very intuitive extension to the Fourier transform. It intercepts a short section of the signal around the moment of concern, that is, making the signal weighting process with a window function to analyze the spectrum distribution of the signal at this time. However, due to the introduction of the time window, the characteristics of the signal are disturbed. When we intend to increase the time resolution and shorten the time window, the resulting spectrum is not related to the characteristics of the original signal; it is meaningless, and vice versa. That is, people say that the localization of time and frequency localization cannot be achieved at the same time. This is mainly caused by the short-time Fourier transform method itself, so people began to study better performance time-frequency analysis methods. In this and the following sections, we give a brief introduction to Wigner distribution and wavelet analysis and discuss their application in random signal analysis.

5.3.2 Wigner distribution and the Wigner–Ville spectrum

5.3.2.1 Wigner distribution of the deterministic signal

Set the continuous time signal $x(t)$ defined in the entire time domain and its complex value, that is, $x(t) \in C, t \in R$. Define the signal from of the Wigner distribution (WD) as:

$$W_x(t,\omega) \overset{\text{def}}{=} \int_{-\infty}^{\infty} x\left(t+\frac{\tau}{2}\right) x^*\left(t-\frac{\tau}{2}\right) \mathrm{e}^{-\mathrm{j}\omega\tau}\,\mathrm{d}\tau \tag{5.52}$$

It can be seen as the Fourier transform of τ for the function of

$$R_{xx}(t,\tau) = x(t+\tau/2)\,x^*(t-\tau/2)$$

The interconnection of Wigner about two consecutive time signals $x(t)$ and $y(t)$ is defined as:

$$W_{x,y}(t,\omega) \overset{\text{def}}{=} \int_{-\infty}^{\infty} x\left(t+\frac{\tau}{2}\right) y^*\left(t-\frac{\tau}{2}\right) \mathrm{e}^{-\mathrm{j}\omega\tau}\,\mathrm{d}\tau \tag{5.53}$$

When there is no confusion, the above two functions are often referred to as Wigner distribution. The Wigner distribution of the discrete-time signal has not been universally accepted so far, but a widely used definition is:

$$W_x(n,\omega) \overset{\text{def}}{=} 2\sum_{k=-\infty}^{\infty} x(t+k)x^*(t-k)\mathrm{e}^{-\mathrm{j}2k\omega} \tag{5.54}$$

Here we examine WD (Wigner Distribution) of some typical signals.

Example 5.3.1. Sliding rectangular window signal:

$$x(t) = \begin{cases} 1, |t| < T \\ 0, |t| > T \end{cases}$$

Through a simple calculation, we can see that WD:

$$W_x(t,\omega) = \begin{cases} \frac{2}{\omega}\sin[2\omega(T-|t|)]\,, & |t| < T \\ 0\,, & |t| > T \end{cases}$$

This kind of WD has the form of sinx/x for the frequency ω, its main lobe width follows the $|t|$ increase and widening. Its WD has the maximum value for $W_x(0,0) = 4T$ when $(t,w) = (0,0)$. In some areas of the plane about (t,ω), $W_x(t,w)$ is negative.

Example 5.3.2. The slide index signal:

$$x(t) = \begin{cases} \mathrm{e}^{\mathrm{j}\omega_0 t}, & |t| < T \\ 0, & |t| > T \end{cases}$$

Its WD by a simple calculation is:

$$W_x(t,w) = \begin{cases} \frac{2}{\omega-\omega_0}\sin 2(\omega-\omega_0)(T-|t|)\,, & |t| < T \\ 0\,, & |t| > T \end{cases}$$

5.3.2.2 Wigner–Ville spectra of nonstationary random signals

Similar to the Wigner distribution of the signal, define its Wigner–Ville (WV) spectrum for nonstationary random signals in order to analyze its time-frequency characteristics. It is widely used in communications, radar and biomedical applications. It is a useful tool for the time-frequency analysis of nonstationary random signals.

First, define the WV spectrum of nonstationary continuous random signals. As mentioned earlier, the symmetric autocorrelation function of nonstationary continuous random signals is:

$$R_X(t,\tau) = E[X(t+\tau/2)X^*(t-\tau/2)] \tag{5.55}$$

Similarly to the WD definition of a deterministic continuous random signal, the WV spectrum of a nonstationary continuous random signal is defined as:

$$W_X(t,\omega) = \int_{-\infty}^{\infty} E[X(t+\tau/2)X^*(t-\tau/2)]\mathrm{e}^{-\mathrm{j}\omega\tau}\,\mathrm{d}\tau \tag{5.56}$$

It is compared with the WD definition of deterministic continuous signals, but only to use mathematical expectations in the above formula. Substituting $R_X(t,\tau)$ into eq. (5.56) yields:

$$W_X(t,\omega) = \int_{-\infty}^{\infty} R_X(t,\tau)\mathrm{e}^{-\mathrm{j}\omega\tau}\,\mathrm{d}\tau \tag{5.57}$$

The WV spectrum of nonstationary stochastic continuous signals is the Fourier transform of τ for the symmetric autocorrelation $R_X(t,\tau)$. The WV spectrum has many properties of the deterministic continuous signal WD and is not required to be limited in the energy of the process and is not limited by the analysis time, as compared to other spectral representations (such as periodic diagrams). Therefore, it is particularly suitable for nonstationary random signal analysis.

Secondly, define the WV spectra of nonstationary discrete random signals. The symmetric autocorrelation function for nonstationary random sequences is:

$$R_X(n,k) = E[X(n+k)X^*(n-k)] \tag{5.58}$$

Similar to the WD definition of a deterministic discrete signal, the WV of a nonstationary discrete random signal (or a WV of a nonstationary random sequence) is defined as:

$$W_X(n,\theta) = 2\sum_{k=-\infty}^{\infty} \mathrm{e}^{-\mathrm{j}2k\theta}R_X(n,k) = 2\sum_{k=-\infty}^{\infty} \mathrm{e}^{-\mathrm{j}2k\theta}E[X(n+k)X^*(n-k)] \tag{5.59}$$

Among these $R_X(n,k)$ denotes a discrete symmetric autocorrelation function. Likewise, it is only necessary to use the mathematical expectations in the above formula comparing with the WD definition of the deterministic discrete signal.

5.3.3 Examples of WD applications

Wigner distribution is widely used. Here, we mention as an example its application in spectral analysis and compare it with the results obtained from other spectral analysis in order to better understand its performance.

With a frequency of ω_1 and when the phase of the single sinusoidal signal is φ, in spectral analysis, it is desirable not to consider the phase φ, and the frequency ω_1 is determined by the peak. In the following, discuss the spectral analysis results of the real sinusoidal signal and the complex sinusoidal signal (or sinusoidal resolution signal) by the usual method. Then use the WD to analyze the spectrum of the sinusoidal analysis signal and, finally, compare the performance of the spectrum analysis.

5.3.3.1 Power spectrum analysis of real sinusoidal signals

A real sinusoidal signal is:

$$x(t) = \cos(\omega_1 t + \varphi) \tag{5.60}$$

We use a rectangular window function $h(x)$ with length T to add a window for $x(t)$:

$$x_h(t) = x(t)h(t) \tag{5.61}$$

Among these:

$$h(t) = \begin{cases} 1\,, & -T/2 \le t \le T/2 \\ 0\,, & \text{otherwise} \end{cases} \tag{5.62}$$

The Fourier spectrum of $x(t)$ is:

$$X(\omega) = \pi[\mathrm{e}^{\mathrm{j}\varphi}\delta(\omega - \omega_1) + \mathrm{e}^{-\mathrm{j}\varphi}\delta(\omega + \omega_1)] \tag{5.63}$$

So:

$$X_h(\omega) = \frac{T}{2}[\mathrm{e}^{\mathrm{j}\varphi}Sa((\omega - \omega_1)T/2) + \mathrm{e}^{\mathrm{j}\varphi}Sa((\omega + \omega_1)T/2)] \tag{5.64}$$

Because the spectrum $X_h(\omega)$ is complex, analysis is usually carried out with its power spectrum, that is:

$$\begin{aligned} |X_h(\omega)|^2 &= X_h(\omega)X_h^*(\omega) \\ &= \left(\frac{T}{2}\right)^2 \Big[Sa^2(\omega - \omega_1)T/2 + Sa^2(\omega + \omega_1)T/2 \\ &\quad + 2\cos(2\varphi)Sa((\omega - \omega_1)T/2)Sa((\omega + \omega_1)T/2)\Big] \end{aligned} \tag{5.65}$$

The first two terms are the parts associated with the positive and negative frequencies of the sinusoidal signal, and they have respective peak values at the real frequency. The third term is the intersection term formed by the interaction of these two parts, which is related to the phase φ, and the peak value of the power spectrum changes with φ. When $2\varphi = n\pi/2$ or $\varphi = n\pi/4$, and n is odd, the third term is zero; only the negative frequency part impacts the spectral value at ω_1. At this time, the power spectrum at ω_1 becomes:

$$|X_h(\omega_1)|^2 = \left(\frac{T}{2}\right)^2 \left[1 + Sa^2(\omega_1 T)\right] \tag{5.66}$$

It is always positive. The general expression of $|X_h(\omega)|^2$ in ω_1 is:

$$|X_h(\omega_1)|^2 = \left(\frac{T}{2}\right) \left[1 + Sa^2(\omega_1 T) + 2\cos(2\varphi) Sa(\omega_1 T)\right] \tag{5.67}$$

In this case, the latter two are associated with the negative frequency part, where Sa square items are small and negligible, but Sa is not always positive, and the item is related to φ.

5.3.3.2 Power spectrum analysis of complex sinusoidal signals

Set the complex sinusoidal signal (or the resolution signal of the real sinusoidal signal) as:

$$x_a(t) = \mathrm{e}^{\mathrm{j}(\omega_1 t + \varphi)} \tag{5.68}$$

The Fourier spectrum is:

$$X_a(\omega) = 2\pi \mathrm{e}^{\mathrm{j}\varphi} \delta(\omega - \omega_1) \tag{5.69}$$

The Fourier spectrum after adding the rectangular window is:

$$X_{ah}(\omega) = X_a(\omega) \otimes H(\omega) = T\mathrm{e}^{\mathrm{j}\varphi} Sa((\omega - \omega_1)T/2) \tag{5.70}$$

The above equation is only related with sinusoidal frequency ω_1. The power spectrum in this case is:

$$|X_{ah}(\omega)|^2 = T^2 Sa^2((\omega - \omega_1)T/2) \tag{5.71}$$

Its power spectrum peak is always at the position of ω_1. Thus, the frequency can be accurately determined without regarding the phase of the signal and the length of the window, and it changes by Sa^2.

5.3.3.3 Analysis of Wigner distribution of complex sine signals

For real sinusoidal resolution signals, the WD is:

$$\begin{aligned} W_{x_a}(t,\omega) &= \int_{-\infty}^{\infty} \mathrm{e}^{\mathrm{j}(\omega_1 t + \varphi + \omega_1 x/2)} \mathrm{e}^{-\mathrm{j}(\omega_1 t + \varphi - \omega_1 x/2)} \mathrm{e}^{-\mathrm{j}\omega\tau}\, \mathrm{d}\tau \\ &= 2\pi\delta(\omega - \omega_1) \end{aligned} \tag{5.72}$$

As can be seen from the above, W_{x_a} is irrelevant with t, nothing and is only the function for ω. Therefore, only consider WD of a rectangular sliding window $h(t)$ with a length T, at $t = 0$:

$$W_h(0,\omega) = \int_{-\infty}^{\infty} h(0+\tau/2)h(0-\tau/2)\mathrm{e}^{-\mathrm{j}\omega\tau}\,\mathrm{d}\tau$$
$$= \int_{-\infty}^{\infty} \mathrm{e}^{-\mathrm{j}\omega\tau}\,\mathrm{d}\tau = 2TSa(\omega T) \tag{5.73}$$

Adding the rectangular window, the WD of the real sinusoidal resolution signal is:

$$W_{x_a h}(t,\omega) = W_{x_a}(t,\omega) \otimes W_h(0,\omega) = 2TSa[(\omega - \omega_1)T] \tag{5.74}$$

The above equation shows that the peak of this WD spectrum is always at the position of ω_1 and φ completely irrelevant to φ, it can be used to determine the frequency accurately, but the spectrum will appear negative, and because it changes with ω by Sa, side lobe attenuation is relatively small.

5.3.3.4 Comparison of spectral analysis performance

As shown in Fig. 5.4, the normal spectrum analysis of the real sinusoidal signal cannot accurately determine the frequency of the sinusoidal signal by its peak, and the main lobe is wide, the resolution is poor, but the power spectrum is always positive and the side lobe changes little. Normal spectrum analysis of the complex sinusoidal signal can accurately determine its frequency, and other performance is the same as that of the real sinusoidal signal. The WD spectrum analysis of the real sinusoidal resolution signal can not only accurately determine its frequency by its peak, but also the main lobe is narrow, the resolution is high, but the spectrum will appear negative, and side lobe amplitude changes are also large.

5.3.4 WV spectra of linear time-varying system outputs

The Wigner distribution of the output of the linear time-varying system in the case of deterministic signal input has been studied. For linear time-varying systems, if the input is a stationary random signal, the output is nonstationary random signal. The following focuses on the WV spectrum of such nonstationary random signals.

5.3.4.1 Two-dimensional Wigner distribution

In order to study the Wigner distribution of linear time-varying systems, we need to know the two-dimensional Wigner distribution first. The two-dimensional Wigner dis-

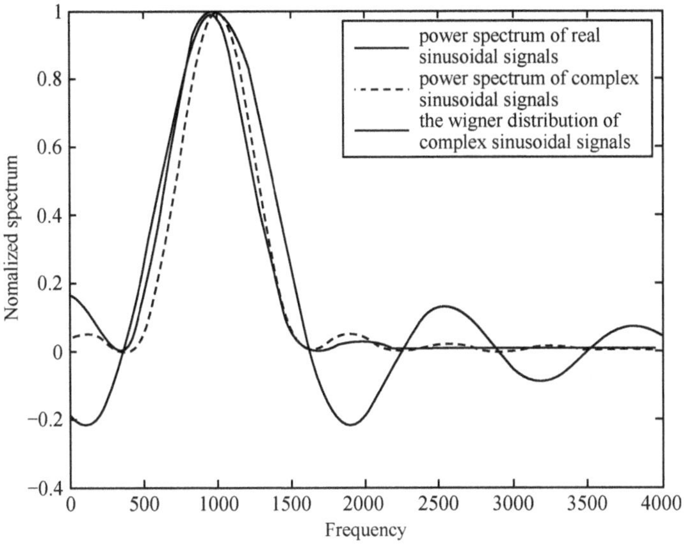

Fig. 5.4: $f = 1$ kHz, $\varphi = 0$, $T = 1$ ms single sine signal spectrum analysis performance.

tribution of signal $f(x, y)$ is defined as:

$$W_f(x, y, u, v) \overset{\text{def}}{=} \int_{-\infty}^{\infty}\int_{-\infty}^{\infty} f\left(x + \frac{\alpha}{2}, y + \frac{\beta}{2}\right) f^*\left(x - \frac{\alpha}{2}, y - \frac{\beta}{2}\right) \exp[-\mathrm{j}(\alpha u + \beta v)]\,\mathrm{d}\alpha\,\mathrm{d}\beta \tag{5.75}$$

The two-dimensional Wigner distribution maintains many properties of one-dimensional Wigner distribution and has been successfully applied in image processing, pattern recognition and so on.

5.3.4.2 Wigner distribution of linear time-varying system outputs

Setting $x(t)$ with the Wigner distribution $W_x(t, \omega)$ through linear time-varying system $h(t, t')$, the Wigner distribution of output $y(t)$ is $W_y(t, \omega)$, both of which have the following relationship:

$$W_y(t, \omega) = \frac{1}{2\pi} \int_{-\infty}^{\infty}\int_{-\infty}^{\infty} W_k(t, t', \omega, \omega') W_x(t', -\omega')\,\mathrm{d}t'\,\mathrm{d}\omega' \tag{5.76}$$

In the formula, $W_h(t, t', \omega, \omega')$ is the two-dimensional Wigner distribution for $h(t, t')$, i.e.,

$$W_h(t, t', \omega, \omega') = \int_{-\infty}^{\infty}\int_{-\infty}^{\infty} h\left(t + \frac{\tau}{2}, t' + \frac{\tau'}{2}\right) h\left(t - \frac{\tau}{2}, t' - \frac{\tau'}{2}\right) \exp[-\mathrm{j}(\omega\tau + \omega'\tau')]\,\mathrm{d}\tau\,\mathrm{d}\tau' \tag{5.77}$$

Test:

$$\frac{1}{2\pi}\int_{-\infty}^{\infty}\int_{-\infty}^{\infty} W_n(t,t',\omega,\omega')W_x(t',-\omega')\,\mathrm{d}t'\,\mathrm{d}\omega'$$

$$=\frac{1}{2\pi}\int_{-\infty}^{\infty}\int_{-\infty}^{\infty}\left[\int_{-\infty}^{\infty}\int_{-\infty}^{\infty} h(t+\tau_1/2,t'+\tau_2/2)h^*(t-\tau_1/2,t'-\tau_2/2)\mathrm{e}^{-\mathrm{j}(\omega\tau_1+\omega'\tau_2)}\,\mathrm{d}\tau_1\,\mathrm{d}\tau_2\right]$$

$$\cdot\left[\int_{-\infty}^{\infty} x(t'+\tau_3/2)x^*(t'-\tau_3/2)\mathrm{e}^{\mathrm{j}\omega'\tau}\,\mathrm{d}\tau_3\right]\mathrm{d}t'\,\mathrm{d}\omega'$$

$$=\frac{1}{2\pi}\int_{-\infty}^{\infty}\int_{-\infty}^{\infty}\int_{-\infty}^{\infty}\int_{-\infty}^{\infty}\left[\int_{-\infty}^{\infty}\mathrm{e}^{-\mathrm{j}\omega'(\tau_2-\tau_3)}\,\mathrm{d}\omega'\right]h(t+\tau_1/2,t'+\tau_2/2)h^*(t-\tau_1/2,t'-\tau_2/2)$$

$$\cdot\, x(t'+\tau_3/2)x^*(t'-\tau_3/2)\mathrm{e}^{-\mathrm{j}\omega\tau_1}\,\mathrm{d}\tau_1\,\mathrm{d}\tau_2\,\mathrm{d}\tau_3\,\mathrm{d}t'$$

$$=\int_{-\infty}^{\infty}\int_{-\infty}^{\infty}\int_{-\infty}^{\infty}\int_{\infty}^{\infty}\delta(\tau_2-\tau_3)h(t+\tau_1/2,t'+\tau_2/2)x(t'+\tau_3/2)$$

$$\cdot\, h^*(t-\tau_1/2,t'-\tau_2/2)x^*(t'-\tau_3/2)\mathrm{e}^{-\mathrm{j}\omega\tau_1}\,\mathrm{d}\tau_1\,\mathrm{d}\tau_2\,\mathrm{d}\tau_3\,\mathrm{d}t'$$

$$=\int_{-\infty}^{\infty}\int_{-\infty}^{\infty}\int_{-\infty}^{\infty} h(t+\tau_1/2,t'+\tau_2/2)x(t'+\tau_2/2)h^*$$

$$\cdot\,(t-\tau_1/2,t'-\tau_2/2)x^*(t'-\tau_2/2)\mathrm{e}^{-\mathrm{j}\omega\tau_1}\,\mathrm{d}\tau_1\,\mathrm{d}\tau_2\,\mathrm{d}t'$$

Let

$$t'+\tau_2/2=\alpha_1\,,\quad t'-\tau_2/2=\alpha_2$$

then

$$t'=(\alpha_1+\alpha_2)/2\,,\quad \tau_2=\alpha_1-\alpha_2$$

The Jacobi formula is:

$$J\begin{vmatrix}\frac{\partial t'}{\partial\alpha_1} & \frac{\partial t'}{\partial\alpha_2}\\ \frac{\partial\tau_2}{\partial\alpha_1} & \frac{\partial\tau_2}{\partial\alpha_2}\end{vmatrix}=\begin{vmatrix}\frac{1}{2} & \frac{1}{2}\\ 1 & -1\end{vmatrix}=-1$$

Therefore, $|J|=1$. Thus we have:

$$\int_{-\infty}^{\infty}\int_{-\infty}^{\infty}\int_{-\infty}^{\infty} h(t+\tau_1/2,\alpha_1)x(\alpha_1)h^*(t-\tau_1/2,\alpha_2)x^*(\alpha_2)\mathrm{e}^{-\mathrm{j}\omega\tau_1}\,\mathrm{d}\alpha_1\,\mathrm{d}\alpha_2\,\mathrm{d}\tau_1$$

$$=\int_{-\infty}^{\infty}\left[\int_{-\infty}^{\infty} h(t+\tau_1/2,\alpha_1)x(\alpha_1)\,\mathrm{d}\alpha_1\right]\left[\int_{-\infty}^{\infty} h^*(t-\tau_1/2,\alpha_2)x(\alpha_2)\,\mathrm{d}\alpha_2\right]\mathrm{e}^{-\mathrm{j}\omega\tau_1}\,\mathrm{d}\tau_1$$

$$=\int_{-\infty}^{\infty} y(t+\tau_1)y^*(t-\tau_1/2)\mathrm{e}^{-\mathrm{j}\omega\tau_1}\,\mathrm{d}\tau_1=W_y(t,\omega)$$

Therefore,

$$W_y(t,\omega) = \frac{1}{2\pi}\int_{-\infty}^{\infty}\int_{-\infty}^{\infty} W_h(t,t',\omega,\omega')W_x(t',-\omega')\,\mathrm{d}t'\,\mathrm{d}\omega$$

If $x(t)$ is the real signal:

$$W_y(t,\omega) = \frac{1}{2\pi}\int_{-\infty}^{\infty}\int_{-\infty}^{\infty} W_h(t,t',\omega,\omega')W_x(t',\omega')\,\mathrm{d}t'\,\mathrm{d}\omega'$$

Proof finished.

5.3.4.3 Relationship between the input and output WV spectra of linear time-varying systems

For linear time-varying systems, when input $X(t)$ is a stationary random signal, its WV spectrum is:

$$WV_X(t,\omega) = \int_{-\infty}^{\infty} E\left[X(t+\tau/2)X^*(t-\tau/2)\right]\mathrm{e}^{-\mathrm{j}\omega\tau}\,\mathrm{d}\tau \tag{5.78}$$

Linear time-varying system output $Y(t)$ is a nonstationary random signal whose WV spectrum is

$$WV_Y(t,\omega) = \int_{-\infty}^{\infty} E\left[Y(t+\tau/2)Y^*(t-\tau/2)\right]\mathrm{e}^{-\mathrm{j}\omega\tau}\,\mathrm{d}\tau \tag{5.79}$$

The relationship between the input and output WV spectra of the linear time-varying system is obtained as:

$$WV_Y(t,\omega) = \frac{1}{2\pi}\int_{-\infty}^{\infty}\int_{-\infty}^{\infty} W_h(t,t',\omega,\omega')WV_X(t',-\omega')\,\mathrm{d}t'\,\mathrm{d}\omega' \tag{5.80}$$

If $X(t)$ is real stationary random signal, the above formula is:

$$WV_Y(t,\omega) = \frac{1}{2\pi}\int_{-\infty}^{\infty}\int_{-\infty}^{\infty} W_h(t,t',\omega,\omega')WV_X(t',\omega')\,\mathrm{d}t'\,\mathrm{d}\omega' \tag{5.81}$$

5.4 Wavelet analysis of nonstationary random signals

5.4.1 Continuous wavelet transform

Setting function $\psi(t) \in L^2(R)$ meets:

$$\int_R \psi(t)\,\mathrm{d}t = 0 \tag{5.82}$$

This $\psi(t)$ is the mother wavelet function. We carry out stretching and translation transformation on the mother wavelet $\psi(t)$; if set the shrink factor is a, the translation factor is b, $a, b \in R$, $a \neq 0$. We obtain the following function:

$$\psi_{a,b}(t) = |a|^{-1/2} \psi\left(\frac{t-b}{a}\right) \tag{5.83}$$

$\psi_{a,b}(t)$ is called the analysis of the wavelet.

The continuous wavelet transform (CWT) of function $f(t) \in L^2(R)$ is defined as:

$$W_{a,b} \overset{\text{def}}{=} \int_R f(t)\psi^*_{a,b}(t)\,\mathrm{d}t \tag{5.84}$$

where $\psi^*_{a,b}(t)$ is the conjugate function of $\psi_{a,b}(t)$. If the mother wavelet function is a real function, $\psi^*_{a,b}(t) = \psi_{a,b}(t)$, then the continuous wavelet transform of $f(t)$ is:

$$W_{a,b} = \int_R f(t)\psi_{a,b}(t)\,\mathrm{d}t = |a|^{-1/2} \int_R f(t)\psi\left(\frac{t-b}{a}\right)\mathrm{d}t \tag{5.85}$$

As can be seen from the above definition, the continuous wavelet transform is the mapping of $f \to W_{a,b}$, if the inverse transformation exists, and it restores the original signal $f(t)$ by $W_{a,b}$. This requires adding the following constraints to the mother wavelets:

$$\int_R \frac{\left|\hat{\psi}(\omega)\right|^2}{|\omega|}\,\mathrm{d}\omega < \infty \tag{5.86}$$

where $\widehat{\psi}(\omega)$ is the Fourier transform for $\psi(t)$. The condition is called the allowable condition, and the mother wavelet $\psi(t)$ satisfying the allowable condition is called the admissible wavelet. In this case, the original signal can be restored by $W_{a,b}$:

$$f(t) = C_\psi^{-1} \int_R \frac{\mathrm{d}a}{a^2} \int_R W_{a,b}\psi_{a,b}(t)\,\mathrm{d}b \tag{5.87}$$

Proof of the above formula is described in the regeneration formula described later, where:

$$C_\psi = \int_R \frac{\left|\hat{\psi}(\omega)\right|^2}{|\omega|}\,\mathrm{d}\omega \tag{5.88}$$

In fact, if the mother wavelet $\psi(t)$ in the infinite meets certain attenuation conditions, namely:

$$|\psi(t)| \le c(1+|t|)^{1+\varepsilon} \tag{5.89}$$

the formula is established. When $|t|$ is big enough and $\varepsilon > 0$, the allowable condition $\int_R \frac{|\psi(\omega)|^2}{|\omega|}\,\mathrm{d}\omega < \infty$ is equivalent to $\int_R \psi(t)\,\mathrm{d}t = 0$ and $\int_R \psi(t)\,\mathrm{d}t = 0$, and then the mother wavelet becomes an admissible wavelet, and the inverse transform of continuous wavelet transform exists.

5.4.2 Two-dimensional phase space

In order to illustrate the time-frequency characteristics of the wavelet transform, we need to start from the time-frequency characteristic of wavelet window function in the phase space. For this reason, we first discuss two-dimensional phase space. The two-dimensional phase space is a Euclidean space with time as the abscissa and the frequency as the ordinate. The finite area in the phase space is called the time-frequency window. The role of the phase space is different from the general coordinate plane; it is not used to describe the relationship of the function but is used to describe a certain physical state or to describe the time-frequency localization characteristics of a system.

The window in the phase space may be a single-window function or it may be a double-window function. The single-window function is defined as follows.

There is a function $g(t) \in L^2(R)$, and $tg(t) \in L^2(R)$ and the point (t_0, ω_0) in the phase space is determined by the following equation:

$$\left.\begin{aligned} t_0 &= \int_R t|g(t)|^2 \,\mathrm{d}t \Big/ \|g\|^2 \\ \omega_0 &= \int_R \omega|\hat{g}(\omega)|^2 \,\mathrm{d}\omega \Big/ \|\hat{g}\|^2 \end{aligned}\right\} \tag{5.90}$$

This is called the center of a single-window function $g(t)$, where:

$$\left.\begin{aligned} \|g\|^2 &= \int_R |g(t)|^2 \,\mathrm{d}t \\ \|\hat{g}\|^2 &= \int_R |g(\omega)|^2 \,\mathrm{d}\omega \end{aligned}\right\} \tag{5.91}$$

and the amount

$$\left.\begin{aligned} \Delta g &= \left[\int_R (t - t_0)^2 |g(t)|^2 \,\mathrm{d}t\right]^{1/2} \Big/ \|g\| \\ \Delta\hat{g} &= \left[\int_R (\omega - \omega_0)^2 |\hat{g}(\omega)|^2 \,\mathrm{d}\omega\right]^{1/2} \Big/ \|\hat{g}\| \end{aligned}\right\} \tag{5.92}$$

This is called the time width and bandwidth of a single-window function $g(t)$. The rectangle with the center (t_0, ω_0) and the sides $2\Delta g$ and $2\Delta\hat{g}$ in the phase space is called the time-frequency window determined by $g(t)$, and $g(t)$ is called a single-window function. In practice, it is usual to make the center of $g(t)$ locate $(0, 0)$ and $\|g\| = 1$; this window function is called the standard single-window function. The double-window function in phase space is defined as follows: you use the following formula to determine the point $(t_0, \omega_0^{\pm})$:

$$\left.\begin{aligned} t_0 &= \int_R t|g(t)|^2 \, dt \Big/ \|g\|^2 \\ \omega_0^- &= \int_{-\infty}^{0} \omega|\hat{g}(\omega)|^2 \, d\omega \Big/ \|\hat{g}\|^2 \\ \omega_0^+ &= \int_{0}^{+\infty} \omega|\hat{g}(\omega)|^2 \, d\omega \Big/ \|\hat{g}\|^2 \end{aligned}\right\} \tag{5.93}$$

Use the following formula to determine $(\Delta g, \Delta_\omega^{\pm})$:

$$\left.\begin{aligned} \Delta g &= \left[\int_R (t - t_0)^2 |g(t)|^2 \, dt\right]^{1/2} \Big/ \|g\| \\ \Delta_\omega^- &= \left[\int_{-\infty}^{0} (\omega - \omega_0^-)^2 |\hat{g}(\omega)|^2 \, d\omega\right]^{1/2} \Big/ \|\hat{g}\| \\ \Delta_\omega^+ &= \left[\int_{0}^{\infty} (\omega - \omega_0^+)^2 |\hat{g}(\omega)|^2 \, d\omega\right]^{1/2} \Big/ \|\hat{g}\| \end{aligned}\right\} \tag{5.94}$$

With the center of (t_0, ω_0^-) and (t_0, ω_0^+) in phase space, the two rectangles with length $2\Delta g$, $2\Delta_\omega^-$ and $2\Delta g$, $2\Delta_\omega^+$ are called the negative-positive time-frequency window determined by $g(t)$, and $g(t)$ is called the double-window function. If we take that t_0 is zero, $\|g\| = 1$ and this double-window function is called the standard double-window function.

The window function is localized in both the time domain and the frequency domain, so it can be used as a basic function for time-frequency localization analysis of the signal. The window it defines describes its local characteristics; that window function itself can use its window size to characterize its local characteristics; the smaller the amount of Δg and $\Delta g'$, indicating the localization degree of the window function in the time domain, the higher frequency domain. However, there is a certain restraining relationship between Δg with $\Delta \hat{g}$, both cannot be arbitrarily small at the same time, and by the uncertainty principle of signal, we have:

$$\Delta g \Delta \hat{g} \geq 1/2 \tag{5.95}$$

When $g(t)$ is the Gaussian window function, the equal sign in the above equation holds.

The single-window function is generally used for time-frequency localized analysis of signals with lowpass characteristics, and the double-window function is generally used as a time-frequency localized analysis of the bandpass characteristics. On the wavelet function $\psi(t)$, because $\int_R \psi(t)\,\mathrm{d}t = 0$ is established, it can be launched that:

$$\hat{\psi}(0) = \int_R \psi(t)\,\mathrm{d}t = 0 \tag{5.96}$$

As can be seen from the above equation, $\psi(t)$ is a bandpass function. Thus, the wavelet function $\psi(t)$ is associated with the double-window function in the phase space.

5.4.3 Time-frequency characteristics of the double-window function of $\psi_{a,b}(t)$

With the concept of the phase space window function and known $\psi(t)$ corresponding to the double-window function in the phase space, the following is a detailed study of the time-frequency characteristics of the double-window function. The double-window function is determined by the analytical wavelet $\psi_{ab}(t)$ derived from the mother wavelet $\psi(t)$.

In order to simplify the problem analysis, set the mother wavelet function $\psi(t)$ as a standard double-window function; $\psi_{a,b}(t)$ derived from it is:

$$\psi_{a,b}(t) = |a|^{-1/2}\psi\left(\frac{t-b}{a}\right) \tag{5.97}$$

The following is a solution to the time-domain center, the frequency-domain center, the time width and the bandwidth about $\psi_{a,b}(t)$.

Time-domain center $t_{a,b}$:

$$t_{a,b} = \int_R t|\psi_{a,b}(t)|^2\,\mathrm{d}t = \int_R t|a|^{\ 1}|\psi\left(\frac{t-b}{a}\right)|^2\,\mathrm{d}t \tag{5.98}$$

Make $(t-b)/a = m$, $t = am + b$, $dt{=}adm$, substitute this into eq. (5.4.17), and then

$$t_{a,b} = \int_R (am+b)|a|^{-1}\psi(m)|^2 a\,\mathrm{d}m \tag{5.99}$$

When $a > 0$, $|a|^{-1} = a^{-1}$,

$$\begin{aligned} t_{a,b} &= \int_R (am+b)|\psi(m)|\,\mathrm{d}m \\ &= a\int_R m|\psi(m)|^2\,\mathrm{d}m + b\int_R |\psi(m)|^2\,\mathrm{d}m \\ &= at_0 + b \end{aligned} \tag{5.100}$$

When $t_0 = 0$, we have

$$t_{a,b} = b \tag{5.101}$$

Frequency-domain center $\omega^{\pm}_{a,b}$:

$$\omega^{-}_{a,b} = \int_{-\infty}^{0} |\omega\hat{\psi}_{a,b}(\omega)|^2 \, \mathrm{d}\omega \tag{5.102}$$

Because:

$$\hat{\psi}_{a,b}(\omega) = \int_{R} |a|^{-1/2} \psi\left(\frac{a-b}{a}\right) \mathrm{e}^{-\mathrm{j}\omega t} \, \mathrm{d}t \tag{5.103}$$

Take $(t-b)/a = m$, $\mathrm{d}t = a\,\mathrm{d}m$ and substitute it into eq. (5.103). Then:

$$\hat{\psi}_{a,b}(\omega) = \int_{R} |a|^{-1/2} \psi(m) \mathrm{e}^{-\mathrm{j}\omega(am+b)} a \, \mathrm{d}m \tag{5.104}$$

When $a > 0$, $|a|^{-1} = a^{-1}$:

$$\hat{\psi}_{a,b}(\omega) = a^{1/2} \int_{R} \psi(m) \mathrm{e}^{-\mathrm{j}a\omega m} \mathrm{e}^{-\mathrm{j}\omega b} \, \mathrm{d}m = a^{1/2} \hat{\psi}(a\omega) \mathrm{e}^{-\mathrm{j}\omega b} \tag{5.105}$$

and:

$$\omega^{-}_{a,b} = \int_{-\infty}^{0} \omega |a^{1/2} \hat{\psi}(a\omega) \mathrm{e}^{-\mathrm{j}\omega b}|^2 \, \mathrm{d}\omega = \int_{-\infty}^{0} \omega a |\psi(a\omega)|^2 \, \mathrm{d}\omega \tag{5.106}$$

Take $a\omega = \omega'$, $\mathrm{d}\omega = 1/a \, \mathrm{d}\omega'$ and substitute into the equation. Then:

$$\omega^{-}_{a,b} = \frac{1}{a} \int_{-\infty}^{0} \omega' |\hat{\psi}(\omega')|^2 \, \mathrm{d}\omega' = \frac{1}{a} \omega^{-}_{0} \tag{5.107}$$

Similarly:

$$\omega^{+}_{a,b} = \frac{1}{a} \omega^{+}_{0} \tag{5.108}$$

Time width:

According to the above derivation method, it is:

$$\Delta_{\psi_{a,b}} = \left[\int_{R} (t - t_{a,b})^2 |\psi_{a,b}(t)|^2 \, \mathrm{d}t \right]^{1/2} = a \Delta_{\psi} \tag{5.109}$$

Bandwidth:

$$\begin{aligned} \Delta^{-}_{\omega_{a,b}} &= \left[\int_{-\infty}^{0} \left(\omega - \omega^{-}_{a,b}\right)^2 \left|\hat{\psi}_{a,b}(\omega)\right| \mathrm{d}\omega \right]^{1/2} = \frac{1}{a} \Delta^{-}_{\omega} \\ \Delta^{+}_{\omega_{a,b}} &= \frac{1}{a} \Delta^{+}_{\omega} \end{aligned} \tag{5.110}$$

From the above results, the two windows determined by $\psi_{a,b}(t)$ in the phase space are:

$$[b - a\Delta_\psi, b + a\Delta_\psi] \cdot \left[\frac{\omega_0^-}{a} - \frac{1}{a}\Delta_\omega^-, \frac{\omega_0^-}{a} + \frac{1}{a}\Delta_\omega^-\right] \tag{5.111}$$

$$[b - a\Delta_\psi, b + a\Delta_\psi] \cdot \left[\frac{\omega_0^+}{a} - \frac{1}{a}\Delta_\omega^+, \frac{\omega_0^+}{a} + \frac{1}{a}\Delta_\omega^+\right] \tag{5.112}$$

In addition, regardless of how a and b change, the following two values of the time-frequency window are unchanged.

The area of the window:

$$2a\Delta_\psi \cdot \frac{2}{a}\Delta_\omega^+ = 4\Delta_\psi\Delta_\psi^+ \tag{5.113}$$

The ratio of the frequency-domain center and the bandwidth:

$$1/a\omega_0^+ \Big/ \frac{1}{a}\Delta_\omega^+ = \omega_0^+/\Delta_\omega^- \tag{5.114}$$

To sum up, the rectangular shape of the time-frequency window determined by $\psi(t)$ changes as the expansion factor a. When the value of a is small, the window is a high-frequency window with a narrow time-bandwidth and a high-frequency bandwidth; when the value of a is large, it is a low-frequency window with a large timeband width and a low-frequency bandwidth, as shown in Fig. 5.5. Because the wavelet function $\psi(t)$ is a bandpass function, through the aforementioned analysis, it is known that variable a of the expansion factor changes, and $\psi_{a,b}(t)$ corresponds to a series of the bandpass system with different bandwidth and central frequency.

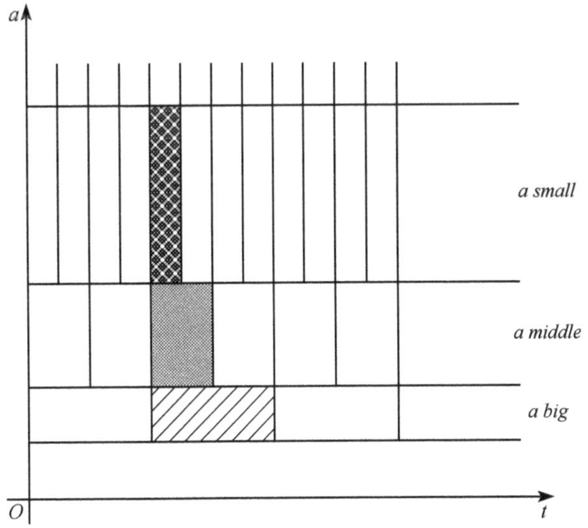

Fig. 5.5: $\psi_{a,b}(t)$ The rectangular shape of the time window.

5.4.4 Physical meaning of the continuous wavelet transform

The continuous wavelet transform is a mathematical transformation method, but in the signal analysis and processing it is of clear and important physical significance. Now, we discuss the wavelet transform from the perspective of system response. The continuous wavelet transform is:

$$W_{a,b} = \int_R |a|^{-1/2} f(t)\psi\left(\frac{t-b}{a}\right)\mathrm{d}t \tag{5.115}$$

If the input signal is $f(t)$ when the unit impulse response is $h_a(t) = |a|^{1/2}\psi(-t/a)$, then the output of the system is:

$$f(t) \otimes |a|^{-1/2}\psi\left(-\frac{t}{a}\right) = \int_R |a|^{-1/2} f(\tau)\psi\left(\frac{\tau-t}{a}\right)\mathrm{d}\tau \tag{5.116}$$

When $t = b$, the output of the system is $W_{a,b}$, which is:

$$f(t) \otimes |a|^{-1/2}\psi\left(-\frac{t}{a}\right) = \int_R |a|^{-1/2} f(t)\psi\left(\frac{t-b}{a}\right)\mathrm{d}t = W_{a,b} \tag{5.117}$$

Therefore, the continuous wavelet transform of $f(t)$ is equivalent to the output of the system with unit impulse response function $h_a(t) = |a|^{1/2}\psi(-t/a)$ or the transfer function $H_a(\omega) = |a|^{1/2}\hat{\psi}(a\omega)$. Through the time-frequency characteristic analysis of the double-window function of the phase space, $\psi(t)$ is a bandpass system; when a changes, $H_a(\omega)$ corresponds to a series of the bandpass system with different bandwidth and central frequency. So, the following conclusions can be drawn:

The continuous wavelet transform of the signal $f(t)$ is the output of a series of bandpass filters on $f(t)$ after filtering, the a in $W_{a,b}$ reflects the bandwidth and central frequency of the bandpass filter, and b is the time parameter of the filtered output.

A series of bandpass filters formed by the change of the scaling factor a are constant Q filters, and Q is the quality factor of the filter, which is equal to the ratio of the central frequency to the bandwidth of the filter. It can be understood by the analysis of the two window functions in the phase space, because $Q = \omega_0/\Delta\omega$, among them ω_0, $\Delta\omega$, respectively, are the central frequency and the bandwidth of the mother wavelet $\psi(t)$.

The scaling factor variable a changes, and the bandwidth and central frequency of the bandpass filter also change. When a is small, the central frequency is large and the bandwidth is also wide; if a large central frequency is small, the bandwidth is also narrow. Through such a bandpass filter to filter the signal $f(t)$, the local characteristics are of great value. The signal changes slowly, and it has mainly low-frequency components. The frequency range is relatively narrow, and at this time the wavelet transform bandpass filter should be equivalent to a situation with a large one. On the other hand, with the signal mutation, which is mainly high-frequency components, the frequency range is also wide, and the wavelet transform of the bandpass filter is equivalent to the situation with variable small a. So, the scaling factor a changes from large to small, and the range of filtering changes from low to high frequency, which is the zoom of the wavelet transform.

In summary, the continuous wavelet transform $W_{a,b}$ of signal $f(t)$ is a transformation in a mathematical form from the definition point of view, and from the perspective of system analysis, it is essentially the output of a signal $f(t)$ after a series of bandpass filters In addition, from the spectral analysis point of view, the wavelet transform is the signal decomposition to a series of selective same-frequency bands.

5.4.5 Application of the wavelet transform in random signal analysis

The use of the wavelet transform in random signal processing has just begun, and it has yet to be researched and developed further. Here, we only take the wavelet-based Karhunen–Loeve expansion (KL expansion) as an example and specify its application.

Letting $X(t)$ be zero the mean stationary random continuous signal, it can be developed into the following series in a finite interval $[0, T]$:

$$X(t) = \sum_{n=1}^{\infty} X_n \varphi_n(t) \tag{5.118}$$

Among these $\{X_n\}$ is an irrelevant random variable sequence, which can be determined as:

$$X_n = \int_0^T X(t)\varphi_n^*(t)\,\mathrm{d}t \tag{5.119}$$

$\{\varphi_n(t)\}$ is the orthogonal basis set, which is the solution of the following characteristic integral equation:

$$\int_0^T R_{XX}(t-s)\varphi_n(s)\,\mathrm{d}s = \lambda_n \varphi_n(t) \tag{5.120}$$

$\varphi_n(t)$ is a characteristic function of eq. (5.4.39), and λn is the eigenvalue. This is the stationary random signal KL unfolds. The important unilateral solution of KL expansion is transforming the analysis of continuous-time stochastic process problems into the analysis of irrelevant random variable sequences.

In practice, the application of KL expansion is very limited, because in addition to the presence of white noise $R_{XX}(t-s) = \sigma^2\delta(t-s)$, it is generally difficult to solve the problem of the characteristic integral equation.

5.5 Non-Gaussian processing and higher-order statistics

5.5.1 Non-Gaussian signal processing

Non-Gaussian signal processing is a new field of signal processing that has devolved rapidly in recent years. The traditional theory and techniques of signal processing are basically based on Gaussian and second-order statistics, because the Gaussian model is relatively simple and applicable in many applications. The signal processing algorithm designed on this model is easy to carry out in theoretical analysis. Since the higher-order cumulates of the third order and above of the Gaussian process are all zero, it is natural to use the second-order statistic to analyze and deal with the signal and noise. In essence, the use of such a Gaussian distribution in the traditional signal processing field is mainly due to the lack of tools for the analysis and processing of more complex models of signal noise. This deficiency includes two aspects. On the one hand, the computational speed and ability of computers is not sufficient to support a large number of computations of complex signal models. On the other hand, the theory and algorithms of signal analysis have not yet provided powerful tools for the analysis and processing of non-Gaussian signal noise. This way, people tend to encounter the various signals and noise as a Gaussian distribution to deal with in practical application.

With the rapid development of computer technology and signal processing theory, people have the ability to use more complex algorithms to analyze and process more complex signal noise models. This way, since the mid-1980s, the theory and technology of non-Gaussian signal processing has received increasing attention from the academic field of signal processing and has been rapidly developed and widely used over a short span of ten years. Unlike traditional signal processing based on second-order statistics, non-Gaussian signal processing mainly uses high-order statistics and fractional lower-order statistics as a tool for signal analysis and processing. High-order statistics and fractional lower-order statistics are very important in areas such as non-Gaussian, nonlinear, noncausal, nonminimum phase and blind signal processing, and so on. Their applications include radar, sonar, geophysics, speech and biomedical areas such as signal processing, image reconstruction, texture analysis, mechanical fault diagnosis, harmonic recovery, array processing, blind deconvolution, blind equalization, blind estimation, blind signal separation and nonlinear waveform analysis. The rapid development of non-Gaussian signal processing and the emergence of a large number of new theoretical new algorithms enable people to solve problems that could not be solved in the past and effectively promote the progress of various fields.

The high-order statistical theory itself is a traditional mathematical field, with roughly 100 years of history from Pearson's study [11] of moments in 1900. In 1920 or so, with the emergence of the maximum likelihood method proposed by Fisher [11], the high-order statistic method was neglected. In the early 1960s, a group of statisticians at the University of California in the United States began to study the application problems of high-order statistics again, but researchers in the field of signal processing did not notice the progress of this research. In 1980, Mendel and his collaborators at the University of Southern California began to study system identification techniques based on high-order statistics and applied the results to the deconvolution of seismic signals and achieved remarkable results. However, at the time, in order to achieve this technology in real time is not possible. In addition, the shorter observation data length is also a major problem affecting the toughness of the algorithm. In the field of signal processing research, the rapid development of high-order statistics began in the mid-1980s. With the development of computer technology and digital signal processing technology, especially the progress of DSP technology, the theory and technology of high-order statistics signal processing have developed rapidly. Since then, almost all academic conferences on signal processing deal with research papers on high-order statistics. Since 1989, the IEEE (Institute of Electrical and Electronics Engineers) has been holding a symposium on high-end statistics every 2 years, and many of the world's leading journals have published special titles for high-end statistics.

Over the past 15 years, the theory and application of high-order statistics have seen a very big development. With respect to MA system parameter identification, scholars have obtained many valuable research results. For example, Giannakos and Mendel put forward GM-RC, method of using one-dimensional diagonal slices of autocorrelation and third-order cumulate to recursively estimate the MA model parameters, and Swami put forward the SM-RC method, etc.; further, there is a class of high-order statistical methods called closed solutions. Giannakos and Mendel established the well-known GM equations by the relationship between autocorrelation function and cumulate. Mugnai constructed the T equation, which avoids the inherent parameter estimation error of the GM equation. Fonollosa and Vrjal proposed the cumulant slice construction equation by selecting arbitrarily and determining the cumulant slice combination method of the model parameters. The method can be realized by singular value decomposition (SVD), and the numerical robustness is strong. Pan and Nikias proposed a double cestrum method, by determining the minimum phase component and the maximum phase component of the impulse response, and then determining the parameters of the MA. The above method forms a linear algebraic solution. Another method about higher-order statistics used in MA system identification is called the optimization solution, including the thorough search method and its optimization algorithm proposed by Lie and Rosenblatt. Wang and Mendel combined the idea of neural networks. Similarly, high-order statistics are widely used in AR system identification. Based on the high-order cumulate Yule–Walker equation, SVD is used to determine the AR parameters and their order. Tugnait determined noncausal AR parameters, the statistic of non-Gaussian inputs, and Gaussian noise statistics by minimizing the objective function. In addition, high-order statistics in ARMA system identification, signal reconstruction with Gaussian colored noise, signal detection with Gaussian noise background, and other issues have also seen significant progress. As the author of the high-level statistics research station pointed out, the problems using autocorrelation research can use high-order statistics for renewed study.

5.5.2 Higher-order moments and high-order cumulates

5.5.2.1 Higher-order moments and high-order cumulates of random variables

The characteristic function of the random variable X was given in the previous section:

$$\Phi(u) = E[\mathrm{e}^{\mathrm{j}uX}] = \int_{-\infty}^{\infty} \exp[\mathrm{j}uX]p(X)\,\mathrm{d}X \tag{5.121}$$

It is called the first feature function. Taking the natural logarithm for $\Phi(u)$, we obtain the second characteristic function:

$$\Psi(u) = \ln[\Phi(u)] \tag{5.122}$$

$\Phi(u)$ is expanded into Taylor series at u=0. We have:

$$\Phi(u) = \sum_{k=0}^{\infty} \Phi^{(k)}(0)\frac{u^k}{k!} \tag{5.123}$$

Among these $\Phi^{(k)}(0) = \frac{\mathrm{d}^k\Phi(u)}{\mathrm{d}u}\Big|_{u=0}$, further simplified to:

$$\begin{aligned}\Phi^{(k)}(0) &= \left[\int_{-\infty}^{\infty} \mathrm{e}^{\mathrm{j}uX}p(X)\,\mathrm{d}X\right]^{(k)}\Bigg|_{u=0}\\ &= \int_{-\infty}^{\infty} (\mathrm{j}X)^k p(X)\,\mathrm{d}X\\ &= \mathrm{j}^k E[X^k]\end{aligned} \tag{5.124}$$

Substitute the above formula $\Phi(u)$ to obtain:

$$\begin{aligned}\Phi(u) &= \sum_{k=0}^{\infty} \frac{\Phi^{(k)}(0)}{\mathrm{j}^k}\frac{(\mathrm{j}u)^k}{k!}\\ &= \sum_{k=0}^{\infty} E[X^k]\frac{(\mathrm{j}u)^k}{k!}\\ &= \sum_{k=0}^{\infty} m_k\frac{(\mathrm{j}u)^k}{k!}\end{aligned} \tag{5.125}$$

Where:

$$m_k \triangleq E[X^k] = \frac{\Phi^{(k)}(0)}{\mathrm{j}^k} = (-\mathrm{j})^k\Phi^{(k)}(0) \tag{5.126}$$

$m_k, k = 1, 2, \ldots$ is called the k-order moment of random variable X. Thus, the first characteristic function is also called the moment generation function of the random variable.

Make the second characteristic function $\Psi(u)$ in the $u = 0$ point to expand into Taylor series. We have:

$$\begin{aligned}\Psi(u) &= \Psi(0) + \sum_{k=1}^{\infty} \frac{\Psi^{(k)}(u)}{k!} \cdot u^k \\ &= 0 + \sum_{k=1}^{\infty} \frac{\Psi^{(k)}(0)/\mathrm{j}^k}{k!} \cdot (\mathrm{j}u)^k \\ &= \sum_{k=1}^{\infty} (-\mathrm{j})^k \Psi^{(k)}(0) \frac{(\mathrm{j}u)^k}{k!} \\ &= \sum_{k=1}^{\infty} \frac{c_k}{k!} \cdot (\mathrm{j}u)^k \end{aligned} \tag{5.127}$$

$$c_k \triangleq \frac{\Psi^{(k)}(0)}{\mathrm{j}^k} = \frac{1}{\mathrm{j}^k} \frac{\mathrm{d}^k}{\mathrm{d}t^k} \ln \Phi(t) \bigg|_{t=0} \tag{5.128}$$

It is called the k-order cumulant of the random variable X. Thus, the second characteristic function is also called the cumulative variable generating function of the random variable.

5.5.2.2 Higher-order moments and higher-order cumulates of random vectors

Set k real random variables $X_1, X_2, \ldots, X_k$; their joint probability density function is $p(X_1, X_2, \ldots, X_k)$. Their first joint characteristic function is:

$$\begin{aligned}\Phi(u_1, u_2, \ldots, u_k) &\triangleq E\{\mathrm{e}^{\mathrm{j}(u_1X_1+u_2X_2+\cdots+u_kX_k)}\} \\ &= \int_{-\infty}^{\infty}\int_{-\infty}^{\infty} \cdots \int_{-\infty}^{\infty} p(X_1, X_2, \ldots, X_k)\mathrm{e}^{\mathrm{j}(u_1X_1+u_2X_2+\cdots+u_kX_k)}\,\mathrm{d}X_1\,\mathrm{d}X_2 \ldots \mathrm{d}X_k\end{aligned} \tag{5.129}$$

The $r = r_1 + r_2 + \cdots + r_k$-order derivative of this function about $u_1, u_2, \ldots, u_k$ is:

$$\frac{\partial \Phi(u_1, u_2, \ldots, u_k)}{\partial u_1^{r_1} \partial u_2^{r_2} \ldots \partial u_k^{r_k}} = \mathrm{j}^r E\{X_1^{r_1} X_2^{r_2} \ldots X_k^{r_k} \mathrm{e}^{\mathrm{j}(u_1X_1+u_2X_2+\cdots+u_kX_k)}\} \tag{5.130}$$

Similarly to the origin of the random variable, the r-order union moment of k random variables $X_1, X_2, \ldots, X_k$ is:

$$m_{r_1 r_2 \ldots r_k} \triangleq E[X_1^{r_1} X_2^{r_2} \ldots X_k^{r_k}] = (-\mathrm{j})^r \frac{\partial^r \Phi(u_1, u_2, \ldots, u_k)}{\partial u_1^{r_1} \partial u_2^{r_2} \ldots \partial u_k^{r_k}} \bigg|_{u_1=u_2=\cdots=u_k=0} \tag{5.131}$$

The second joint characteristic function of the k random variables is defined as the natural logarithm of their first joint characteristic function:

$$\Psi(u_1, u_2, \ldots, u_k) = \ln[\Phi(u_1, u_2, \ldots, u_k)] \tag{5.132}$$

Therefore, the r-order cumulate of the k random variables $X_1, X_2, \ldots, X_k$ is:

$$c_{r_1 r_2 \ldots r_k} \triangleq \mathrm{cum}[X_1^{r_1}, X_2^{r_2}, \ldots, X_k^{r_k}] = (-\mathrm{j})^r \left. \frac{\partial^r \Psi(u_1, u_2, \ldots, u_k)}{\partial u_1^{r_1} \partial u_2^{r_2} \ldots \partial u_k^{r_k}} \right|_{u_1=u_2=\cdots=u_k=0} \tag{5.133}$$

5.5.2.3 *k*-order moments and *k*-order cumulates for stationary continuous random signals

The k-order moments of the stationary continuous random signal $X(t)$ is:

$$m_{kX}(\tau_1, \ldots, \tau_{K-1}) \triangleq E\left[X(t)X(t+\tau_1)\ldots X(t+\tau_{K-1})\right] \tag{5.134}$$

Among these, $\tau_i, i = 1, 2, \ldots, k-1$ is any real number.

The k-order cumulate of the stationary continuous random signal $X(t)$ is:

$$c_{kX}(\tau_1, \ldots, \tau_{K-1}) \triangleq \mathrm{cum}\left[X(t), X(t+\tau_1), \ldots, X(t+\tau_{K-1})\right] \tag{5.135}$$

Among these, $\tau_i, i = 1, 2, \ldots, k-1$ is any real number.

It is necessary to know the relationship between the higher-order moments and the higher-order cumulates for the stationary random signals.

First-order cumulate (mean function):

$$c_{1X} = m_{1X} = E[X(t)] \tag{5.136}$$

Second-order moment (autocorrelation function):

$$m_{2X}(\tau_1) = E[X(t)X(t+\tau_1)] \tag{5.137}$$

Second-order cumulate (covariance function):

$$c_{2X}(\tau_1) = m_{2X}(\tau_1) - (m_{1X})^2 = c_{2X}(-\tau_1) \tag{5.138}$$

Third-order cumulate:

$$c_{3X}(\tau_1, \tau_2) = m_{3X}(\tau_1, \tau_2) - m_{1X}[m_{2X}(\tau_1) + m_{2X}(\tau_2) + m_{2X}(\tau_2 - \tau_1)] + 2(m_{1X})^3 \tag{5.139}$$

Example 5.5.1. The Laplace distribution is a commonly used symmetric distribution. Solve their k-order moments and k-order cumulates ($k = 1, 2, 3, 4$).

Solution: The probability density function of the Laplace distribution is $p(x) = 0.5\mathrm{e}^{-|x|}$. The moment generating function and the cumulative quantity generating function are, respectively,

$$\boldsymbol{\Phi}(u) = 0.5 \int_{-\infty}^{\infty} \mathrm{e}^{\mathrm{j}ux-|x|}\,\mathrm{d}x = \frac{1}{1-u^2}$$

$$\boldsymbol{\Psi}(u) = \ln\left[\boldsymbol{\Phi}(u)\right] = -\ln(1-u^2)$$

The $k(k = 1, 2, 3, 4)$-order moments and k-order cumulates are:

$$m_1 = \Phi'(u)|_{u=0} = \left.\frac{2u}{(1-u^2)^2}\right|_{u=0} = 0$$
$$c_1 = \Psi'(u)|_{u=0} = \left.\frac{2u}{1-u^2}\right|_{u=0} = 0$$
$$m_2 = \Phi''(u)|_{u=0} = \left.\frac{2-10u^2}{(1-u^2)^3}\right|_{u=0} = 2$$
$$c_2 = \Psi''(u)|_{u=0} = \ln''[\Phi(u)]|_{u=0} = \left.\frac{2-4u^2}{(1-u^2)^2}\right|_{u=0} = 2$$
$$m_3 = \Phi'''(u)|_{u=0} = \left.\left(\frac{2-10u^2}{(1-u^2)^3}\right)'\right|_{u=0} = 0$$
$$c_3 = \Psi'(u)|_{u=0} = \ln'''[\Phi(u)]|_{u=0} = 0$$
$$m_4 = \Phi^{(4)}(u)|_{u=0} = 24$$
$$c_4 = \Psi^{(4)}(u)|_{u=0} = 12$$

5.5.3 The higher-order spectrum

It is well known that the power spectral density of a signal is defined as the Fourier transform of its autocorrelation function. Similarly, spectra of higher-order moments and higher-order cumulates can be defined. In general, we classify the moment and cumulative spectra whose order is greater than 2 as high-order spectra.

5.5.3.1 Definition of higher-order moments

If the k-order moment function $m_{kX}(\tau_1, \ldots, \tau_{k-1})$ of the random process $X(n)$ meets the absolute condition, namely:

$$\sum_{\tau_1=-\infty}^{\infty}\sum_{\tau_2=-\infty}^{\infty}\cdots\sum_{\tau_{k-1}=-\infty}^{\infty}|m_{kX}(\tau_1, \tau_2, \ldots, \tau_{k-1})| < \infty \tag{5.140}$$

then k-order moments of $X(n)$ exist and are defined as $k-1$-dimensional discrete Fourier transforms of k-order moments:

$$M_{kX}(\omega_1, \omega_2, \ldots, \omega_{k-1}) = \sum_{\tau_1=-\infty}^{\infty}\sum_{\tau_2=-\infty}^{\infty}\cdots\sum_{\tau_{k-1}=-\infty}^{\infty} m_{kX}(\tau_1, \tau_2, \ldots, \tau_{k-1}) \exp\{-\mathrm{j}(\omega_1\tau_1 + \omega_2\tau_2 + \cdots + \omega_{k-1}\tau_{k-1})\} \tag{5.141}$$

Among these, $|\omega_i| \leq \pi$, $i = 1, 2, \ldots, k-1$, $|\omega_1, \omega_2, \ldots, \omega_{k-1}| \leq \pi$.

5.5.3.2 Definition of the higher-order cumulative spectrum

If the k-order cumulate function $c_{kX}(\tau_1, \ldots, \tau_{k-1})$ of the random process $X(n)$ meets the absolute condition, it is:

$$\sum_{\tau_1=-\infty}^{\infty} \sum_{\tau_2=-\infty}^{\infty} \cdots \sum_{\tau_{k-1}=-\infty}^{\infty} |c_{kX}(\tau_1, \tau_2, \ldots, \tau_{k-1})| < \infty\,, \quad i = 1, 2, \ldots, k-1 \tag{5.142}$$

Or:

$$\sum_{\tau_1=-\infty}^{\infty} \sum_{\tau_2=-\infty}^{\infty} \cdots \sum_{\tau_{k-1}=-\infty}^{\infty} (1 + |\tau_i|)|c_{kX}(\tau_1, \tau_2, \ldots, \tau_{k-1})| < \infty\,, \quad i = 1, 2, \ldots, k-1 \tag{5.143}$$

Then, the k-order cumulative spectrum $C_{nX}(\omega_1, \ldots, \omega_{k-1})$ of $X(n)$ exists, and it is continuous and defined as the $k-1$-dimensional Fourier transform of k-order cumulates for $X(n)$:

$$C_{kX}(\omega_1, \ldots, \omega_{k-1}) = \sum_{\tau_1=-\infty}^{\infty} \sum_{\tau_2=-\infty}^{\infty} \cdots \sum_{\tau_{k-1}=-\infty}^{\infty} c_{kX}(\tau_1, \tau_2, \ldots, \tau_{k-1}) \exp\{-\mathrm{j}(\omega_1\tau_1 + \omega_2\tau_2 + \cdots + \omega_{k-1}\tau_{k-1})\} \tag{5.144}$$

Among these, $|\omega_i| \le \pi$, $i = 1, 2, \ldots, k-1$, $|\omega_1, \omega_2, \ldots, \omega_{k-1}| \le \pi$.

Normally, $M_{kX}(\omega_1, \ldots, \omega_{k-1})$ and $C_{kX}(\omega_1, \ldots, \omega_{k-1})$ are plural and can be expressed as:

$$M_{kX}(\omega_1, \ldots, \omega_{k-1}) = |M_{kX}(\omega_1, \ldots, \omega_{k-1})| \exp\{\mathrm{j}\varphi_{kX}(\omega_1, \omega_2, \ldots, \omega_{k-1})\} \tag{5.145}$$

$$C_{kX}(\omega_1, \ldots, \omega_{k-1}) = |C_{kX}(\omega_1, \ldots, \omega_{k-1})| \exp\{\mathrm{j}\varphi_{kX}(\omega_1, \omega_2, \ldots, \omega_{k-1})\}\varphi \tag{5.146}$$

In addition, the higher-order moments and the cumulative spectrum are periodic functions with 2π cycles:

$$M_{kX}(\omega_1, \ldots, \omega_{k-1}) = M_{kX}(\omega_1 + 2\pi l_1, \omega_2 + 2\pi l_2, \ldots, \omega_{k-1} + 2\pi l_{k-1}) \tag{5.147}$$

$$C_{kX}(\omega_1, \ldots, \omega_{k-1}) = C_{kX}(\omega_1 + 2\pi l_1, \omega_2 + 2\pi l_2, \ldots, \omega_{k-1} + 2\pi l_{k-1}) \tag{5.148}$$

Among these, $l_1, l_2, \ldots, l_{k-1}$ is an integer. Sometimes, the high-order spectrum (or multispectrum) is a special high-order cumulative spectrum; the most commonly used is the third-order spectrum, also known as the double spectrum, because it is the function of two independent variables ω_1, ω_2.

5.5.3.3 Special cases of the cumulative spectrum

The power spectrum ($k = 2$):

$$C_{2X}(\omega) = \sum_{\tau_1=-\infty}^{\infty} c_{2X}(\tau) \exp\{-\mathrm{j}(\omega\tau)\} \tag{5.149}$$

Among these $|\omega_i| \le \pi$, if the random process $X(n)$ is the zero mean, then the above equation is a Wiener–Xining equation, and:

$$c_{2X}(\tau) = c_{2X}(-\tau) \tag{5.150}$$

$$C_{2X}(\omega) = C_{2X}(-\omega) \tag{5.151}$$

$$C_{2X}(\omega) \ge 0 \tag{5.152}$$

The double spectrum ($k = 3$):

$$C_{3X}(\omega_1, \omega_2) = \sum_{\tau_1=-\infty}^{\infty} \sum_{\tau_2=-\infty}^{\infty} c_{3X}(\tau_1, \tau_2) \exp\{-\mathrm{j}(\omega_1\tau_1 + \omega_2\tau_2)\} \tag{5.153}$$

We have the following symmetry:

$$\begin{aligned} C_{3X}(\omega_1, \omega_2) &= C_{3X}(\omega_2, \omega_1) \\ &= C_{3X}(-\omega_2, -\omega_1) = C_{3X}(-\omega_1 - \omega_2, \omega_2) \\ &= C_{3X}(\omega_1, -\omega_1 - \omega_2) = C_{3X}(-\omega_1 - \omega_2, \omega_1) \\ &= C_{3X}(\omega_2, -\omega_1 - \omega_2) \end{aligned} \tag{5.154}$$

For the real random process, double spectrums have 12 symmetric regions. Triple spectrums ($k = 4$):

$$C_{4X}(\omega_1, \omega_2, \omega_3) = \sum_{\tau_1=-\infty}^{\infty} \sum_{\tau_2=-\infty}^{\infty} c_{4X}(\tau_1, \tau_2, \tau_3) \exp\{-\mathrm{j}(\omega_1\tau_1 + \omega_2\tau_2 + \omega_3\tau_3)\} \tag{5.155}$$

Triple spectrums also have symmetry:

$$\begin{aligned} C_{4X}(\omega_1, \omega_2, \omega_3) &= C_{4X}(\omega_2, \omega_1, \omega_3) = C_{4X}(\omega_3, \omega_2, \omega_1) = C_{4X}(\omega_1, \omega_3, \omega_2) \\ &= C_{4X}(\omega_2, \omega_3, \omega_1) = C_{4X}(\omega_3, \omega_1, \omega_2) = \cdots \end{aligned} \tag{5.156}$$

5.5.3.4 Variance, slope and the peak state expressed by the cumulative spectrum

The Fourier transform of the definition of the higher-order cumulate spectrum is:

$$\begin{aligned} c_{kX}(\tau_1, \tau_2, \ldots, \tau_{k-1}) = \frac{1}{2\pi} \sum_{\tau_1=-\infty}^{\infty} \sum_{\tau_2=-\infty}^{\infty} \cdots \sum_{\tau_{k-1}=-\infty}^{\infty} C_{kX}(\omega_1, \omega_2, \ldots, \omega_{k-1}) \\ \exp\{\mathrm{j}(\omega_1\tau_1 + \omega_2\tau_2 + \cdots + \omega_{k-1}\tau_{k-1})\} \end{aligned} \tag{5.157}$$

The variance expressed by the cumulative spectrum can be defined as:

$$c_{2X}(0) = \frac{1}{2\pi} \int_{-\pi}^{\pi} C_{2X}(\omega)\,\mathrm{d}\omega \tag{5.158}$$

The slope is:

$$c_{3X}(0, 0) = \frac{1}{(2\pi)^2} \int_{-\pi}^{\pi} \int_{-\pi}^{\pi} C_{3X}(\omega_1, \omega_2)\,\mathrm{d}\omega_1\,\mathrm{d}\omega_2 \tag{5.159}$$

The peak state is:

$$c_{4X}(0,0,0) = \frac{1}{(2\pi)^3}\int_{-\pi}^{\pi}\int_{-\pi}^{\pi}\int_{-\pi}^{\pi} C_{4X}(\omega_1,\omega_2,\omega_3)\,\mathrm{d}\omega_1\,\mathrm{d}\omega_2\,\mathrm{d}\omega_3 \tag{5.160}$$

The variance describes the degree of the probability distribution dispersing from the mean, and the slope describes the asymmetry of the probability distribution. The peak state describes the degree of probability distribution curve or flatness.

5.5.3.5 One-dimensional slices of the cumulative and the cumulative spectrum

Since the high-order cumulate spectrum is a multidimensional function, it is computationally complex. Nagata [11] suggests using one-dimensional segments of multidimensional cumulates and their one-dimensional Fourier transforms as useful means to extract information from the high-order statistical information of non-Gaussian stationary signals. Considering a non- Gaussian stationary process $X(n)$, its third-order cumulate is:

$$c_{3X}(\tau_1,\tau_2) = \mathrm{cum}[X(n), X(n+\tau_1), X(n+\tau_2)] \tag{5.161}$$

A one-dimensional slice of $c_{3X}(\tau_1,\tau_2)$ is defined as:

$$r_{2,1}^X(\tau) \triangleq \mathrm{cum}[X(n), X(n), X(n+\tau)] = c_{3X}(0,\tau) \tag{5.162}$$

$$r_{1,2}^X(\tau) \triangleq \mathrm{cum}[X(n), X(n+\tau), X(n+\tau)] = c_{3X}(\tau,\tau) \tag{5.163}$$

$$r_{1,2}^X(\tau) \triangleq r_{1,2}^X(-\tau) \tag{5.164}$$

Taking $r_{2,1}^X(\tau)$ as an example, the one-dimensional spectrum is defined as:

$$R_{2,1}^X(\omega) \triangleq \sum_{\tau=-\infty}^{\infty} r_{2,1}^X(\tau)\mathrm{e}^{-\mathrm{j}\omega\tau} \tag{5.165}$$

It can be shown that the one-dimensional spectrum has the following relationship with the double spectrum:

$$R_{2,1}^X(\omega) = \frac{1}{2\pi}\int_{-\pi}^{+\pi} C_3^X(\omega,\sigma)\,\mathrm{d}\sigma \tag{5.166}$$

It represents the integration of the double spectrum along a certain frequency axis.

A one-dimensional slice of the fourth-order cumulate about a non-Gaussian stationary process $X(n)$ can be defined as:

$$\begin{aligned} r_{3,1}^X(\tau) &\triangleq \mathrm{cum}[X(n), X(n), X(n), X(n+\tau)] \\ &= c_{3X}(0,0,\tau) = c_{3X}(0,\tau,0) = c_{3X}(\tau,0,0) \end{aligned} \tag{5.167}$$

$$\begin{aligned} r_{2,2}^X(\tau) &\triangleq \mathrm{cum}[X(n), X(n), X(n+\tau), X(n+\tau)] \\ &= c_{3X}(0,\tau,\tau) = c_{3X}(0,\tau,\tau) = c_{3X}(\tau,0,\tau) \end{aligned} \tag{5.168}$$

$$r_{1,3}^X(\tau) \triangleq \mathrm{cum}[X(n), X(n+\tau), X(n+\tau), X(n+\tau)] = c_{3X}(\tau,\tau,\tau) \tag{5.169}$$

A one-dimensional spectrum is defined as:

$$R^X_{3,1}(\omega) \triangleq \sum_{\tau=-\infty}^{\infty} r^X_{3,1}(\tau)\mathrm{e}^{-\mathrm{j}\omega\tau} \tag{5.170}$$

$$R^X_{2,2}(\omega) \triangleq \sum_{\tau=-\infty}^{\infty} r^X_{2,2}(\tau)\mathrm{e}^{-\mathrm{j}\omega\tau} \tag{5.171}$$

$$R^X_{3,1}(\omega) = \frac{1}{(2\pi)^2}\int_{-\pi}^{+\pi}\int_{-\pi}^{+\pi} C_{4X}(\omega, \sigma_1, \sigma_2)\frac{\mathrm{d}\sigma_1}{\mathrm{d}\sigma_2} \tag{5.172}$$

$$R^X_{2,2}(\omega_1+\omega_2) = \frac{1}{2}\int_{-\pi}^{+\pi}\int_{-\pi}^{+\pi} C_{4X}(\omega_1, \omega_2, \omega_3)\,\mathrm{d}(\omega_1-\omega_2)\,\mathrm{d}\omega_3 \tag{5.173}$$

In random signal processing, the cumulative spectrum is much more applied than the moment, mainly for the following reasons.

First, the cumulative spectrum of Gaussian processes greater than the second order is zero, and a non-zero cumulative spectrum provides a non-Gaussian measure. Second, the cumulative amount provides a suitable measure of the independence of the time series statistics. Third, the cumulative amount of the sum of the two independent stationary processes with zero mean is equal to the sum of their respective cumulative spectrums, and the moments do not have this property. Fourth, the assumptions of each state are easier to satisfy when estimating the cumulative spectrum than when estimating the moment spectrum. The moment is mainly used in the analysis process of the determined signal.

5.5.4 Non-Gaussian processes and linear systems

The non-Gaussian process is a process that the probability density distribution is a non-normal distribution. From the analysis of the previous sections, it is known that the high-order cumulate of the Gaussian process is zero, and the non-Gaussian process must have the characteristic that the high-order cumulate is not zero.

5.5.4.1 The Non-Gaussian white noise process

Setting $W(n)$ as a stationary non-Gaussian process, $E[W(n)] = 0$, the k-order cumulative amount is:

$$\begin{aligned} c_{kw}(\tau_1, \tau_2, \ldots, \tau_{k-1}) &= \mathrm{cum}[W(n), W(n+\tau_1), \ldots, W(n+\tau_{k-1})] \\ &= \gamma_{kw}\delta(\tau_1, \tau_2, \ldots, \tau_{k-1}) \end{aligned} \tag{5.174}$$

Among these, γ_{kw} is a constant, $\delta(\tau_1, \tau_2, \ldots, \tau_{k-1})$ is the $k-1$-dimensional impulse function, and $W(n)$ is called a k-order non-Gaussian white noise process. An independent non-Gaussian white noise process is called high-order white noise. Of course, we do not have to assume all γ_{kw} are bounded. For $k \geq 2$, γ_{kw} cannot be all zero. $W(n)$'s k-order cumulative spectrum is:

$$C_{kw}(\omega_1, \omega_2, \ldots, \omega_{k-1}) = \gamma_{kw} \tag{5.175}$$

The white non-Gaussian process is a flat spectrum at all frequencies. When $k = 2, 3, 4$, the power spectrum, the double spectrum and the triple spectrum corresponding to the white noise are:

$$C_{2w}(\omega) = \gamma_{2w} \tag{5.176}$$

$$C_{3w}(\omega_1, \omega_2) = \gamma_{3w} \tag{5.177}$$

$$C_{4w}(\omega_1, \omega_2, \omega_3) = \gamma_{4w} \tag{5.178}$$

Among these, γ_{2w} is the variance of $W(n)$, and γ_{3w}, γ_{4w} are its slope and peak state, respectively.

5.5.4.2 The Non-Gaussian white noise process and linear systems

Non-Gaussian white noise $E(n)$ passes through a the linear time-invariant system shown in Fig. 5.6; the additive noise $V(n)$ is Gaussian colored noise, $V(n)$ is statistically independent of $E(n)$ and also statistically independent of $X(n)$. We have:

$$Y(n) = X(n) + V(n) \tag{5.179}$$

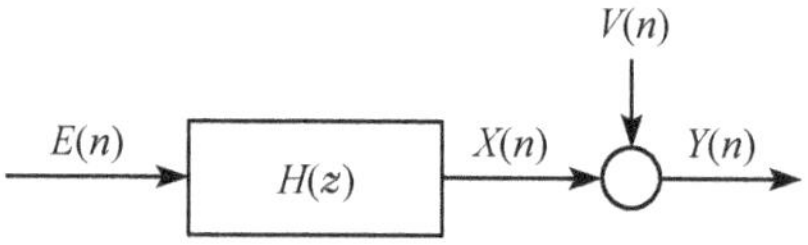

Fig. 5.6: Schematic of a linear time-invariant system.

According to the nature of the cumulative, output the k-order cumulative amount of $Y(n)$ as:

$$\begin{aligned} c_{kY}(\tau_1, \tau_2, \ldots, \tau_{k-1}) &= c_{kX}(\tau_1, \tau_2, \ldots, \tau_{k-1}) + c_{kV}(\tau_1, \tau_2, \ldots, \tau_{k-1}) \\ &= c_{kX}(\tau_1, \tau_2, \ldots, \tau_{k-1}) \end{aligned} \tag{5.180}$$

$$X(n) = E(n) \cup h(n) = \sum_{i=-\infty}^{\infty} h(i)E(n-i) \tag{5.181}$$

$$\begin{aligned} &c_{kX}(\tau_1, \tau_2, \ldots, \tau_{k-1}) = \operatorname{cum}[X(n), X(n+\tau_1), \ldots, X(n+\tau_{k-1})] \\ &= \operatorname{cum}\left(\sum_{i=-\infty}^{\infty} h(i_1)E(n-i_1), \sum_{i=-\infty}^{\infty} h(i_2)E(n+\tau_1-i_2), \ldots, \sum_{i=-\infty}^{\infty} h(i_k)E(n+\tau_{k-1}-i_k) \right) \\ &= \sum_{i_1=-\infty}^{\infty} h(i_1) \sum_{i_2=-\infty}^{\infty} h(i_2) \cdots \sum_{i_k=-\infty}^{\infty} h(i_k) \\ &\quad \cdot \operatorname{cum}[E(n-i_1), E(n+\tau_1-i_2), \ldots, E(n+\tau_{k-1}-i_k)] \end{aligned} \tag{5.182}$$

According to the definition of cumulative quantity: $E(n)$

$$c_{kE}(\tau_1,\tau_2,\ldots,\tau_{k-1}) = \text{cum}[E(n), E(n+\tau_1),\ldots,E(n+\tau_{k-1})] \tag{5.183}$$

We have:

$$\begin{aligned} &c_{kY}(\tau_1,\tau_2,\ldots,\tau_{k-1}) = c_{kX}(\tau_1,\tau_2,\ldots,\tau_{k-1}) \\ &= \sum_{i_1=-\infty}^{\infty}\sum_{i_2=-\infty}^{\infty}\sum_{i_k=-\infty}^{\infty} h(i_1)h(i_2)\ldots h(i_k)\cdot c_{kE}(\tau_1+i_1-i_2,\tau_2+i_1-i_3,\ldots,\tau_{k-1}+i_1-i_k) \end{aligned} \tag{5.184}$$

The above equation describes the relationship between the cumulative amount of the system output signal and the cumulative amount of input noise and the system impulse response. For the above formula use the $k-1$-order Fourier transform:

$$\begin{aligned} C_{kY}(\omega_1,\omega_2,\ldots,\omega_{k-1}) = H(\omega_1)H(\omega_2)\ldots H(\omega_{k-1})H^*(\omega_1+\omega_2+\cdots+\omega_{k-1}) \\ \cdot C_{kE}(\omega_1,\omega_2,\ldots,\omega_{k-1}) \end{aligned} \tag{5.185}$$

Among these, $H(\omega_i) = \sum_{i_k=-\infty}^{\infty} h(i)\mathrm{e}^{-j\omega i}$. The above equation describes the relationship between the higher-order spectrum of the system output signal and the higher-order spectrum of the input signal and the system transfer function. The transformation form about z is:

$$\begin{aligned} C_{kY}(z_1,z_2,\ldots,z_{k-1}) = H(z_1)H(z_2)\ldots H(z_{k-1})H^*(z_1+z_2+\cdots+z_{k-1}) \\ \cdot C_{kE}(z_1,z_2,\ldots,z_{k-1}) \end{aligned} \tag{5.186}$$

In particular, when the system input $E(n)$ is high-order white noise, the k-order cumulates and cumulative spectra of the output signal are, respectively:

$$c_{kY}(\tau_1,\tau_2,\ldots,\tau_{k-1}) = \gamma_{kE}\sum_{i=-\infty}^{\infty} h(i)h(i+\tau_1)\ldots h(i+\tau_{k-1}) \tag{5.187}$$

$$C_{kY}(\omega_1,\omega_2,\ldots,\omega_{k-1}) = \gamma_{kE}H(\omega_1)H(\omega_2)\ldots H(\omega_{k-1})H^*(\omega_1+\omega_2+\cdots+\omega_{k-1}) \tag{5.188}$$

When $k = 2, 3, 4$, the cumulative and accumulated spectrum of the output signal are as follows:

$$c_{2Y}(\tau) = R(\tau) = \sigma_E^2\sum_{i=-\infty}^{\infty} h(i)h(i+\tau) \tag{5.189}$$

$$c_{3Y}(\tau_1,\tau_2) = \gamma_{3E}\sum_{i=-\infty}^{\infty} h(i)h(i+\tau_1)h(i+\tau_2) \tag{5.190}$$

$$c_{4Y}(\tau_1,\tau_2,\tau_3) = \gamma_{4E}\sum_{i=-\infty}^{\infty} h(i)h(i+\tau_1)h(i+\tau_2)h(i+\tau_3) \tag{5.191}$$

$$C_{2Y}(\omega) = P_X(\omega) = \sigma_E^2 H(\omega)H^*(\omega) = \sigma_E^2 |H(\omega)|^2 \tag{5.192}$$

$$C_{3Y}(\omega_1, \omega_2) = B_Y(\omega_1, \omega_2) = \gamma_{3E} H(\omega_1) \cdot H(\omega_2) H^*(-\omega_1 - \omega_2) \tag{5.193}$$

$$C_{4Y}(\omega_1, \omega_2, \omega_3) = \gamma_{4E} H(\omega_1) \cdot H(\omega_2) \cdot H(\omega_3) H^*(-\omega_1 - \omega_2 - \omega_3) \tag{5.194}$$

In addition, because of the formula $C_{kY}(\omega_1, \omega_2, \ldots, \omega_{k-1})$, there is a constant relationship between the k-order cumulate spectrum and the $k-1$-order of the cumulative spectrum as follows:

$$C_{kY}(\omega_1, \omega_2, \ldots, \omega_{k-2}, 0) = C_{k-1,Y}(\omega_1, \omega_2, \ldots, \omega_{k-2}) \cdot H(0) \cdot \frac{\gamma_{kY}}{\gamma_{k-1,Y}} \tag{5.195}$$

Therefore, the power spectrum of the non-Gaussian linear process can be reconstructed by its double spectrum and a constant term. The reconstruction expression is as follows:

$$C_{3Y}(\omega, 0) = C_{2Y}(\omega) H(0) \frac{\gamma_{3Y}}{\gamma_{2Y}} \tag{5.196}$$

Exercises

5.1 Use MATLAB to generate a constant sinusoidal combination process in white noise, and estimate the mean, variance, correlation function and power spectrum by using multiple samples.

5.2 Program with MATLAB to generate a segmented stationary Gaussian process and analyze its time-varying power spectrum and short-term power spectrum.

5.3 Use MATLAB to generate a white noise process and generate a Wiener process on this basis. Then, analyze the mean and correlation function.

5.4 If the linear discrete system is a causal system, its Green function is:

$$g(n, m) = \begin{cases} \frac{e^{-m^2}}{2n+1}, & n \geq m \\ 0, & n < m \end{cases}$$

Try to solve the generalized transfer function.

5.5 Seek the Wigner distribution of the signal $x(t) = Ae^{j\omega_0 t}$.

5.6 Generate any stationary random signal with MATLAB and input it into a linear time-varying system to obtain a nonstationary random signal; analyze its WV spectrum.

5.7 Set the Gaussian distribution probability density function to be $f(x) = \frac{0.5}{\sqrt{2\pi\sigma^2}} e^{x^2/2\sigma^2}$. Find the k-order moments and k-order cumulates ($k = 1, 2, 3, 4$).

5.8 Set the probability density function of uniform distribution be

$$f(x) = \begin{cases} \frac{1}{2c}, & |x| \leq c \\ 0, & \text{elsewhere} \end{cases}$$

Find the k-order moments and k-order cumulates ($k = 1, 2, 3, 4$).

A Appendix

A.1 Power spectral density of nonstationary random processes

For nonstationary random processes $X(t)$, the correlation function is related to both time intervals τ and time t:

$$R_X(\tau, t) = R_X(t, t+\tau) = E[X(t)X(t+\tau)] \tag{A.1}$$

This is called the instantaneous correlation function. Its Fourier transform is:

$$G_X(\omega, t) = \int_{-\infty}^{\infty} R_X(\tau, t) e^{-j\omega\tau}\, d\tau \tag{A.2}$$

$G_X(\omega, t)$ also relates to time t and is called the instantaneous power spectral density. It is necessary to take an average time to make the power spectral density a single-valued function of frequency, so the power spectral density of the nonstationary random process is defined as:

$$\tilde{G}_X(\omega) = \lim_{T\to\infty} \frac{1}{2T} \int_{-T}^{T} G_X(\omega, t)\, dt \tag{A.3}$$

Or:

$$\tilde{G}_X(\omega) = \lim_{T\to\infty} \frac{1}{2T} \int_{-T}^{T} \left[\int_{-\infty}^{\infty} R_X(\tau, t) e^{-j\omega\tau}\, d\tau \right] dt \tag{A.4}$$

With the above formula, since the order of time-average and integral operations can be interchanged, there are two algorithms.

(1) Algorithm 1
First, Fourier transform:

$$G_X(\omega, t) = \int_{-\infty}^{\infty} R_X(\tau, t) e^{-j\omega\tau}\, d\tau$$

and then do the time average:

$$\tilde{G}_X(\omega) = \lim_{T\to\infty} \frac{1}{2T} \int_{-T}^{T} G_X(\omega, t)\, dt$$

(2) Algorithm 2
First, take the time average:

$$\langle R_X(\tau) \rangle = \lim_{T\to\infty} \frac{1}{2T} \int_{-T}^{T} R_X(\tau, t)\, dt \tag{A.5}$$

https://doi.org/10.1515/9783110593808-007

and then do the Fourier transform:

$$\tilde{G}_X(\omega) = \int_{-\infty}^{\infty} \langle R_X(\tau) \rangle e^{-j\omega\tau}\, d\tau \tag{A.6}$$

In general, Algorithm 2 is a little easier to calculate.

Example A.1.1. It is known that noise is a narrowband amplitude modulation oscillation $X(t) = n(t)\cos\omega_0 t$, where constant ω_0 is the angular frequency of the carrier, $n(t)$ is the zero mean stationary noise, and $R_n(\tau)$ is the correlation function. Find the power spectral density $\tilde{G}_X(\omega)$ of the nonstationary stochastic process $X(t)$.

Solution:

$$\begin{aligned} R_X(\tau, t) &= E[X(t)X(t+\tau)] \\ &= E[n(t)n(t+\tau)\cos\omega_0 t\cos\omega_0(t+\tau)] \\ &= \frac{1}{2}R_n(\tau)[\cos\omega_0\tau + \cos\omega_0(2t+\tau)] \end{aligned} \tag{A.7}$$

Using Algorithm 2, the previous equation is substituted into eq. (A.5) to obtain:

$$\langle R_X(\tau) \rangle = \frac{1}{2}R_n(\tau)\cos\omega_0\tau \tag{A.8}$$

So, we substitute that into eq. (A.6):

$$\begin{aligned} \tilde{G}_X(\omega) &= \int_{-\infty}^{\infty} \frac{1}{2}R_n(\tau)\cos\omega_0\tau e^{-j\omega\tau}\, d\tau \\ &= \int_{0}^{\infty} R_n(\tau)\cos\omega_0\tau\cos\omega\tau\, d\tau \\ &= \int_{0}^{\infty} \frac{1}{2}R_n(\tau)[\cos(\omega-\omega_0)\tau + \cos(\omega+\omega_0)\tau]\, d\tau \end{aligned}$$

Equation 1.176 can be calculated to give:

$$\tilde{G}_X(\omega) = \frac{1}{4}[G_n(\omega-\omega_0) + G_n(\omega+\omega_0)] \tag{A.9}$$

The above equation shows that, like the amplitude modulation oscillation of the known signal, the power spectrum of the noise amplitude modulation oscillation is also distributed in the carrier frequency ω_0 in the two nearby sidebands, and its shape is the same as the low-frequency noise power spectrum $G_n(\omega)$, but the modulus is reduced to a quarter of the original, as is shown in Fig. A.1.

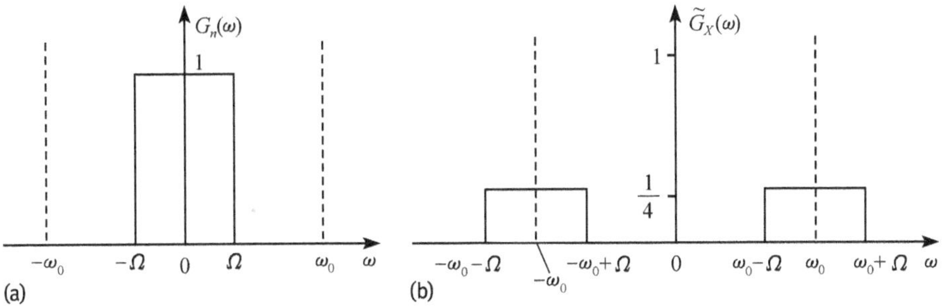

Fig. A.1: The power spectrum of narrowband noise amplitude modulation oscillation.

A.2 Proof of a double integral formula

When $0 \le \varphi \le 2\pi$, we have the integral formula:

$$\int_0^\infty \int_0^\infty Z_1 Z_2 \exp\left[-\left(Z_1^2 + Z_2^2 + 2Z_1Z_2\cos\varphi\right)\right] \mathrm{d}Z_1\,\mathrm{d}Z_2 = \frac{1}{4}\csc^2\varphi(1-\varphi\cot\varphi) \quad \text{(A.10)}$$

Prove: Transform rectangular coordinates Z_1, Z_2 to polar coordinates r, θ, take:

$$Z_1 = r\cos\theta\,, \quad Z_2 = r\sin\theta \quad \text{(A.11)}$$

Then:

$$\int_0^\infty \int_0^\infty Z_1 Z_2 \exp\left[-\left(Z_1^2 + Z_2^2 + 2Z_1Z_2\cos\varphi\right)\right] \mathrm{d}Z_1\,\mathrm{d}Z_2$$

$$= \int_0^{\pi/2} \sin\theta\cos\theta \int_0^\infty r^3 \mathrm{e}^{-r^2(1+\cos\varphi\sin 2\theta)}\,\mathrm{d}r\,\mathrm{d}\theta \quad \text{(A.12)}$$

Because:

$$\int_0^\infty x^{2a+1}\mathrm{e}^{-px^2}\,\mathrm{d}x = \frac{a!}{2p^{a+1}}\,, \quad p > 0 \quad \text{(A.13)}$$

So:

$$\int_0^\infty r^3 \mathrm{e}^{-r^2(1+\cos\varphi\sin 2\theta)}\,\mathrm{d}r = \frac{1}{2(1+\cos\varphi\sin 2\theta)^2} \quad \text{(A.14)}$$

Therefore the right-hand side of eq. (A.12) is differentiated into:

$$\int_0^{\pi/2} \frac{\sin 2\theta}{4(1+\cos\varphi\sin 2\theta)^2}\,\mathrm{d}\theta = \frac{1}{8}\int_0^{\pi/2} \frac{\sin 2\theta}{(1+\cos\varphi\sin 2\theta)^2}\,\mathrm{d}(2\theta) \quad \text{(A.15)}$$

Taking $\cos\varphi = K$, $2\theta = x$, then the double integral of the right-hand side of eq. (A.12) is equivalent to the calculation of a single integral:

$$\frac{1}{8}\int_0^{\pi} \frac{\sin x}{(1+K\sin x)^2}\,dx \tag{A.16}$$

Use the integral formula:

$$\int \frac{\alpha+\beta\sin x}{(a+b\sin x)^n}\,dx = \frac{1}{(n-1)(a^2-b^2)} \cdot \left[\frac{(b\alpha-a\beta)\cos x}{(a+b\sin x)^{n-1}} + \int \frac{(a\alpha-b\beta)(n-1)+(a\beta-b\alpha)(n-2)\sin x}{(a+b\sin x)^{n-1}}\,dx\right] \tag{A.17}$$

to obtain

$$\int_0^{\pi} \frac{\sin x}{(1+K\sin x)^2}\,dx = \frac{1}{1-K^2}\left[\frac{-\cos x}{1+K\sin x}\Big|_0^{\pi} + \int_0^{\pi}\frac{-K}{(1+K\sin x)^2}\,dx\right]$$
$$= \frac{1}{1-K^2}\left[2 - K\int_0^{\pi}\frac{1}{(1+K\sin x)^2}\,dx\right] \tag{A.18}$$

and then use the integral formula:

$$\int \frac{1}{a+b\sin x}\,dx = \frac{2}{\sqrt{a^2-b^2}}\arctan\left(\frac{a\tan\frac{x}{2}+b}{\sqrt{a^2-b^2}}\right), \quad a^2>b^2 \tag{A.19}$$

to obtain:

$$\int_0^{\pi} \frac{\sin x}{(1+K\sin x)^2}\,dx = \frac{1}{1-K^2}\left[2 - \frac{2K}{\sqrt{1-K^2}}\arctan\left(\frac{\tan\frac{x}{2}+K}{\sqrt{1-K^2}}\right)\Big|_0^{\pi}\right]$$
$$= \frac{2}{1-K^2}\left[1-\frac{K}{\sqrt{1-K^2}}\left(\frac{\pi}{2}-\arctan\frac{K}{\sqrt{1-K^2}}\right)\right]$$
$$= \frac{2}{\sin^2\varphi}\left[1-\frac{\cos\varphi}{\sin\varphi}\left(\frac{\pi}{2}-\arctan\frac{\cos\varphi}{\sin\varphi}\right)\right]$$
$$= 2\csc^2\varphi\left[1-\cot\varphi\left(\frac{\pi}{2}-\arctan\frac{1}{\tan\varphi}\right)\right] \tag{A.20}$$

Using the inverse trigonometric function:

$$\arctan x = \frac{\pi}{2} - \arctan\frac{1}{x} \tag{A.21}$$

we obtain:

$$\frac{\pi}{2} - \arctan\frac{1}{\tan\varphi} = \arctan(\tan\varphi) = \varphi \tag{A.22}$$

Therefore:

$$\frac{1}{8}\int_0^{\pi} \frac{\sin x}{(1+K\sin x)^2}\,dx = \frac{1}{4}\csc^2\varphi(1-\varphi\cot\varphi) \tag{A.23}$$

Thus, the equivalent double-integral formula eq. (A.10) is established.

A.3 Derivation of the detector voltage transfer coefficient

When the voltage on the nonlinear circuit load has a negative effect on the nonlinear device, and this effect cannot be ignored, the nonlinear stochastic differential equation should be listed and the nonlinear transformation of the random process should be solved. It is necessary to use this method when it comes to the reaction of the envelope detector load voltage. The typical circuit of the envelope detector is shown in Fig. 4.17. It is known that the voltage–current characteristics of the nonlinear device are:

$$i = f(u) \tag{A.24}$$

The input voltage of the detector is the high-frequency narrowband process is:

$$X(t) = A(t)\cos\left(\omega_0 t + \varphi(t)\right) \tag{A.25}$$

The output voltage of the detector is $Y(t)$. The internal resistance of the input voltage source is ignored. The following relations are obtained from the figure:

$$i = i_1 + i_2 \tag{A.26}$$

$$Y(t) = \frac{1}{C_L}\int i_1\,\mathrm{d}t = i_2 R_L \tag{A.27}$$

$$u = X(t) - Y(t) \tag{A.28}$$

It is not difficult to obtain the nonlinear stochastic differential equation of the circuit:

$$\frac{\mathrm{d}Y(t)}{\mathrm{d}t} + \frac{1}{R_L C_L}Y(t) = \frac{1}{C_L}f[X(t) - Y(t)] \tag{A.29}$$

Taking eq. (A.25) into the above formula, taking $\varphi(t) = \omega_0 t + \varphi(t)$, and eliminating the time variable t:

$$\frac{\mathrm{d}Y}{\mathrm{d}t} + \frac{1}{R_L C_L}Y = \frac{1}{C_L}f[A\cos\varphi - Y] \tag{A.30}$$

It is very difficult to precisely solve this nonlinear differential equation, and usually its approximate solution is obtained.

When the two conditions of the envelope detector are satisfied with time ($\tau \gg T_0$, $\tau \ll \tau_0$), the output voltage $Y(t)$ of the detector in which only a low-frequency voltage instantaneous value $Y_0(t)$ is determined by the envelope $A(t)$ of the narrowband process, where for these two voltages, the detector is equivalent to no inertia, so it can be analyzed with the non-inert slowly varying enveloping method.

It is approximately believed that reaction voltage $Y(t)$ only includes low-frequency components $Y_0(t)$, and this voltage is also the function of the input voltage $A\cos\varphi$, and therefore, it is the same as the slowly varying enveloping method without reaction; the nonlinear function $f\left[A\cos\varphi - Y\right]$ is still a periodic function of non random variable φ and can be expanded into Fourier series. The high-frequency components ($\mathrm{d}Y/\mathrm{d}t$ high-frequency component) are ignored, and the low-frequency components only are extracted at both ends of the above equation, so the following relationship is obtained:

$$Y_0 = \frac{R_L}{\pi} f_0 [A \cos\varphi - Y_0] \tag{A.31}$$

In the formula:

$$f_0[A\cos\varphi - Y_0] = \frac{1}{\pi}\int_0^{\pi} f[A\cos\varphi - Y_0]\,\mathrm{d}\varphi \tag{A.32}$$

I_0 (Including dc) is the low-frequency current component of a nonlinear device. So, the relation of no inertial transformation between the output voltage $Y_0(t)$ and input envelope voltage $A(t)$ is obtained:

$$Y_0 = \frac{R_L}{\pi}\int_0^{\pi} f[A\cos\varphi - Y_0]\,\mathrm{d}\varphi \tag{A.33}$$

After the voltage–current characteristics $i = f(u)$ of nonlinear devices, the output low-frequency voltage $Y_0(t)$ of the detector can be solved by using the above formula.

A.3.1 Half-wave linear detector

Set the volt ampere characteristic:

$$i = f(u) = \begin{cases} bu\,, & u \geq 0 \\ 0\,, & u < 0 \end{cases} \tag{A.34}$$

That is:

$$f[A\cos\varphi - Y_0] = \begin{cases} b[A\cos\varphi - Y_0]\,, & |\varphi| \leq \arccos\frac{Y_0}{A} \\ 0\,, & \text{otherwise} \end{cases} \tag{A.35}$$

We can obtain it from eq. (A.33):

$$Y_0 = \frac{bR_L A}{\pi}\int_0^{\arccos\frac{Y_0}{A}} \left[\cos\varphi - \frac{Y_0}{A}\right]\mathrm{d}\varphi \tag{A.36}$$

If $Y_0/A = K_d$, then:

$$Y_0 = \frac{bR_L A}{\pi}\int_0^{\arccos K_d} [\cos\varphi - K_d]\,\mathrm{d}\varphi = \frac{bR_L A}{\pi}[\sin(\arccos K_d) - K_d \arccos K_d] \tag{A.37}$$

Because $\sin(\arccos K_d) = \sqrt{1 - [\cos(\arccos K_d)]^2} = \sqrt{1 - K_d^2}$, therefore:

$$K_d = \frac{Y_0}{A} = \frac{bR_L}{\pi}\left[\sqrt{1 - K_d^2} - K_d \arccos K_d\right] \tag{A.38}$$

or:

$$bR_L = \frac{K_d \pi}{\sqrt{1 - K_d^2} - K_d \arccos K_d} \tag{A.39}$$

The above equation shows that the voltage coefficient factor of the linear detector is only related to bR_L but has nothing to do with the input envelope voltage $A(t)$.

For diode detectors, $b = 1/R_i$, the internal resistance of the diode R_i changes with the operating point. The relationship between K_d and R_L/R_i according to the formula is shown in Fig. 4.18. When R_L/R_i is known, K_d is determined by the graph, and then the relationship is:

$$Y_0 = K_d A \tag{A.40}$$

The output voltage $Y_0(t)$ of the linear detector is proportional to the input envelope voltage $A(t)$.

A.3.2 Full-wave square-law detector

Set the volt ampere characteristic as:

$$i = f(u) = \begin{cases} bu^2, & u \geq 0 \\ 0, & u < 0 \end{cases} \tag{A.41}$$

That is:

$$f[A\cos\varphi - Y_0] = \begin{cases} b[A\cos\varphi - Y_0]^2, & |\varphi| \leq \arccos\frac{Y_0}{A} \\ 0, & \text{others} \end{cases} \tag{A.42}$$

By imitating the previous equation we can obtain:

$$Y_0 = \frac{bR_L A^2}{\pi} \int_0^{\arccos K_d} [\cos\varphi - K_d]^2 \, d\varphi = \frac{bR_L A^2}{\pi} \int_0^{\arccos K_d} \left[\cos^2\varphi - 2K_d \cos\varphi + K_d^2\right] d\varphi \tag{A.43}$$

After integration, we obtain:

$$Y_0 = \frac{bR_L A^2}{2\pi} \left[\left(1 + 2K_d^2\right) \arccos K_d - 3K_d \sqrt{1 - K_d^2}\right] \tag{A.44}$$

or

$$bR_L A = \frac{2\pi K_d}{\left(1 + 2K_d^2\right) \arccos K_d - 3K_d \sqrt{1 - K_d^2}} \tag{A.45}$$

where $K_d = Y_0/A$. The pattern is shown in Fig. 4.19. This formula shows that the voltage transfer coefficient K_d of the square-law detector is not only related to bR_L but also to the input envelope voltage $A(t)$.

When $bR_LA < 0.1$, we have $K_d \ll 1$, approximating $\cos^{-1} K_d \approx \pi/2$:

$$bR_LA \approx \frac{2\pi K_d}{\pi/2} = 4K_d \tag{A.46}$$

or:

$$K_d \approx \frac{1}{4} bR_LA \tag{A.47}$$

So, the output voltage is:

$$Y_0 = K_d A \approx \frac{1}{4} bR_LA^2 \tag{A.48}$$

It is shown that the output voltage $Y_0(t)$ of the square law detector is the square of the input envelope voltage $A(t)$.

A.4 Derivation of the statistical mean of random processes in Rice distribution

According to the definition of the mean, we have:

$$\begin{aligned}\overline{R} = \int_{-\infty}^{\infty} R \cdot p(R)\,\mathrm{d}R &= \int_{-\infty}^{\infty} \frac{R^2}{\sigma^2} \exp\left[-\frac{R^2 + A^2}{2\sigma^2}\right] I_0\left(\frac{RA}{\sigma^2}\right) \mathrm{d}R \\ &= \frac{1}{\sigma^2} \exp\left[-\frac{A^2}{2\sigma^2}\right] \int_0^{\infty} R^2 \exp\left[-\frac{R^2}{2\sigma^2}\right] I_0\left(\frac{RA}{\sigma^2}\right) \mathrm{d}R\end{aligned} \tag{A.49}$$

The Bessel function has the following integral formula:

$$\int_0^{\infty} I_0(at)\mathrm{e}^{-ht^2}\,\mathrm{d}t = \frac{1}{2}\sqrt{\frac{\pi}{h}}\mathrm{e}^{\frac{a^2}{8h}} I_0\left(\frac{a^2}{8h}\right) \tag{A.50}$$

Take the derivation of parameters h at the two ends of the above equation and use the properties of the Bessel function:

$$\frac{\mathrm{d}}{\mathrm{d}x} I_0(x) = I_1(x) \tag{A.51}$$

Then:

$$\int_0^{\infty} I_0(at)\mathrm{e}^{-ht^2} t^2\,\mathrm{d}t = \frac{\sqrt{\pi}}{4h^{3/2}}\mathrm{e}^{\frac{a^2}{8h}} \left\{ I_0\left(\frac{a^2}{8h}\right) + \frac{a^2}{4h}\left[I_0\left(\frac{a^2}{8h}\right) + I_1\left(\frac{a^2}{8h}\right)\right]\right\} \tag{A.52}$$

Take:

$$a = \frac{A}{\sigma^2}, \quad h = \frac{1}{2\sigma^2}, \quad Q = \frac{A}{\sqrt{2}\sigma} \tag{A.53}$$

Then:

$$\frac{a^2}{8h} = \frac{A^2}{4\sigma^2} = \frac{Q^2}{2} \tag{A.54}$$

So, the formula above is brought into eq. (A.52), and we obtain:

$$\int_0^\infty R^2 \exp\left[-\frac{R^2}{2\sigma^2}\right] I_0\left(\frac{RA}{\sigma^2}\right) \mathrm{d}R = \sqrt{\frac{\pi}{2}}\sigma^3 \mathrm{e}^{-\frac{Q^2}{2}} \left\{I_0\left(\frac{Q^2}{2}\right) + Q^2\left[I_0\left(\frac{Q^2}{2}\right) + I_1\left(\frac{Q^2}{2}\right)\right]\right\} \tag{A.55}$$

Then the formula above is brought into eq. (A.49), and we can conclude that:

$$\overline{R} = \sqrt{\frac{\pi}{2}}\sigma \cdot \mathrm{e}^{-\sigma^2/2}\left[\left(1 + Q^2\right)I_0\left(\frac{Q^2}{2}\right) + Q^2 I_1\left(\frac{Q^2}{2}\right)\right] \tag{A.56}$$

Bibliography

[1] Cohen, 2000. Time-frequency analysis: theory and application, Xi'an: Xi'an Jiaotong University Press.
[2] Ge Y. B., 2005. Probability and mathematical statistics. Beijing: Tsinghua University Press.
[3] He Y. H., Qian W. M., 2004. A concise course of stochastic process. Shangnhai: Tongji University Press.
[4] Li B. B., Ma W. P., Tian H. X., 2006. Random signal analysis. Beijing: Science Press.
[5] Long Y. H., 2001. Probability and mathematical statistics. Beijing: Higher Education Press.
[6] Mao Y. C., Hu Q. Y., 1998. Random process. Xi'an: Xidian University Press.
[7] Papoulis A., 2004. Probability, random variables and stochastic processes. Xi'an: Xi'an Jiaotong University Press.
[8] Qiu T. S., Zhang X. X., Li X. B., et al, 2004. Yongmei Sun. Statistical signal processing: Non-Gaussian signal processing and application. Beijing: Publishing House of Electronics Industry.
[9] Shen H. F., 2003. Probability and mathematical statistics. 4th edn. Beijing: Higher Education Press.
[10] Sheng Z., Xie S. Q., Pan C. Y., 2001. Probability and mathematical statistics. 3rd edn. Beijing: Higher Education Press.
[11] Wang H. Y., Qiu T. S., Chen Z., 2008. Nonstationary random signal analysis and processing. 2nd edn. Beijing: National Defend Industry Press.
[12] Wang J. S., Liu J. K., 2003. Essentials of stochastic processes. Tianjin: Tianjin University Press.
[13] Wang R. X., 2006. Random process. Xi'an: Xi'an Jiaotong University Press.
[14] Wang Y. X., Sun H. X., 2003. Probability and stochastic processes. Beijing: Beijing University of Posts and Telecommunications Press.
[15] Yan Z., Xiao Z. R., Yue C. L., 2003. A new edition of probability theory and mathematical statistics. Hefei: Press of University of Science and Technology of China.
[16] Zhang A. W., 1984. Random signal analysis. Xi'an: Xidian University Press.
[17] Zhang Z. Q., Chen H. J., 2003. Random process. Xi'an: Xidian University Press.
[18] Zhu H., Huang H. N., Li Y. Q., et al, 1990. Random signal analysis. Beijing: Beijing Institute of Technology Press.

https://doi.org/10.1515/9783110593808-008

Index

https://doi.org/10.1515/9783110593808-009

www.ingramcontent.com/pod-product-compliance
Lightning Source LLC
Chambersburg PA
CBHW080928220326
41598CB00034B/5714
* 9 7 8 3 1 1 0 5 9 5 3 6 9 *